Land-Based Greenhouse Gas Mitigation for Carbon Neutrality

Land-Based Greenhouse Gas Mitigation for Carbon Neutrality

Guest Editors

Li'na Liu
Zhenhua Zhang
Rong Wu
Suman Aryal

MDPI

Basel • Beijing • Wuhan • Barcelona • Belgrade • Novi Sad • Cluj • Manchester

Guest Editors
Li'na Liu
Northwest Institute of
Eco–Environment and
Resources
Chinese Academy of Sciences
Lanzhou
China

Zhenhua Zhang
Institute of Green Finance
Lanzhou University
Lanzhou
China

Rong Wu
School of Architecture and
Urban Planning
Guangdong University of
Technology
Guangzhou
China

Suman Aryal
Faculty of Health,
Communities and Education
TAFE Queensland
Toowoomba
Australia

Editorial Office
MDPI AG
Grosspeteranlage 5
4052 Basel, Switzerland

This is a reprint of the Special Issue, published open access by the journal *Land* (ISSN 2073-445X), freely accessible at: https://www.mdpi.com/journal/land/special_issues/MYP5110O4Q.

For citation purposes, cite each article independently as indicated on the article page online and as indicated below:

Lastname, A.A.; Lastname, B.B. Article Title. *Journal Name* **Year**, *Volume Number*, Page Range.

ISBN 978-3-7258-5837-8 (Hbk)
ISBN 978-3-7258-5838-5 (PDF)
https://doi.org/10.3390/books978-3-7258-5838-5

Contents

land

MDPI

Article

The Heterogeneity of the Drivers of Urban Form in China: Perspectives from Regional Disparities and Development Stage Variations

Rong Wu [1,†], Yongli Zhang [1,†], Meilin Dai [1], Qingyin Li [1] and Changlong Sun [2,*]

1 School of Architecture and Urban Planning, Guangdong University of Technology, Guangzhou 510062, China; wurong@gdut.edu.cn (R.W.); 2112110048@mail2.gdut.edu.cn (Y.Z.); 3219008409@mail2.gdut.edu.cn (M.D.); 2112110025@mail2.gdut.edu.cn (Q.L.)

2 School of Economics and Management, Wenzhou University of Technology, Wenzhou 325000, China

* Correspondence: 20200436@wzu.edu.cn

† These authors contributed equally to this work.

Abstract: Although there have been many discussions about the influencing factors of urban expansion, the heterogeneity of the driving mechanisms behind urban form remains poorly understood. Therefore, this paper evaluated the heterogeneous impacts of potential determinants on urban form, considering regional disparities and the stage of development. Based on land use data collected from Landsat ETM and TM scenes, the landscape metrics of urban size, urban centrality, urban shape irregularity, and urban fragmentation were measured to describe the urban form of 265 Chinese cities. We find that the regional disparities and development-stage variations significantly affect urban form. All urban form variables showed a significant stair-stepping difference in cities at various development stages, indicating that as a city upgrades its level of development, the intensity of urban expansion gradually increases, the shape of the urban edge becomes more fragmented and the urban built-up area becomes more compact. Urban form in Chinese cities shows significant geographical heterogeneity in terms of its driving forces. The effect of the socioeconomic factors on urban form also presented changes depending on the development stage. Our results provide helpful references for policymakers within urban spatial structure planning and land resource management.

Keywords: urban form; driving forces; geographical heterogeneity; development stage; China

1. Introduction

Urban populations have witnessed constant growth in recent decades, and urbanization is now a global phenomenon [1]. At present, more than 50% of the global population lives in urban areas, and this proportion is expected to exceed 67% by 2050 [2]. The built-up area of cities around the world has expanded rapidly in line with the increase in urban population [3]. Beyond sheer size, this expansion has also led to changes in other urban form parameters, such as urban compactness and urban shape complexity. Urban areas are the engines of population aggregation and economic growth, upgrade industrial structures, and lead to social prosperity [4]. However, urban areas not only offer opportunities for social and economic development—they are also linked to the creation of severe environmental challenges [5,6]. Aiming to address these challenges, sustainable plans for urban development must be devised; such plans, in turn, require a greater scientific comprehensive of the spatiotemporal patterns and causes of changes in urban form [7].

Urban form, which is the physical arrangement of structures, spaces, and objects that make up cities, plays a pivotal role in society. It refers to the spatial distribution and organization of urban entities, encompassing factors such as building density, land use patterns, infrastructure, and transportation systems. This formation significantly impacts not only how we navigate cities, but also how resources are utilized, and how physical activity within these spaces is facilitated [8–10]. Urban form has been measured, characterized, and evaluated in a variety of ways, sometimes indirectly—for instance, population density [11] and the economy [12]. Recent developments in remote sensing technologies have seen the increasing utilization of landscape metrics to measure urban form, which is often addressed regarding the three aspects of urban expansion, urban shape complexity, and urban compactness [13,14]. Beyond these analyses, urban form has also long been considered as a field of intervention and study within the disciplines of urban management and spatial planning [15,16]. In the above traditions, much of the research has been devoted to investigating the influencing factors and mechanisms at work in urban form, with scholars addressing aspects as diverse as physical factors [17–19], population [20–22], economic growth [23–26], infrastructure [27,28], industry [29], and policy [30–32]. Although increasing attention has focused on the determinants of urban form, previous studies in the field have only focused on urban expansion (or urban growth) [33–35], examining various factors in order to investigate its causes [36,37]. Nevertheless, the term "urban form" describes the form of the spatial distribution of the composition of urban entities, including not only urban growth but also urban shape and fragmentation. There has been little discussion about the determinants of these other aspects of urban form to date.

In addition, although much of the previous literature has focused on the determinants of urban form, they only focused on one single city or region, such as Beijing [38,39], Guangzhou [24,40], Jiangsu [41] or the Yangtze River Delta [42,43]. The strength and direction of the impact of the various determinants on changes in urban form appears to differ between studies. Some studies have analyzed data from different periods and found that the contribution of various determinants to urban expansion changes over time [19,44]. This finding naturally provokes the question: do the driving forces behind changes in urban form vary in space and at different developmental stages? Despite the importance of this question, variations in the influencing factors behind changes in urban form have rarely been studied in relation to their effects at different geographical scales and/or economic development levels. This study will attempt to fill this gap by evaluating the influencing factors that affect urban form in China, considering a range of different economic development stages and different geographical locations.

To sum up, lots of studies exist have explored the influencing factors of urban expansion, and these findings enhance our comprehension of the spatial and temporal characteristics of urban growth and its influencing factors. Nevertheless, previous studies suffer from some drawbacks. Firstly, the research has generally been directed towards the influencing factors behind urban sprawl (or urban growth) [45–47]. Urban form, however, is a more comprehensive term than urban expansion, and knowledge about the driving forces of urban form—which is necessary for urban management and spatial planning—is still lacking. Second, the heterogeneity of urban form factors is rarely discussed within contemporary scholarly discourse. Due to the wide range of natural environments, geophysical conditions, and socioeconomic conditions that characterize urban areas, the driving forces behind urban form can be expected to vary from region to region and stage to stage. This variation has not yet been comprehensively dealt with by the literature. Third, existing analyses of the driving forces at work in urban form have mainly been carried out at the level of individual cities [20,28]. Several studies have produced estimates of the drivers of urban sprawl in some counties or cities [48]. A nationwide survey of the drivers of urban form, however, remains lacking in the Chinese context. Such a study would be essential to the formulation of national policies. Thus, the contribution of this paper is to evaluate the spatial–temporal patterns of the urban form of China's urban areas, and to verify the heterogeneity of the driving forces behind those identified patterns. On the basis of land use data interpreted from Landsat TM scenes and Landsat ETM scenes, landscape metrics are applied to estimate the urban form of 265 cities in China, taking into account the perspectives of urban size, urban centrality, urban shape irregularity, and urban fragmentation. A panel regression model is adopted to evaluate the impacts of the selected potential determinants on urban form with consideration of regional location and development stage. The results of this study constitute a helpful reference for policymakers within urban land-use management and spatial planning.

2. Data and Methodology

2.1. Study Area

Since the economic reforms of 1978, China—the second-largest economy and largest developing country in the world—has witnessed the greatest flow of rural–urban migration in world history, with an urbanization rate increase from 17.6% in 1978 to 57.35% in 2016, at an average annual growth rate of 1.02%. With the total area of urban land expanding from 7438 km^2 in 1981 to 45,566 km^2 in 2015, urban areas in China have also undergone significant expansion, accompanied by considerable changes to the configuration of the urban landscape pattern. With its vast territory and large differences in the level of development of its various regions, China offers scholars an excellent opportunity to study the spatiotemporal pattern of urban form and geographical heterogeneity in the driving mechanisms of urban form. In this paper, we selected all 265 prefecture-level cities in China as study cases (Figure 1). Using Chenery's criteria, which are based on per capita GDP and the most common indicator used to evaluate the stage of economic development [49], the 265 cities were classified into three development categories on the basis of per capita GDP: cities in the primary and middle stages of industrialization (77 cities), cities in the late stage of industrialization (117 cities), and cities in the developed stage (68 cities) (Figure 2).

Figure 1. Study area.

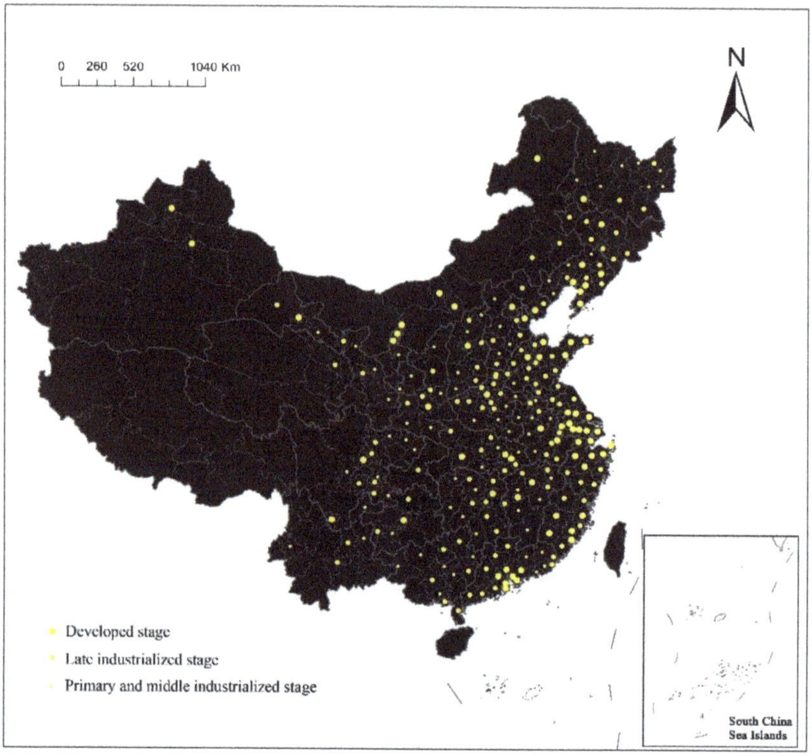

Figure 2. The cities included in this study, and their different stages of development.

2.2. Indicators Quantifying Urban Form

In order to characterize the spatiotemporal dynamics at work in the urban form patterns of China's 265 prefecture cities from 1990 to 2015, urban built-up area boundaries were determined using the global database of annual urban dynamics data from 1985 to 2015 at 30 m resolution, employing a large amount of surface reflectance data provided by Landsat satellite, which were published by Liu et al. (2020) [50].

Landscape metrics, which aim to measure both the regulation and the design of the uses of urban space, have been widely used to represent urban form by describing the pattern and structure of a landscape. According to previous studies, six landscape metrics were adopted to measure urban form and its changes: total area (TA), number of urban patches (NP), the largest patch index (LPI), the landscape shape index (LSI), the percentage of like adjacencies (PLADJ), and the aggregation index (AI). These metrics were considered to characterize four aspects of urban form, namely size, centrality, shape irregularity and fragmentation. Each of these aspects provides a unique perspective on the understanding of urban form. The 'size' is often associated with urban expansion, depicting the extent of a city's growth. 'Centrality' relates to urban compactness, reflecting the efficient use of space within a city. 'Shape irregularity' ties in with urban shape complexity, encapsulating the intricacy of the city's layout, while 'fragmentation' examines the degree of discontinuity in urban spaces. In the context of urban form, these aspects interact in complex ways. For instance, unchecked urban expansion can lead to an increase in shape irregularity and fragmentation, while compromising centrality or urban compactness. The choice of these metrics and an understanding of their interplay provide a more comprehensive view of urban form, addressing not only growth but also shape and fragmentation, and thus enriching our investigation of urban form.

Size is measured by two metrics: TA and NP. TA, which measures the total area of all urban land within all patches, helps to reveal the expansion process behind the built-up area of a particular city. NP measures discrete urban areas throughout the urban landscape, and generally increases with the rapid growth in the urban core area. Nevertheless, NP is expected to decrease if the urban area expands and merges into an overall urban area. Urban centrality is characterized by LPI, which reveals the percentage of the largest patch in the urban landscape area and represents the dominance of the city in the landscape. The extent of shape irregularity was represented by LSI in this study. LSI provides an indication of the geometric complexity of an entire urban area by measuring the perimeter-to area ratio of the whole landscape. As the LSI values increase, the shape of the urban built-up area becomes more complex. Urban fragmentation was described by PLADJ and AI. PLADJ is an absolute indicator of the degree of urban landscape aggregation, and its value ranges from 0 to 100%, with larger values representing a more concentrated urban landscape. AI is calculated as an area-weighted average class clustering index and is used to calculate the probability that urban patches are adjacent to each other throughout a given landscape. A larger AI value represents a more compact urban built-up area—AI is 1 when the urban built-up area is maximally disaggregated and equals 100 when the urban area is aggregated to the maximum extent into a compact, single patch. Table 1 gives a detailed description of these six landscape metrics.

Table 1. The details of the selected landscape metrics.

Landscape Metrics	Equations	Explanation
Total areas (TA)	$\mathrm{TA} = \sum_{j=1}^{n} a_{ij}(\frac{1}{10,000})$	a_{ij} represents the total areas of patch ij
Number of urban patches (NP)	$\mathrm{NP} = n_i$	n_i is the number of patches in the landscape of patch i
Largest patch index (LPI)	$\mathrm{LPI} = \frac{\max\limits_{1 \le j \le n}\left(a_{ij}\right)}{\mathrm{TA}}(100)$	a_{ij} denotes the area of patch ij, TA is the total landscape area
Landscape shape index (LSI)	$\mathrm{LSI} = \frac{0.25\sum_{k=1}^{m} e_{ik}^{*}}{\sqrt{\mathrm{TA}}}$	e_{ik}^{*} is the total length of edge in a landscape between class i and k TA denotes the total landscape areas
Percentage of like adjacencies (PLADJ)	$\mathrm{PLADJ} = \left(\frac{g_{ii}}{\sum_{k=1}^{m} g_{ik}}\right)(100)$	g_{ii} is the number of like adjacencies between pixels of patch type i based on the double-count method, g_{ik} is the number of adjacencies between pixels of patch type i and k based on the double-count method
Aggregation index (AI)	$\mathrm{AI} = \left[\sum_{i=1}^{m}\left(\frac{g_{ii}}{max \rightarrow g_{ii}}\right)\right] \times 100$	gii stands for the number of like adjacencies (joins) between pixels of urban patch

2.3. Panel Regression Model for Influencing Factors Analysis

Whilst the task of evaluating the driving forces of urban extension has garnered considerable attention from scholars, previous studies have mainly emphasized the influencing factors of urban expansion. The influencing factors of other aspects of urban form have not always been examined, nor has the fact that socioeconomic factors may modify these other aspects. Observations of multiple individuals in multiple periods can be treated simultaneously by the panel regression model; recognizing this capacity, this study used panel regression to quantitatively measure the driving mechanisms behind urban form, taking the research period of 1990–2015 into account. On the basis of previous studies and data availability, we selected five variables to identify the underlying mechanisms at work in urban form, namely population, gross domestic product, industrial structure, per capita urban road area, and fixed investment (Figure 3). The five variables were assumed to be linked with urban form by means of the following models:

$$TA_{it} = \alpha_0 + \alpha_1 \mathrm{P} + \alpha_2 \mathrm{GDP} + \alpha_3 \mathrm{IS} + \alpha_4 \mathrm{ROAD} + \alpha_5 \mathrm{IS} + \varepsilon_{\mathrm{it}} \quad (1)$$

$$NP_{it} = \alpha_0 + \alpha_1\text{P} + \alpha_2\text{GDP} + \alpha_3\text{IS} + \alpha_4\text{ROAD} + \alpha_5\text{IS} + \varepsilon_{\text{it}} \tag{2}$$

$$LPI_{it} = \alpha_0 + \alpha_1\text{P} + \alpha_2\text{GDP} + \alpha_3\text{IS} + \alpha_4\text{ROAD} + \alpha_5\text{IS} + \varepsilon_{\text{it}} \tag{3}$$

$$LSI_{it} = \alpha_0 + \alpha_1\text{P} + \alpha_2\text{GDP} + \alpha_3\text{IS} + \alpha_4\text{ROAD} + \alpha_5\text{IS} + \varepsilon_{\text{it}} \tag{4}$$

$$PLADJ_{it} = \alpha_0 + \alpha_1\text{P} + \alpha_2\text{GDP} + \alpha_3\text{IS} + \alpha_4\text{ROAD} + \alpha_5\text{IS} + \varepsilon_{\text{it}} \tag{5}$$

$$AI_{it} = \alpha_0 + \alpha_1\text{P} + \alpha_2\text{GDP} + \alpha_3\text{IS} + \alpha_4\text{ROAD} + \alpha_5\text{IS} + \varepsilon_{\text{it}} \tag{6}$$

where TA_{it} represents total area of city i in year t, NP_{it} represents the number of urban patches of city i in year t, LPI_{it} stands for number of urban patches of city i in year t, LSI_{it} represents landscape shape index of city i in year t, $PLADJ_{it}$ denotes percentage of like adjacencies of city i in year t, AI_{it} stands for the aggregation index of city i in year t, the intercepts for all individuals are denoted by α_0, α_1 to α_7 denote coefficients of the independent variables. P is population, GDP refers to gross domestic product, IS represents industrial structure, ROAD denotes per capita urban road area, FAI stands for fixed asset investment, and ε_{it} is the random error.

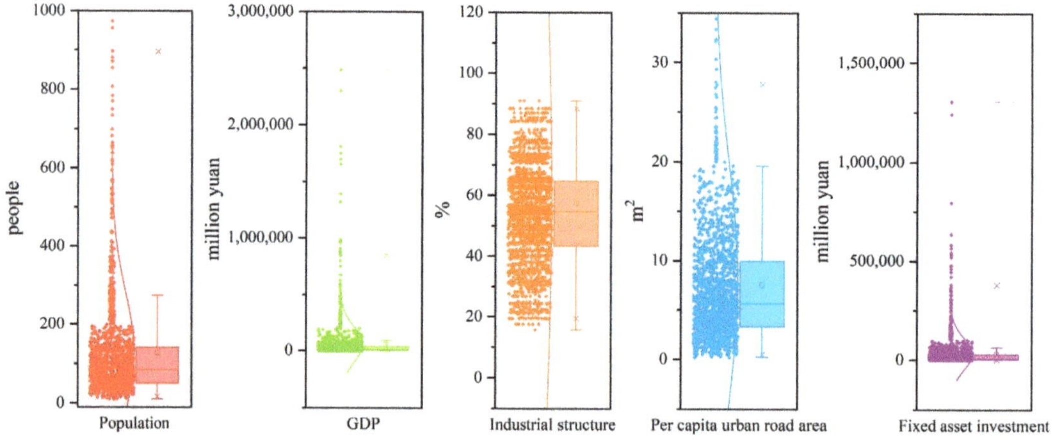

Figure 3. Box charts of the five driving factors.

Population (P) is the foundation of urbanization and is believed to constitute a crucial variable, driving urban expansion and the transmutation of urban form by affecting the urban market, rigid demand, and agglomeration ability [23]. The process of urbanization implies a growing urban population and an increasing demand for urban construction land, which are realized by means of urban expansion and urban renewal, which in turn drive microscopic changes in urban landscapes [45]. Gross domestic product (GDP) is the most commonly used economic variable to characterize macro-level economic development. Industrial structure (IS) can also reflect economic development. The prior literature has found that the power of economic development often determines the urban form itself [1,51]. Per capita urban road area (ROAD) represents transportation infrastructure, an index that has improved in the past 30 years. ROAD is believed to have significant impacts on urban form. On the one hand, urban expansion along particular routes is one of the popular urban growth patterns [46]. On the other hand, transportation infrastructure also directly leads to the fragmentation of urban landscapes. Fixed asset investment (FAI) is assumed to constitute one of the influencing factors in relation to urban form, as this acts as a basic financial security for urban development [45]. All data were from the China City Statistical Yearbook (1996–2016).

3. Results and Discussion

3.1. Dynamic Pattern of Urban Form

Between 1990 and 2015, China experienced accelerating industrialization and urbanization, which together caused a significant expansion of the urban area of cities and brought about dramatic changes in urban form. Figure 4 shows the change characteristics of the four aspects of urban form studied over this period in China. From these figures, significant differences can be seen in the urban form of different regions at different times. Further comparison of urban form metrics between regions allows for a more elaborate observation of the changes in urban form between 1990 and 2015. A significant increase in the urban area (represented by TA and NP) can be observed between 1990 and 2015. In 2015, higher levels of expansion were predominantly observed on the east coast and in Chongqing. It is not surprising that cities located in the eastern region had a larger average urban area than those situated in the central and western regions. Such a finding seems to be consistent with the previous literature findings, showing that the attributes of urban area vary from region to region [45,52]. Such variation reflects, to a great extent, the relatively early development of urban land in the east, as well as its status as a developmental focus area of the "reform and opening" policy [45]. We found a significant decrease in urban centrality from 1990 to 2015; in 1990, LSI was found to have a relatively higher value in the cities along the Yangtze River and the southeastern coastal areas, where high-density hydrological networks exist. This finding reflects the way in which the urban development in these regions has been limited by natural conditions, whilst cities located in more developed regions may have experienced the benefits of a "leap-forward" development model. For these reasons, the geometrical shape of urban areas in those cities may appear more irregular. The LSI of all regions increased in the period from 1990 to 2015. The spatial distribution of LSI was relatively even in 2015, showing that as cities expand, they become restricted by various factors, which makes their urban boundaries more irregular. A significant decrease in the urban compactness (represented by PLADJ and AI) can be observed between 1990 and 2015. This finding reveals a trend showing that China's urban areas are becoming less intensive and compact.

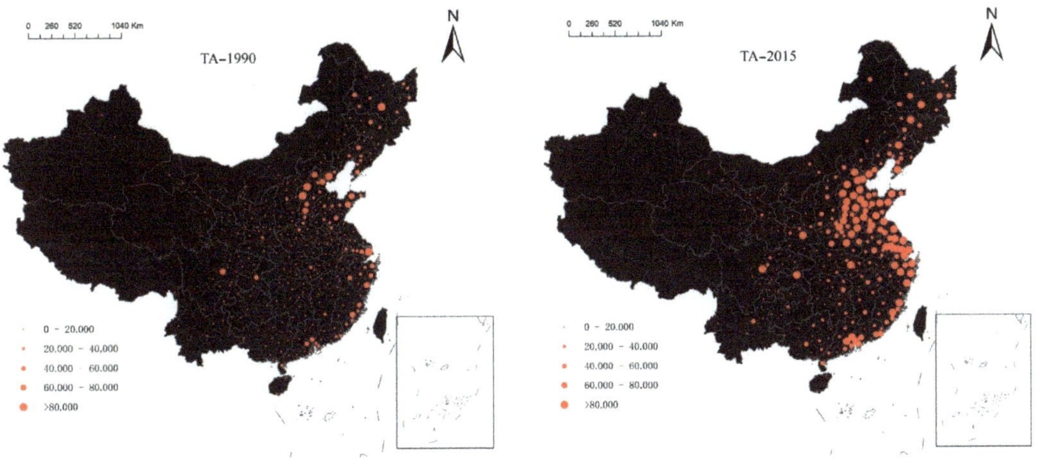

Figure 4. *Cont.*

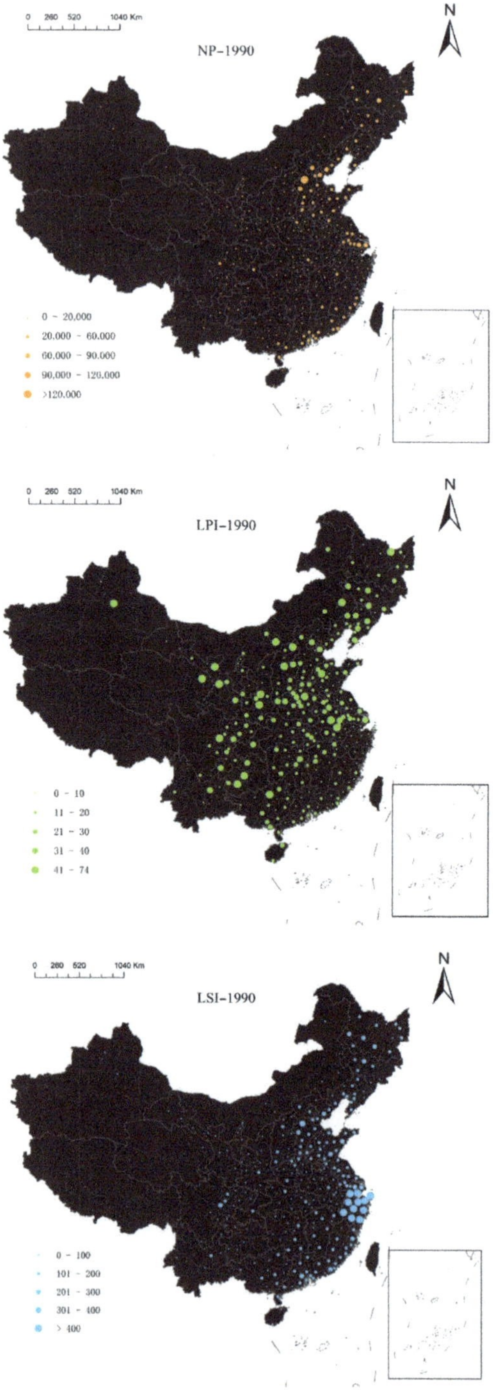

Figure 4. *Cont.*

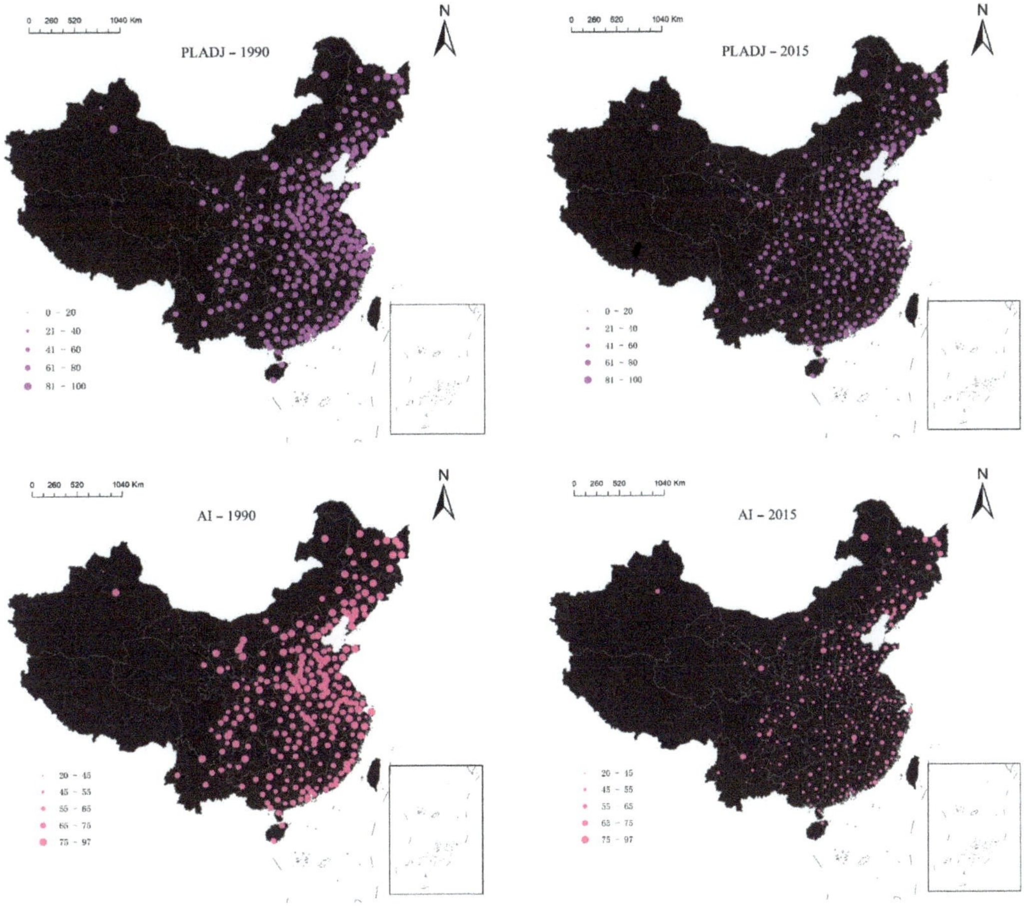

Figure 4. Urban form changes between 1990 and 2015.

As mentioned above, the studied cities were classified into three categories on the basis of their per capita GDP, in accordance with Chenery's economic development stage criterion, namely the primary and middle stages of industrialization, late-stage industrialization, and the developed stage. Figure 5 contains box charts that show the urban form characteristics (measured using TA, NP, LPI, LSI, PLADJ and AI metrics) of cities in the three different economic development categories. It can be seen from Figure 5 that TA and NP showed a significant stair-stepping difference. Meanwhile, the urban construction areas of cities at the developed stage were much larger than those of cities at the other two stages. This phenomenon was most obvious in 2015, when total area evidenced a larger average difference with a greater significance than the figures for 1990. The urban area of the developed-stage cities increased by 236.65%, whilst the cities in a stage of late industrialization and cities in the primary and middle stages of industrialization increased by 186.96% and 157.45%, respectively, between 1990 and 2015. The average intensity of urban expansion was highest for developed cities, followed by cities at a stage of late industrialization, with the lowest intensity being recorded in cities at the primary and middle stages. This indicates that, as cities progress in economic development terms, the scale of new construction land and the intensity of urban expansion gradually increase. LPI showed a significant stair-stepping difference in 1990, but in 2015 there was no such significant difference and the LPI value significantly decreased, indicating that the polarization phenomenon of the urban area has been significantly alleviated. The measures of LSI also showed an apparent stair-stepping difference, revealing that as cities enter more advanced stages of development, the shape of their urban edges becomes more fragmented. The measures of PLADJ and AI in 2015 were significantly smaller than those in 1990, which means that the urban compactness of cities at different stages has declined. These findings clearly indicate that the location and development stage of a city significantly affects its urban form.

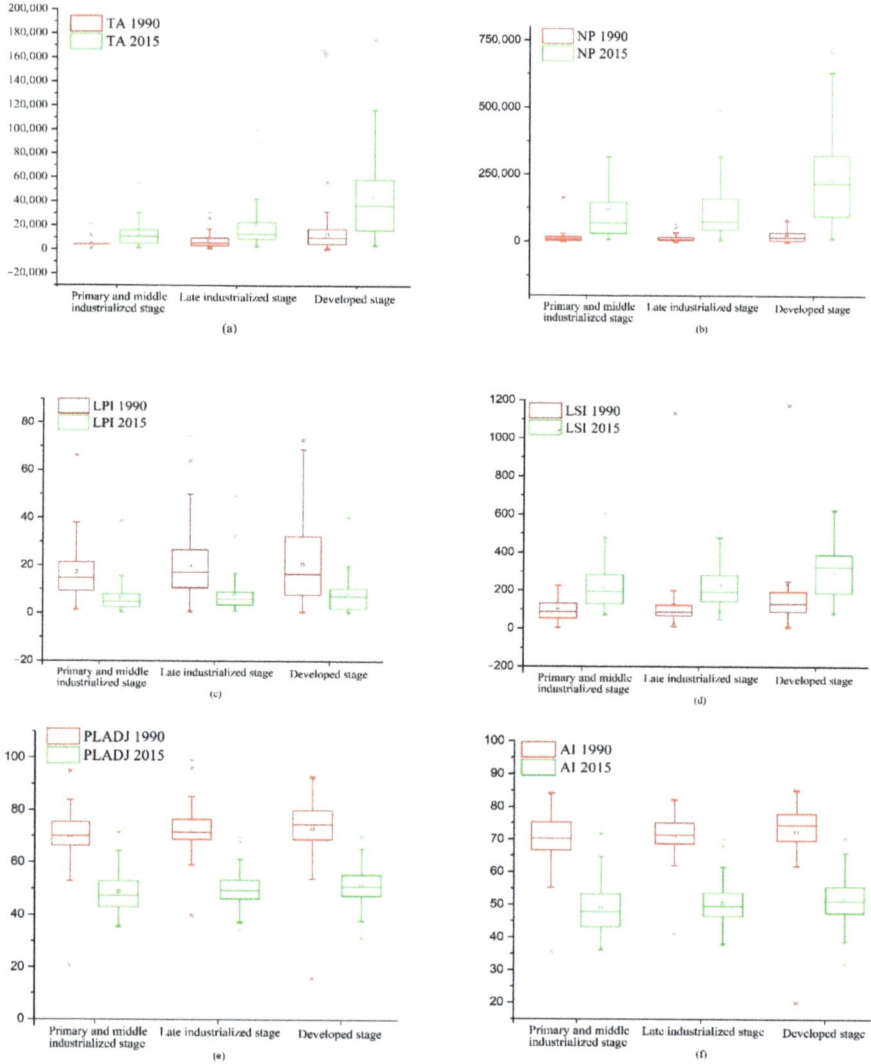

Figure 5. Box charts showing the urban form characteristics of cities at different economic development stages, where the (**a–f**) means the variations of total area (TA), number of urban patches (NP), the largest patch index (LPI), the landscape shape index (LSI), the percentage of like adjacencies (PLADJ), and the aggregation index (AI). The x in the figure means the outliers.

3.2. Influencing Factors of Urban Form

Prior to the panel regression analysis, a variance inflation factor (VIF) test was used to verify the presence of severe multicollinearity between the independent variables. It can be seen from Table 2 that the VIFs of all five variables were less than 10 and the tolerances were greater than 0.1, revealing that the selected five variables are not collinear. Thus, we were able to examine the parameters of the panel data model. The most widely used forms of panel regression model are the random effect (RE) estimator with a large degree of freedom and the fixed effect (FE) estimator with a relatively small degree of freedom. The Hausman test is generally used to determine which estimator is more suitable. In this study, a panel regression model was established to investigate the effects of socioeconomic variables on urban form.

Table 2. Correlation coefficients of the independent variables.

	VIF	Tolerance
P	1.965	0.509
GDP	4.609	0.217
IS	1.060	0.943
ROAD	1.070	0.935
FAI	4.632	0.216

Table 3 displays the estimation results. It can be seen that the factor of population exhibited a significant positive relationship with TA, NP, and LSI, showing that growth in the total population of a given county will result in more complex patterns of urban development and an increase in the urban area. Most existing studies have also drawn similar conclusions, noting that population changes directly affect urban form [45,46]. While population exhibited a negative relationship with LPI, PLADJ and AI, it was indicated that increases in the total population of a given county will lead to a decrease in city centrality and compactness. The economic disparity between urban and rural areas, coupled with increases in urban employment opportunities, lead to the massive migration of people from rural to urban areas, bringing aboutrapid growth in the construction area of the cities receiving these migrants [45]. GDP displayed a positive effect in relation to TA, NP, LPI, PLADJ and AI, while it exerted a negative effect on LSI and AI, that the growth of GDP brings about more centricity, more compact, less complex patterns of urban development and the growth of urban areas. Such results are in accordance with prior studies that found GDP to be a driver of urban expansion [47]. Industrial structure has a positive impact on TA, NP and LSI while demonstrating negative correlations with respect to LPI, PLADJ and A, indicating that an increase in the metric of industrial structure results in an expansion of the urban area and more complex patterns of urban development, a result that reinforces the findings of previous studies [46]. The infrastructure factor, represented by the ROAD metric, was shown to have statistically significant effects on the selected landscape metrics. Of these six variables, ROAD was found to have a positive impact on TA, revealing that ROAD benefits urban development by providing easier transportation access. This finding reflects the gradual evolution of infrastructure investment into an important driving force for urban land expansion. ROAD was found to have negative effects on LPI, PLADJ and AI; it was revealed that the improvement in infrastructure will result in less centrality and less compactness in a given city. In addition, we note that FAI exhibited a positive relationship with respect to TA, NP and LSI, indicating that increases in FAI result in complexity in patterns of urban development and an increase in urban areas.

3.3. Heterogeneity Effect

It can be seen from the above results that significant differences exist between China's regions in terms of urban form, which reflects their diverse economic levels, development policies, and natural conditions. For the purpose of measuring the relationship between different factors and urban form in various regions, three panel models were established to estimate the various effects of the selected driving forces on urban form in the regions of eastern China, central China, and western China.

Table 3. Estimated results of panel data model.

	TA	NP	LPI	LSI	PLADJ	AI
P	0.0440	0.1109	−0.4799	0.0124	−0.2563	−0.2490
	(1.58)	(5.13) ***	(−0.98)	(0.17)	(−3.62) ***	(−3.42) ***
GDP	0.0845	0.0958	0.1068	−0.0312	0.1032	0.1101
	(2.78) ***	(3.28) ***	(2.01) **	(−0.39)	(1.34)	(1.39)
IS	0.0190	0.0202	−0.9599	0.0238	−0.2469	−0.2444
	(5.35) ***	(5.74) ***	(−15.46) ***	(2.53) **	(−27.41) ***	(−26.35) ***
ROAD	0.0955	0.0978	−0.2317	0.2901	−0.5348	−0.5642
	(3.17) ***	(3.32) ***	(−4.43) ***	(3.66) ***	(−7.05) ***	(−7.22) ***
FAI	0.1518	0.1339	−0.4725	0.4619	−0.6006	−0.6276
	(5.04) ***	(4.61) ***	(−8.97) ***	(5.79) ***	(−7.86) ***	(−7.98) ***
Estimation methods	FE	RE	FE	FE	FE	FE
R-squared	0.5417	0.4701	0.8039	0.5144	0.6618	0.5226
Observations	6864	6864	6864	6864	6864	6864

** denotes $p < 0.05$, *** denotes $p < 0.01$.

Tables 4–6 set out the estimated results for these three different regions in China. These findings show that population exerted a negative effect on TA and NP in western China, but had a positive influence on TA and NP in eastern and central China, indicating that increases in the size of the population result in a decrease in urban size in western China and an increase in urban size in eastern and central China. The coefficient of population's effect on urban area was the highest in eastern China, followed by central China and then western China, indicating that population size has the strongest impact on the urban area of cities in eastern China. GDP was observed to have exerted a positive influence on urban size in central and western regions but a negative influence in eastern China. This discovery is inconsistent with the prior literature, which has recognized the role of economic growth as the influencing factor of urban growth [2,25,53]. The industrial structure variable was positively related to the variable of urban size in all three regions, supporting the perspective that industrialization is a significant part of urban extension in China [54]. The coefficient of the effect of industrial structure on the size of cities was the highest in eastern China, lower in central China, and the lowest in western China, indicating that industrial structure has the largest impact on urban size in eastern China. The ROAD variable was found to be positively linked to TA and NP. A number of other relevant studies have drawn a similar conclusion, attributing the positive impact of ROAD on urban size to easier transportation access [47]. The influence of ROAD in central and western China seemed to be more significant than in eastern China, a finding that reveals that traffic has a greater impact on urban size in less developed regions. The impact of FAI on urban expansion also showed regional differences. A positive correlation between FAI and urban size was found in all three regions.

The impact of the five socioeconomic factors on urban centrality demonstrated regional heterogeneity. The results indicated that the population variable negatively correlated with LPI in eastern and central, while it exerted a positive effect in western China. This finding indicated that increases in the size of the population result in a decrease in urban centrality in eastern and central China and an increase in urban centrality in western China. GDP had negative effects on LPI in western China and exerted a positive influence on LPI in eastern and central China. This indicates that increases in GDP result in an increase in urban centrality in eastern and central China, but lead to a decrease in urban centrality in western China. The industrial structure factor was found to correlate to decreases in LPI in all three regions, which demonstrates that industrial structure directly impacts urban centrality. ROAD and FAI maintained a negative correlation with LPI in all three regions, revealing that transportation impacts urban centrality. Moreover, for cities in the eastern region, FAI was identified as being the most influential factor in relation to urban centrality, while in cities in central and western regions, the most influential factor was ROAD.

The impacts of socioeconomic factors on urban shape complexity also varied across different regions. Population was observed to exert a positive impact on LSI in eastern and central China while it showed negative effects in western China, suggesting that increases in the size of the population result in more complex urban expansion shapes in eastern and central China and less complex urban expansion shapes in western China. Surprisingly, no significant influence was observed between GDP and urban shape complexity in eastern and central China; GDP was only correlated to urban shape complexity in western China. Further, the industrial structure factor showed negative effects on LSI in eastern China and exerted a negative influence on LSI in western and central China. The transportation variable, represented by ROAD, was not statistically significant in eastern China, and correlated to increases in LSI in western and central China. This finding corroborates the findings of several previous studies that highlight the role of public transit in contributing to urban form in China [48]. FAI was statistically significant in relation to LSI in eastern and central China. FAI was found to have a positive effect on LSI, indicating that increases in FAI result in increases in urban shape complexity.

The impact of the five socioeconomic factors on urban compactness demonstrated regional heterogeneity. The results indicated that the population variable was statistically significant for PLADH and AI in eastern and central China and was not found to exert a significant influence on western China, which is inconsistent with the findings of previous studies [45]. Population growth stimulates the increase in urban areas and the number of urban areas; as such, increases in the size of a population lead to more complex urban development patterns. GDP showed positive effects on PLADJ and AI in eastern China and exerted a negative influence on PLADJ and AI in western and central China. This indicates that increases in GDP result in an increase in urban compactness in eastern China, but a decrease in urban compactness in western and central China. The industrial structure factor was found to correlate to decreases in PLADJ and AI in all regions, which demonstrates that industrial structure directly impacts urban compactness. ROAD maintained a significant correlation with PLADJ and AI in all regions, revealing that transportation also impacts urban compactness. These results are in line with earlier cases that have shown that the evolution of urban road networks can have an impact on urban form [48]. This finding can be partially illustrated by the way in which urban road development can promote the formation of sub-centers and form a decentralized urban form. FAI exerted a significant impact on PLADJ and AI in eastern China, while this factor seems to have little impact on PLADJ and AI in central and western China.

Table 4. The estimated results for eastern China.

	TA	NP	LPI	LSI	PLADJ	AI
P	0.0497 (1.34)	0.2289 (1.69) *	−0.0885 (−1.60)	0.0089 (0.04)	−0.3946 (−3.29) ***	−0.3691 (−3.36)***
GDP	−0.1115 (−3.94) ***	−0.5130 (−4.97) ***	0.1187 (2.81) ***	−0.3195 (−1.88) *	0.4937 (5.39) ***	0.4889 (5.83) ***
IS	0.0295 (6.82) ***	0.1299 (8.23) ***	−0.0537 (−8.29) ***	−0.0713 (−2.74) ***	−0.2802 (−19.98) ***	−0.2393 (−18.64) ***
ROAD	0.0496 (2.28) ***	0.1904 (2.40) ***	−0.0430 (−1.32)	0.1176 (0.90)	−0.1786 (−2.53) **	−0.1885 (−2.92) ***
FAI	0.4962 (12.18) ***	1.9507 (13.13) ***	−0.4840 (−7.95) ***	1.0115 (4.13) ***	−1.2193 (−9.25) ***	−1.2441 (−10.30) ***
Estimation methods	FE	FE	FE	FE	FE	FE
R-squared	0.8196	0.8081	0.8884	0.3893	0.8283	0.8566
Observations	2366	2366	2366	2366	2366	2366

* denotes $p < 0.1$, ** denotes $p < 0.05$, *** denotes $p < 0.01$.

Table 5. The estimated results for central China.

	TA	NP	LPI	LSI	PLADJ	AI
P	0.0422 (0.53)	0.2541 (2.79) ***	−0.2428 (−2.60) ***	0.3551 (4.80) ***	−0.6780 (−5.37) ***	−0.7263 (−5.36) ***
GDP	0.0479 (0.18)	−0.0144 (−0.05)	0.7258 (2.13) **	−0.0388 (−0.15)	−0.6135 (−1.43)	−0.6904 (−1.50)
IS	0.0102 (1.28)	0.0386 (4.17) ***	−0.0748 (−7.17) ***	0.0477 (6.35) ***	−0.1749 (−13.63) ***	−0.1742 (−12.64) ***
ROAD	0.5605 (2.96) ***	2.0982 (9.61) ***	−2.2168 (−9.05) ***	2.2278 (12.56) ***	−4.5712 (−15.10) ***	−4.7558 (−14.62) ***
FAI	0.0942 (0.54)	0.5841 (2.93) ***	−0.7858 (−3.55) ***	0.3174 (1.96) **	0.0172 (0.06)	0.0328 (0.11)
Estimation methods	FE	FE	RE	FE	FE	FE
R-squared	0.3073	0.5535	0.7279	0.6326	0.7068	0.6971
Observations	2964	2964	2964	2964	2964	2364

** denotes $p < 0.05$, *** denotes $p < 0.01$.

Table 6. The estimated results for western China.

	TA	NP	LPI	LSI	PLADJ	AI
P	−0.0122 (−2.43) **	−0.0569 (−2.40) **	0.1343 (1.60)	−0.0342 (−1.25)	0.1231 (1.14)	0.0315 (0.27)
GDP	0.1345 (6.62) ***	0.5865 (6.11) ***	−0.9665 (−2.71) ***	0.1847 (1.66) *	−0.6409 (−1.47)	−0.6070 (−1.29)
IS	0.0032 (3.78) ***	0.0142 (3.58) ***	−0.0926 (−6.26) ***	0.0313 (6.80) ***	−0.1578 (−8.72) ***	−0.1933 (−9.88) ***
ROAD	0.1731 (9.21) ***	0.8682 (9.78) ***	−2.1512 (−6.52) ***	1.6132 (15.69) ***	−2.8050 (−6.95) ***	−3.3333 (−7.64) ***
FAI	0.0416 (3.31) ***	0.0562 (0.95)	−0.1724 (0.78)	0.1008 (1.46)	−0.0085 (−0.03)	0.0411 (0.14)
Estimation methods	FE	FE	RE	FE	FE	FE
R-squared	0.9590	0.7428	0.8175	0.8318	0.7481	0.7202
Observations	1820	1820	1820	1820	1820	1820

* denotes $p < 0.1$, ** denotes $p < 0.05$, *** denotes $p < 0.01$.

Although China generally entered the middle stage of industrialization in 2010, given the vast territory of China, regional natural resources, economic foundations, and policy differences, different levels of socioeconomic development can be seen in different regions. For the purpose of measuring the associations between various factors and urban form in cities at different economic development levels, we also established three panel regression models to test the various effects of selected driving forces on urban form in cities at the primary and middle stages of industrialization, late stage of industrialization, and the developed stage.

Tables 7–9 review the estimated results with respect to the different economic development levels in China. They reveal that population was only statistically significant in relation to its effects on TA and NP for cities at the developed stage. Moreover, the influence of population on TA was the greatest in cities at the developed stage, indicating that with an increase in a city's level of development, population becomes more important in relation to urban expansion. GDP is shown to have exerted a positive influence on TA at all stages. Industrial structure was statistically significant for urban size at all stages and the influence of industrial structure on urban size was the highest in developed cities. The influence of ROAD on cities in the late stage of industrialization and within the developed stage seemed to be more significant than in cities at the primary and middle stages. This finding reveals that traffic has a greater impact on urban expansion in more developed periods. The coefficient of ROAD's influence on the size of urban areas was highest in cities in the late stage of industrialization, followed by cities in the developed stage, and was lowest in cities in the primary and middle stages, indicating that transportation development has the most significant influence on urban expansion in the later stages of industrialization. In addition, FAI was shown to exert significant effects on urban size at the developed stage. Moreover, for cities at the primary and middle stages, GDP was identified as the most influential factor in relation to urban expansion, while in cities experiencing a late industrialization or developed stage, the most influential factor was ROAD.

Table 7. The estimated results for cities at primary and middle industrialization stages.

	TA	NP	LPI	LSI	PLADJ	AI
P	0.0519	0.1964	−0.0552	0.1893	−0.2475	−0.2646
	(1.94) *	(1.85) *	(−0.63)	(2.34) **	(−1.70) *	(−1.72) *
GDP	1.3079	5.9918	−1.1411	3.4037	−6.1397	−6.1034
	(4.20) ***	(4.93) ***	(−1.13)	(3.65) ***	(−3.63) ***	(−3.40) ***
IS	0.0121	0.0583	−0.1137	0.0745	−0.2477	−0.2616
	(3.59) ***	(4.21) ***	(−10.18) ***	(7.12) ***	(−13.60) ***	(−13.50) ***
ROAD	0.0189	0.0927	−0.0493	0.0818	−0.1302	−0.1406
	(1.35)	(1.57)	(−1.05)	(1.84) *	(−1.70) *	(−1.73) *
FAI	−0.2174	−0.9677	−0.4742	−0.3930	0.9313	0.8029
	(−1.30)	(−1.45)	(−0.86)	(−0.77)	(1.03)	(0.83)
Estimation methods	FE	RE	RE	RE	FE	FE
R-squared	0.8353	0.5681	0.7961	0.6622	0.7415	0.7194
Observations	2236	2236	2236	2236	2236	2236

* denotes $p < 0.1$, ** denotes $p < 0.05$, *** denotes $p < 0.01$.

Table 8. The estimated results for cities at the late industrialization stage.

	TA	NP	LPI	LSI	PLADJ	AI
P	−0.0022	0.0115	0.0752	0.0488	−0.1280	−0.0916
	(−0.04)	(0.15)	(0.96)	(0.59)	(−1.29)	(−0.89)
GDP	0.2704	0.6933	−0.1573	0.3624	−1.2827	−1.5204
	(0.76)	(1.60)	(−0.34)	(0.65)	(−2.30) **	(−2.63) ***
IS	0.0041	0.0194	−0.0569	−0.0001	−0.1648	−0.1504
	(0.51)	(1.97) **	(−5.41) ***	(−0.00)	(−13.04) ***	(−11.45) ***
ROAD	0.8874	3.2804	−2.4576	2.7891	−4.8847	−5.0637
	(4.67) ***	(14.24) ***	(−9.99) ***	(9.26) ***	(−16.51) ***	(−16.47) ***
FAI	−0.0130	0.1442	−0.2200	0.0908	0.4140	0.5036
	(−0.07)	(0.62)	(−0.89)	(0.30)	(1.38)	(1.62)
Estimation methods	FE	FE	RE	RE	FE	FE
R−squared	0.3817	0.6680	0.7878	0.3591	0.6713	0.6588
Observations	2938	2938	2938	2938	2938	2938

** denotes $p < 0.05$, *** denotes $p < 0.01$.

The effect of the five socioeconomic factors on urban centrality also demonstrated heterogeneity in terms of the development level of the cities being studied. Our results indicate that the population variable brought about negative effects on LSI in cities at the primary and middle stages of industrialization as well as the developed stage, and had a positive influence on LSI in cities in the late stage. In addition, GDP did not appear to exert a significant effect on urban shape complexity in cities at the primary and middle stages of industrialization, or in cities at the late stage of industrialization; it only has a significant positive effect on LSI in the developed stage. The industrial structure was significantly correlated with LSI, revealing that industrial structure impacts urban centrality. ROAD was found to maintain a significant correlation with respect to LPI in late industrialized and developed cities, revealing that the influence of transportation on urban compactness increased advances in development stage. FAI exerted a negative impact on LPI in cities at all stages of development.

Table 9. The estimated results for cities at the developed stage.

	TA	NP	LPI	LSI	PLADJ	AI
P	0.1090 (3.68) ***	0.3705 (3.38) ***	−0.1530 (−1.90) *	0.0242 (0.13)	−0.4095 (−3.67) ***	−0.3972 (−3.79) ***
GDP	0.0475 (2.23) **	0.1346 (1.71) *	0.1971 (3.41) ***	−0.0561 (−0.42)	0.2253 (2.81) ***	0.2231 (2.97) ***
IS	0.0146 (3.77) ***	0.0626 (4.37) ***	−0.0572 (−5.43) ***	−0.0753 (−3.10) ***	−0.1888 (−12.89) ***	−0.1801 (−13.11) ***
ROAD	0.3496 (5.66) ***	1.3981 (6.13) ***	−0.5823 (−3.48) ***	0.8692 (2.26) **	−1.7035 (−7.32) ***	−1.8171 (−8.33) ***
FAI	0.1606 (6.27) ***	0.6429 (6.79) ***	−0.5604 (−8.06) ***	0.4789 (2.99) ***	−0.6426 (−6.65) ***	−0.6496 (−7.17) ***
Estimation methods	FE	FE	FE	FE	FE	FE
R-squared	0.8130	0.5687	0.8941	0.4175	0.8205	0.8399
Observations	2002	2002	2002	2002	2002	2002

* denotes $p < 0.1$, ** denotes $p < 0.05$, *** denotes $p < 0.01$.

The impact of the five socioeconomic factors on urban shape complexity also showed heterogeneity in relation to the development level of cities. Population showed significant positive effects at each stage of development. GDP showed positive effects in relation to LSI in cities within the primary and middle stages of industrialization, as well as in cities at a stage of late industrialization, and was found to exert a negative influence on LSI in developed cities. This indicates that the influence of GDP on urban shape complexity first increases and then decreases as development progresses. The industrial structure was observed to demonstrate positive correlations with respect to LSI in cities within the primary and middle stages of industrialization, and exerted a negative influence on LSI in late-industrialized cities and developed cities. The transportation variable, represented by ROAD, was positive in cities at all stages. The positive effects of ROAD on urban shape complexity can probably be attributed to interaction effects with other factors. In addition, FAI exerted positive effects on LSI in cities in the late industrialization stage and the developed stage, and had a negative influence on LSI in cities in the primary and middle stages of industrialization.

The impact of the five socioeconomic factors on urban compactness also demonstrated heterogeneity in terms of the development level of the cities being studied. Our results indicate that the population variable brought about negative effects on PLADJ and AI in cities at each stage of development. In addition, GDP had negative effects in relation to PLADJ and AI in cities at the primary and middle stages of industrialization as well as in cities at a stage of late industrialization and was observed to demonstrate positive correlations with respect to PLADJ and AI in developed cities. The industrial structure was significantly correlated with PLADJ and AI, revealing that industrial structure impacts urban compactness. ROAD was found to maintain a significant correlation with respect to PLADJ and AI in late-industrialized and developed cities, revealing that the influence of transportation on urban compactness increased advances in development stage. FAI did not appear to have a significant influence with respect to urban compactness in cities at the primary and middle stages of industrialization, or in cities at the late stage of industrialization or the developed stage.

4. Conclusions and Policy Implications

This paper aimed to measure the impacts of a range of selected potential determinants on urban form in 265 prefecture-level cities in China by considering regional disparities and variations in the stage of development evidenced by a given city. Landscape metrics were employed to quantitatively address urban form from the perspectives of urban size, urban centrality, urban shape irregularity, and urban fragmentation. We also selected five potential influencing factors that constituted the independent variables for this paper. A panel regression model was utilized to measure the association between the three urban form variables and the five driving forces.

The main results can be concluded as follows: first, significant differences existed in the urban form of cities in different regions at different times. All three of the urban form variables addressed in the study showed a significant stair-stepping difference in cities at various development stages. Second, urban form in China showed significant geographical heterogeneity in terms of the driving forces bringing about change. Most of the explanatory factors explored in this study presented significant regional differences. Population, industrial structure, and FAI were found to exert the greatest impact on urban size in eastern China, revealing that these factors have a greater impact on urban size in more developed regions. Third, five socioeconomic factors regarding urban form also had a different impact between cities at different stages of development. For example, GDP showed positive effects in relation to LSI in cities in the primary and middle stages of industrialization, as well as in cities at a stage of late industrialization, and was observed to demonstrated negative correlations with respect to LSI in developed cities. This indicates that the influence of GDP on urban shape complexity first increases and then decreases as development progresses.

In recent decades, urban areas in China have also undergone significant expansion, accompanied by considerable changes to the configuration of the urban landscape pattern. China is a vast territory, with large differences in development between various regions [55–58]. The results of this paper contain significant policy suggestions for the Chinese government. The variations that we found in the influence exerted by a range of factors on urban form reveal the extent of regional disparities and development differences within China. To achieve a sustainable urban form, future policymakers should consider differences in the developmental stage of a city, as well as regional differences, avoiding a "one size fits all" approaches to policy formulation. In addition, based on sound analyses of the determinants of urban form changes, decision-makers working with urban landscapes should be aware of the importance of accurately differentiating driving mechanisms. Regional differences and development stage changes need to be considered when formulating urban development policies in China. In particular, the results of this study suggest that population plays an important role in rapid urban expansion in eastern China and in cities at the developed stage. China should strengthen the policy of "population access", especially in the developed, large eastern cities, implementing strict population access in order to control disorderly and unplanned urban expansion. Combining population policy with urbanization policy can solve the problem of urban expansion management in China [59–61]. In addition, the ROAD variable was observed to have a significant impact on urban form, suggesting that urban expansion and morphological evolution are related to improved transportation infrastructure. Therefore, optimizing existing road networks and rationally planning road development may effectively guide the direction and speed of urban growth. Furthermore, the close connection between urban form and industrial structure shows that secondary industries in China's development heavily rely on resource inputs, and development cannot be achieved with land resources. Urgent acceleration in the upgrading and transformation of Chinese industry is thus required if the country is to go from relying on extensive resource use and labor-intensive growth methods to a capital-intensive, knowledge-intensive mode of growth.

Despite the valuable findings presented in this paper, some limitations must be acknowledged. Firstly, this study conceptualizes urban form in terms of urban expansion, urban shape irregularity, and urban compactness, represented by certain landscape metrics. However, urban form is a complex phenomenon that is shaped by a multitude of factors, extending beyond the aspects discussed in this paper. Moreover, while landscape metrics provide a useful tool for quantifying urban form, they do not represent the only possible measure. Secondly, the multitude of variables that influence urban form poses another limitation to this study. Although this research considers a number of key determinants, it cannot encompass all possible variables that might affect urban form [62–64]. Therefore, while the factors studied here are important, the complexity of urban form could be further influenced by additional factors not included in this study.

Future research in this field could explore alternative or additional metrics to quantify urban form, possibly capturing aspects not represented by the landscape metrics used in this study. Furthermore, given the multifaceted nature of urban form, studies that incorporate a wider range of influencing variables, particularly ones unique to different socio-cultural or geopolitical contexts, could enhance our understanding of urban form. Such endeavors will contribute to a more comprehensive understanding of urban form and its determinants, leading to more effective urban planning and management strategies.

Author Contributions: Conceptualization, R.W. and Y.Z.; methodology, R.W. and Y.Z.; software, M.D. and Q.L.; validation, M.D. and Q.L.; formal analysis, Y.Z.; investigation, C.S.; resources, C.S.; data curation, Q.L.; writing—original draft preparation, R.W. and Y.Z.; supervision, R.W.; project administration, R.W.; funding acquisition, R.W. All authors have read and agreed to the published version of the manuscript.

Funding: This work was supported by the National Natural Science Foundation of China (No. 42001147), Guangdong Province Natural Science Fund (No. 2022A1515011728).

Institutional Review Board Statement: Not applicable.

Informed Consent Statement: Not applicable.

Data Availability Statement: Data are contained within the article.

Conflicts of Interest: The authors declare no conflict of interest.

References

1. Zhang, Z.H.; Su, S.L.; Xiao, R.; Jiang, D.W.; Wu, J.P. Identifying determinants of urban growth from a multi-scale perspective: A case study of the urban agglomeration around Hangzhou Bay, China. *Appl. Geogr.* **2013**, *45*, 193–202. [CrossRef]
2. Zhang, Q.; Su, S. Determinants of urban expansion and their relative importance: A comparative analysis of 30 major metropolitans in China. *Habitat Int.* **2016**, *58*, 89–107. [CrossRef]
3. Ramachandra, T.V.; Aithal, B.H.; Sanna, D.D. Insights to urban dynamics through landscape spatial pattern analysis. *Int. J. Appl. Earth Obs. Geoinf.* **2012**, *18*, 329–343.
4. Seto, K.C.; Sanchez-Rodriguez, R.; Fragkias, M. The New Geography of Contemporary Urbanization and the Environment. *Annu. Rev. Environ. Resour.* **2010**, *35*, 167–194. [CrossRef]
5. Liu, J.Y.; Deng, X.Z. Impacts and mitigation of climate change on Chinese cities. *Curr. Opin. Environ. Sustain.* **2011**, *3*, 188–192. [CrossRef]
6. Pauchard, A.; Aguayo, M.; Pena, E.; Urrutia, R. Multiple effects of urbanization on the biodiversity of developing countries: The case of a fast-growing metropolitan area (Concepcion, Chile). *Biol. Conserv.* **2006**, *127*, 272–281. [CrossRef]
7. Schneider, A.; Chang, C.; Paulsen, K. The changing spatial form of cities in Western China. *Landsc. Urban Plan.* **2015**, *135*, 40–61. [CrossRef]
8. Duque, J.C.; Lozano-Gracia, N.; Patino, J.E.; Restrepo, P. Urban form and productivity: What shapes are Latin-American cities? *Environ. Plan. B Urban Anal. City Sci.* **2022**, *49*, 131–150. [CrossRef]
9. Grimm, N.B.; Faeth, S.H.; Golubiewski, N.E.; Redman, C.L.; Wu, J.; Bai, X.; Briggs, J.M. Global change and the ecology of cities. *Science* **2008**, *319*, 756–760. [CrossRef]
10. Tsai, Y.H. Quantifying urban form: Compactness versus 'Sprawl'. *Urban Stud.* **2005**, *42*, 141–161. [CrossRef]
11. Li, F.; Zhou, T. Effects of urban form on air quality in China: An analysis based on the spatial autoregressive model. *Cities* **2019**, *89*, 130–140. [CrossRef]
12. Zhang, C.Q.; Chen, P.Y. Industrialization, urbanization, and carbon emission efficiency of Yangtze River Economic Belt-empirical analysis based on stochastic frontier model. *Environ. Sci. Pollut. Res.* **2021**, *28*, 66914–66929. [CrossRef] [PubMed]
13. Shi, K.; Xu, T.; Li, Y.; Chen, Z.; Gong, W.; Wu, J.; Yu, B. Effects of urban forms on CO_2 emissions in China from a multi-perspective analysis. *J. Environ. Manag.* **2020**, *262*, 110300. [CrossRef]
14. Zhou, C.S.; Li, S.J.; Wang, S.J. Examining the Impacts of Urban Form on Air Pollution in Developing Countries: A Case Study of China's Megacities. *Int. J. Environ. Res. Public Health* **2018**, *15*, 1565. [CrossRef]
15. Schneider, A.; Woodcock, C.E. Compact, dispersed, fragmented, extensive? A comparison of urban growth in twenty-five global cities using remotely sensed data, pattern metrics and census information. *Urban Stud.* **2008**, *45*, 659–692. [CrossRef]
16. Wang, S.J.; Wang, J.Y.; Fang, C.L.; Li, S.J. Estimating the impacts of urban form on CO_2 emission efficiency in the Pearl River Delta, China. *Cities* **2019**, *85*, 117–129. [CrossRef]

17. Luo, J.; Wei, Y.H.D. Modeling spatial variations of urban growth patterns in Chinese cities: The case of Nanjing. *Landsc. Urban Plan.* **2009**, *91*, 51–64. [CrossRef]
18. Muller, K.; Steinmeier, C.; Kuchler, M. Urban growth along motorways in Switzerland. *Landsc. Urban Plan.* **2010**, *98*, 3–12. [CrossRef]
19. Vermeiren, K.; Van, R.A.; Loopmans, M.; Serwajja, E.; Mukwaya, P. Urban growth of Kampala, Uganda: Pattern analysis and scenario development. *Landsc. Urban Plan.* **2012**, *106*, 199–206. [CrossRef]
20. Braimoh, A.K.; Onishi, T. Spatial determinants of urban land use change in Lagos, Nigeria. *Land Use Policy* **2007**, *24*, 502–515. [CrossRef]
21. Dewan, A.M.; Yamaguchi, Y. Land use and land cover change in Greater Dhaka, Bangladesh: Using remote sensing to promote sustainable urbanization. *Appl. Geogr.* **2009**, *29*, 390–401. [CrossRef]
22. Hernández-Flores, M.D.L.L.; Otazo-Sánchez, E.M.; Galeana-Pizaña, M.; Roldán-Cruz, E.I.; Razo-Zárate, R.; González-Ramírez, C.A.; Galindo-Castillo, E.; Gordillo-Martínez, A.J. Urban driving forces and megacity expansion threats. Study case in the Mexico City periphery. *Habitat Int.* **2017**, *64*, 109–122. [CrossRef]
23. Liu, J.Y.; Zhan, J.Y.; Deng, X.Z. Spatio-temporal patterns and driving forces of urban land expansion in China during the economic reform era. *AMBIO J. Hum. Environ.* **2005**, *34*, 450–455. [CrossRef]
24. Ma, Y.L.; Xu, R.S. Remote sensing monitoring and driving force analysis of urban expansion in Guangzhou City, China. *Habitat Int.* **2010**, *34*, 228–235. [CrossRef]
25. Maimaitijiang, M.; Ghulam, A.; Sandoval, J.S.O.; Maimaitiyiming, M. Drivers of land cover and land use changes in St. Louis metropolitan area over the past 40 years characterized by remote sensing and census population data. *Int. J. Appl. Earth Obs. Geoinf.* **2015**, *35*, 161–174. [CrossRef]
26. Shao, Z.F.; Ding, L.; Li, D.R.; Altan, O.; Huq, M.E.; Li, C.M. Exploring the Relationship between Urbanization and Ecological Environment Using Remote Sensing Images and Statistical Data: A Case Study in the Yangtze River Delta, China. *Sustainability* **2020**, *12*, 5620. [CrossRef]
27. Aljoufie, M.; Brussel, M.; Zuidgeest, M.; van Maarseveen, M. Urban growth and transport infrastructure interaction in Jeddah between 1980 and 2007. *Int. J. Appl. Earth Obs. Geoinf.* **2013**, *21*, 493–505. [CrossRef]
28. Chen, J.F.; Chang, K.T.; Karacsonyi, D.; Zhang, X.L. Comparing urban land expansion and its driving factors in Shenzhen and Dongguan, China. *Habitat Int.* **2014**, *43*, 61–71. [CrossRef]
29. Jones, C.; Henderson, D. Broadband and uneven spatial development: The case of Cardiff City-Region. *Local Econ.* **2019**, *34*, 228–247. [CrossRef]
30. Feng, J.; Lichtenberg, E.; Ding, C.R. Balancing act: Economic incentives, administrative restrictions, and urban land expansion in China. *China Econ. Rev.* **2015**, *36*, 184–197. [CrossRef]
31. Xiao, J.; Shen, Y.; Ge, J.; Tateishi, R.; Tang, C.; Liang, Y.; Huang, Z. Evaluating urban expansion and land use change in Shijiazhuang, China, by using GIS and remote sensing. *Landsc. Urban Plan.* **2006**, *75*, 69–80. [CrossRef]
32. Ye, C.; Liu, Z.J.; Cai, W.B.; Chen, R.S.; Liu, L.L.; Cai, Y.L. Spatial Production and Governance of Urban Agglomeration in China 2000-2015: Yangtze River Delta as a Case. *Sustainability* **2019**, *11*, 1343. [CrossRef]
33. Li, H.; Wei, Y.D.; Liao, F.H.; Huang, Z.J. Administrative hierarchy and urban land expansion in transitional China. *Appl. Geogr.* **2015**, *56*, 177–186. [CrossRef]
34. Li, H.; Wei, Y.D.; Korinek, K. Modelling urban expansion in the transitional Greater Mekong Region. *Urban. Stud.* **2018**, *55*, 1729–1748. [CrossRef]
35. Wei, Y.D.; Ewing, R. Urban expansion, sprawl and inequality. *Landsc. Urban Plan.* **2018**, *177*, 259–265. [CrossRef]
36. Wei, Y.D.; Ye, X.Y. Urbanization, urban land expansion and environmental change in China. *Stoch. Environ. Res. Risk Assess.* **2014**, *28*, 757–765. [CrossRef]
37. Wei, W.; Wang, X.; Zhu, H.; Li, J.; Zhou, S.; Zou, Z. Carbon emissions of urban power grid in Jing-Jin-Ji region: Characteristics and influential factors. *J. Clean. Prod.* **2017**, *168*, 428–440. [CrossRef]
38. Li, X.M.; Zhou, W.Q.; Ouyang, Z.Y. Forty years of urban expansion in Beijing: What is the relative importance of physical, socioeconomic, and neighborhood factors? *Appl. Geogr.* **2013**, *38*, 1–10. [CrossRef]
39. Wu, W.J.; Zhao, S.Q.; Zhu, C.; Jiang, J.L. A comparative study of urban expansion in Beijing, Tianjin and Shijiazhuang over the past three decades. *Landsc. Urban Plan.* **2015**, *134*, 93–106. [CrossRef]
40. Sun, C.; Wu, Z.F.; Lv, Z.Q.; Yao, N.; Wei, J.B. Quantifying different types of urban growth and the change dynamic in Guangzhou using multi-temporal remote sensing data. *Int. J. Appl. Earth Obs. Geoinf.* **2013**, *21*, 409–417. [CrossRef]
41. Xiong, G.P.; Cao, X.; Hamm, N.A.S.; Lin, T.; Zhang, G.Q.; Chen, B.H. Unbalanced Development Characteristics and Driving Mechanisms of Regional Urban Spatial Form: A Case Study of Jiangsu Province, China. *Sustainability* **2021**, *13*, 3121. [CrossRef]
42. Haas, J.; Ban, Y.F. Urban growth and environmental impacts in Jing-Jin-Ji, the Yangtze, River Delta and the Pearl River Delta. *Int. J. Appl. Earth Obs. Geoinf.* **2014**, *30*, 42–55. [CrossRef]
43. Tian, G.J.; Jiang, J.; Yang, Z.F.; Zhang, Y.Q. The urban growth, size distribution and spatio-temporal dynamic pattern of the Yangtze River Delta megalopolitan region, China. *Ecol. Model.* **2011**, *222*, 865–878. [CrossRef]
44. Aspinall, R. Modelling land use change with generalized linear models—A multi-model analysis of change between 1860 and 2000 in Gallatin Valley, Montana. *J. Environ. Manag.* **2004**, *72*, 91–103. [CrossRef] [PubMed]

45. Fang, C.; Li, G.; Wang, S. Changing and Differentiated Urban Landscape in China: Spatiotemporal Patterns and Driving Forces. *Environ. Sci. Technol.* **2016**, *50*, 2217–2227. [CrossRef]
46. Li, G.; Sun, S.; Fang, C. The varying driving forces of urban expansion in China: Insights from a spatial-temporal analysis. *Landsc. Urban Plan.* **2018**, *174*, 63–77. [CrossRef]
47. Thapa, R.B.; Murayama, Y. Drivers of urban growth in the Kathmandu valley, Nepal: Examining the efficacy of the analytic hierarchy process. *Appl. Geogr.* **2010**, *30*, 70–83. [CrossRef]
48. Li, S.; Zhou, C.; Wang, S.; Gao, S.; Liu, Z. Spatial Heterogeneity in the Determinants of Urban Form: An Analysis of Chinese Cities with a GWR Approach. *Sustainability* **2019**, *11*, 479. [CrossRef]
49. Qi, Y.; Yang, Y.; Jin, F. China's economic development stage and its spatio-temporal evolution: A prefectural-level analysis. *J. Geogr. Sci.* **2013**, *23*, 297–314. [CrossRef]
50. Liu, X.; Huang, Y.; Xu, X.; Li, X.; Li, X.; Ciais, P.; Lin, P.; Gong, K.; Ziegler, A.D.; Chen, A.; et al. High-spatiotemporal-resolution mapping of global urban change from 1985 to 2015. *Nat. Sustain.* **2020**, *3*, 564–570. [CrossRef]
51. Seto, K.C.; Kaufmann, R.K. Modeling the drivers of urban land use change in the Pearl River Delta, China: Integrating remote sensing with socioeconomic data. *Land Econ.* **2003**, *79*, 106–121. [CrossRef]
52. Kuang, W.H.; Chi, W.F.; Lu, D.S.; Dou, Y.Y. A comparative analysis of megacity expansions in China and the U.S.: Patterns, rates and driving forces. *Landsc. Urban Plan.* **2014**, *132*, 121–135. [CrossRef]
53. Angel, S.; Parent, J.; Civco, D.L.; Blei, A.; Potere, D. The dimensions of global urban expansion: Estimates and projections for all countries, 2000–2050. *Prog. Plann.* **2011**, *75*, 53–107. [CrossRef]
54. Tian, L. Land use dynamics driven by rural industrialization and land finance in the pen-urban areas of China: "The examples of Jiangyin and Shunde". *Land Use Policy* **2015**, *45*, 117–127. [CrossRef]
55. Wu, R.; Wang, J.Y.; Wang, S.J.; Feng, K.S. The drivers of declining CO_2 emissions trends in developed nations using an extended STIRPAT model: A historical and prospective analysis. *Renew. Sust. Energ. Rev.* **2021**, *149*, 111328. [CrossRef]
56. Zhou, C.S.; Wang, S.J.; Feng, K.S. Examining the socioeconomic determinants of CO_2 emissions in China: A historical and prospective analysis. *Resour Conserv Recycl.* **2018**, *130*, 1–11. [CrossRef]
57. Wang, S.J.; Huang, Y.Y.; Zhou, Y.Q. Spatial spillover effect and driving forces of carbon emission intensity at the city level in China. *J. Geogr. Sci.* **2019**, *29*, 231–252. [CrossRef]
58. Wang, S.J.; Cui, Z.T.; Lin, J.J.; Xie, J.Y.; Su, K. The coupling relationship between urbanization and ecological resilience in the Pearl River Delta. *J. Geogr. Sci.* **2022**, *32*, 44–64. [CrossRef]
59. Xie, Z.H.; Wu, R.; Wang, S.J. How technological progress affects the carbon emission efficiency? Evidence from national panel quantile regression. *J. Clean. Prod.* **2021**, *307*, 127133. [CrossRef]
60. Wang, S.J.; Zeng, J.Y.; Liu, X.P. Examining the multiple impacts of technological progress on CO_2 emissions in China: A panel quantile regression approach. *Renew. Sust. Energ. Rev.* **2019**, *103*, 140–150. [CrossRef]
61. Wang, S.J.; Shi, C.Y.; Fang, C.L.; Feng, K.S. Examining the spatial variations of determinants of energy-related CO_2 emissions in China at the city level using Geographically Weighted Regression Model. *Appl. Energy* **2019**, *235*, 95–105. [CrossRef]
62. Wang, S.J.; Liu, X.P.; Zhou, C.S.; Hu, J.C.; Ou, J.P. Examining the impacts of socioeconomic factors, urban form, and transportation networks on CO_2 emissions in China's megacities. *Appl. Energy* **2017**, *185*, 189–200. [CrossRef]
63. Fang, C.L.; Wang, S.J.; Li, G.D. Changing urban forms and carbon dioxide emissions in China: A case study of 30 provincial capital cities. *Appl. Energy* **2015**, *158*, 519–531. [CrossRef]
64. Wang, S.J.; Fang, C.L.; Guan, X.L.; Pang, B.; Ma, H.T. Urbanisation, energy consumption, and carbon dioxide emissions in China: A panel data analysis of China's provinces. *Appl. Energy* **2014**, *136*, 738–749. [CrossRef]

land

MDPI

Article

How Does Intensive Land Use Affect Low-Carbon Transition in China? New Evidence from the Spatial Econometric Analysis

Xiao Ling [1], Yue Gao [2] and Guoyong Wu [3,*]

[1] Business School, Hubei University, Wuhan 430062, China; lingxiao@stu.hubu.edu.cn
[2] Business School, Zhengzhou University, Zhengzhou 450001, China; gaoyue980913@163.com
[3] Western Modernization Research Institute, Guizhou University, Guiyang 550025, China
* Correspondence: wgyong@126.com

Abstract: Anthropogenic land cover change is one of the primary sources of increasing carbon emissions and affects the potential of terrestrial ecosystems to store carbon and act as carbon sinks. As a necessary means to reduce land expansion, land-use intensification significantly impacts greenhouse gas emission reduction and the low-carbon transition of the economy. This paper constructs a framework for the relationship between intensive land use (*ILU*) and low carbon transition (*LCT*), considering direct and spatially driven effects. First, this paper constructs a multidimensional indicator to measure intensive land use and documents the spatial pattern of intensive land use levels in China. Second, this paper assesses the spatial driving effect of intensive land use on China's economic low-carbon transition. Based on data from 283 Chinese cities from 2006–2019 and using a spatial Durbin model, the study provides empirical evidence that intensive land use can significantly promote low-carbon transition in neighboring and economically linked cities (especially in eastern cities, large and medium-sized cities, and veteran economic circles). Tests introducing exogenous policy shocks further confirm the robustness of the findings. In addition, industrial structure transformation and technology spillovers are identified as the dual mechanism channels of intensive land use for low-carbon transition in China, and the spatial driving effect on neighboring cities attenuating with geographic distance is also confirmed.

Keywords: intensive land use; low carbon transformation; industrial structure transformation; technology spillovers; national and regional policy; land space planning

1. Introduction

Global warming caused by greenhouse gas (GHG) emissions seriously threatens the natural and social environments on which human beings depend for survival [1,2]. The series of chain reactions across ecosystems triggered by greenhouse gases has become a massive challenge for all humanity [3,4]. The International Energy Agency (IEA) estimates that global energy-related carbon dioxide (CO_2) emissions will grow by 0.9% in 2022, reaching a record high of over 36.8 Gt [5]. Among them, carbon dioxide emissions from energy combustion and industrial processes account for 89% of total energy-related greenhouse gas emissions; methane from energy combustion, leakage, and venting accounts for 10%. They are all mainly from onshore oil and gas field operations and the production of coal for power. Compared to 1880, 2022 is also the fifth hottest year globally, fraught with extreme weather events [6,7]. As the country with the most rapid economic development in the 20th century, China has become the world's largest emitter of carbon dioxide since 2007, with carbon emissions rising from 8.83 billion tons in 2011 to 9.90 billion tons in 2020 [8]. As the urbanization rate of the population rises (to 64.72% in 2021), large-scale migration and the concentration of human activities will result in continued land expansion and land carbon emissions. China is actively taking on the corresponding obligations to mitigate global warming, noting that it should effectively promote carbon peaking and carbon-neutral actions. In this context, China urgently needs to find breakthrough solutions to accelerate economic activities' low-carbon transition to control the increasingly severe climate problem [9].

Since the 19th century, land use has influenced terrestrial ecosystem carbon balance through changes in land cover status and the human activities it hosts [10]. It has been recognized as an essential factor influencing regional carbon source/sink patterns [11]. From 1750 to 2011, an estimated 180 Gt has been released globally due to deforestation and other land use changes [12]; more than 66% of energy and 80% of carbon emissions may be related to the scale and productivity of land use [13], and the loss of carbon stocks in terrestrial ecosystems due to the occupation of forest resources worldwide is the second largest source of carbon emissions [14]. The loss of ecosystem carbon stocks will be exacerbated by unintentional land expansion, and the overconcentration of human activities brought about by expansion will also generate high consumption and emissions [15]. As the world's largest carbon emitter, land use carbon emissions have become an essential source of carbon emissions in China, reaching 3.2×10^9 t in 2015, an increase of about 2.45 times compared with 1999. As of 2020, China's land-use carbon emissions will remain high (see Figure 1). Assuming that the 1.5 °C global temperature control target of the Paris Agreement is to be achieved, further attention needs to be paid to the critical role of intensive land use in the low-carbon transition of the economy. China is implementing policy elements of intensive land utilization to promote a low-carbon transition in economic development. In particular, since the promulgation of the Regulations on the Economical and Intensive Utilization of Land in 2014, land regulation for carbon emission reduction has become an important means of promoting grassroots efforts to achieve carbon neutrality targets. The regulation encourages small-scale centralized and intensive land use and emphasizes green and livable land use [16]. Zhao (2021) assessed the carbon emission reduction contribution of the Outline of China's Overall Land Use Plan (2005–2020). Based on 2005, optimizing the land use structure will contribute 27.6% to the achievement of the target of carbon emission reduction of 40% to 45% per unit of GDP in 2020 [17]. In the context of the country's emphasis on coordinated "economic-ecological" development, intensive land use is crucial for China to achieve the goal of "carbon neutrality" and low-carbon transition.

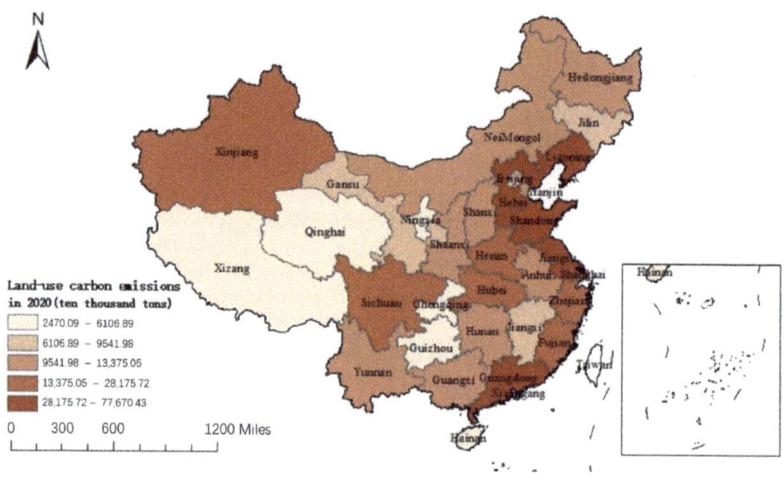

Figure 1. Land use carbon emissions for 31 provinces in China in 2020.

In China, fiscal revenue from land concessions has long been local governments' primary income source. To solve the fiscal balance gap, some local governments have been keen to attract industrial investment by taking advantage of their resource endowment and geographic location [18], dramatically expanding industrial scale while promoting development. This process has not only resulted in massive waste of urban land and rapid urbanization but also led to differences in the spatial pattern of land use carbon emissions [19]. With China's coordinated economic development strategy deepening, the flow of technology, personnel, capital, and other factors between regions has further accelerated. The spatial correlation of economic development, energy consumption, and agricultural activities has broken through the limitations of geographic location. The spatial correlation of land-use carbon emissions will also be further complicated. Under the spatial differences in population distribution and economic resources, the differences in land carbon emissions of each province in China are apparent; carbon emissions are high in the eastern coastal areas (Figure 1). In this context, it is significant to carry out a study on the spatial differences in carbon emissions from land use for the synergistic emission reduction of regional land use.

As land is essential for population, industry, transportation, buildings, and energy use, can improved intensive land use help promote China's low-carbon transition? Moreover, in what ways does it achieve "economic-ecological" synergistic development? Given China's goal of achieving carbon neutrality by 2050, these questions' conceptual and applied consequences are critical to assess. Existing studies have intensely discussed the relationship between land use and carbon emissions. First, some scholars have expressed the hidden concern that land expansion may lead to increased carbon emissions in earlier studies [20,21], arguing that these crude features are an essential cause of higher environmental costs and unsustainable problems [22,23]. Some scholars have also tested the relationship between land use structure and carbon emissions [24]. It is pointed out that the structural imbalance of ecological land, agricultural land, and urban land [25], as well as the increase in the proportion of urban built-up areas [26,27], will lead to a more obvious greenhouse effect. Then, some scholars explored the environmental benefits of optimizing land use. For example, Xie et al. (2018) pointed out that improving industrial land use efficiency at the national level contributes 37.52% to the total CO_2 emission reduction, in which R&D investment in intensive land use is the most effective way to promote emission reduction [28]. Additionally, Goh et al. (2018) concluded that decarbonizing land use can be an effective method of reducing carbon emissions [29]. These conclusions are also recognized by scholars such as Peng et al. (2022) [30] and Zhang et al. (2023) [31]. However, some scholars, such as Zhu et al. (2022), believe that China's current land use optimization has caused a greater degree of carbon emissions [32]. Scholars have only analyzed the impact of optimized land use on the intensity and efficiency of carbon emissions in isolation, and their views have not yet reached a consensus. Therefore, despite the intense academic discussion and research on these issues so far, there still exists a vast research space, such as the characteristics of spatial distribution, direct and spillover effects, and impact mechanisms, which constitute the initial motivation of this study.

Based on the consideration of breaking through the limitations of the existing literature, we decided to identify and assess the driving effect of land intensification on China's low-carbon transition from a spatial perspective and evaluate the mechanism of its action in terms of both green upgrading of industries and clean technology spillovers (Figure 2). The possible marginal contribution consists of the following three points. Firstly, we constructed a framework for the relationship between land use intensification and low carbon transition considering both direct and spatial driving effects and numericized land use intensification and low carbon transition in the form of multiple composite indicators. Compared with traditional studies, we provide a more comprehensive analysis from a spatiotemporal perspective (i.e., spatial distribution, spatial autocorrelation, evolution, spillover effects, spatial decay, and spatial heterogeneity). Secondly, compared with the traditional single research method, we adopt the exogenous policy shock test to support the conclusion of the driving effect and examine the dual channels of influence of land intensification and low-carbon transition from the dual perspectives of green transformation of industrial structure and clean technology spillovers, which expands the empirical research in related fields. Finally, we provide practical policy recommendations for policymakers regarding the efficiency of low-carbon economies and the focus on green, livable, and efficient living and production environments in emerging countries such as China.

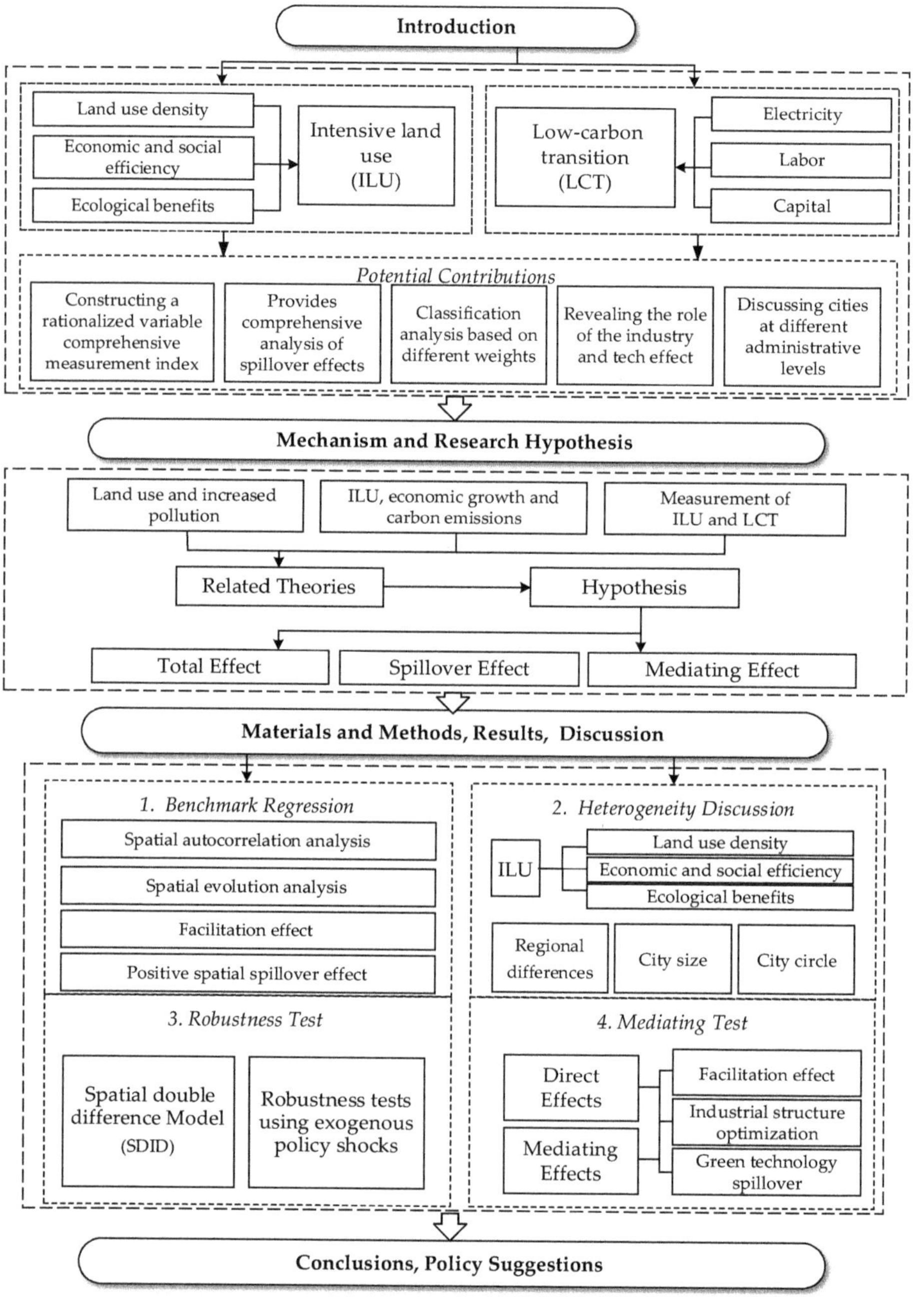

Figure 2. Outline of the research framework.

2. Mechanism and Research Hypothesis

ILU is a critical way to build an urban ecological civilization, focusing on conservation, efficiency, and ecology. Economic green transformation is the primary way to realize the goal of sustainable development, focusing on economic efficiency and carbon reduction [33]. Compared with traditional land use, *ILU* embodies the concept of scientific and clean development. Whether it is to reduce emission sources or increase carbon sink absorption, *ILU* plays an important role [34]. The mechanism between *ILU* and China's low-carbon transition can be analyzed in economic and ecological systems.

2.1. Promoting Effect of Intensive Land Use on Low-Carbon Transition

First, we consider the economic growth effect. *ILU* is to increase the input of factors such as capital, labor, and technology on the urban land stock and improve land use efficiency through rational layout and optimization of land use structure to promote sustainable development [35]. According to the law of increasing and decreasing land remuneration, before reaching the highest point of remuneration, the more capital and labor force invested in the land per unit area, the higher the economic output obtained, that is, the higher the intensity of land use and development, the higher the contribution to economic growth [36].

Second, we consider the carbon emission reduction effect. At the ecosystem level, *ILU* corresponds to the impact of land use change on soil carbon stock and vegetation carbon stock. According to the land use classification, construction land is the primary carbon source, while ecological and agricultural land are essential sources of carbon sinks. *ILU* effectively reduces the conversion of agricultural and ecological land, such as garden land, forest land, and grassland, to construction land and increases carbon sink absorption within the ecosystem [37]. In addition, at the economic system level, with the strengthening of land use constraints, low energy-consuming technologies, enterprises, and industrial chains will be "squeezed out", thus reducing carbon emissions. Compact land space pattern facilitates public transportation use and reduces infrastructure construction waste [38]. It helps to improve the efficiency of centralized energy supply and utilization and reduces the growth of carbon sources from land construction.

2.2. Spatial Spillover Mechanism

According to the theory of agglomeration effect, the increased density and spatial proximity of economic activities on land contribute to the economies of scale in production and transactions and resulting spillovers on a local scale [39]. Considering that carbon emissions are mainly influenced by socioeconomic drivers such as the stage of economic development, energy resource endowment, and consumption patterns. Land use mode, scale, structure, and intensity are closely related to industrial development status and technological progress [40]. Therefore, intensive land use affects the low-carbon transformation of neighboring cities mainly by influencing the socioeconomic constitutive factors reflected in industrial development and technological progress [41,42].

The mediating role of industrial structure transformation. On the one hand, as the land use pattern constrains the industrial layout, the crude land approach is challenging to promote the automatic transformation of industry. The theoretical connotation of *ILU* includes the principles of "structural optimization" and "market allocation." Conversely, *ILU* encourages the development of resource-saving and environment-friendly industries, which helps cities establish a modernized industrial system characterized by low energy consumption and emissions [40]. On the other hand, due to market competition and the price mechanism, industrial agglomeration forces enterprises to improve production technology and techniques and promotes upgrading the industrial chain layout [43]. As a result, with the gradual withdrawal of resource-intensive enterprises, the utility of energy utilization has been enhanced. The booming development of high-tech enterprises and service industries is conducive to easing resource dependence and environmental pressure. This transformation of industrial structure promotes the diversified division of labor among enterprises, which helps the low-carbon production links and industrial clusters among cities to generate symbiosis, complement each other, and promote the low-carbon transformation and development of neighboring cities [44].

The intermediary role of technology spillover. The agglomeration economy formed by *ILU* makes cities accumulate innovative resources and also promotes technological overflow from cities. The theory of *ILU* includes the principles of "conservation first" and "reform and innovation," which help accelerate the R&D, innovation, and application of production, environmental protection, and energy-saving technologies and generate technological factor overflows from the region to neighboring or related regions. The overflow accelerates the learning and disseminating of green technologies and innovation among cities. Neighboring cities at this time can realize imitation and secondary innovation with the help of technological overflow, thus promoting green transformation.

Overall, spatial spillovers from economic systems and ecosystems diffuse the effects of NTUs on local industrial restructuring and technological spillovers to neighboring regions, which ultimately manifests itself in the diffusion of local *ILUs*' driving effects on neighboring regions' *LCT*. The above spillover mechanisms accumulate layer by layer, forming the total effect of *ILU* on the overall regional *LCT*.

Hypothesis (H1). *Intensive land use drives low-carbon transition and can promote neighboring cities to achieve low-carbon transition goals.*

Hypothesis (H2). *Intensive land use can promote low-carbon transition through industrial structure transformation and technology spillover.*

2.3. Spatial Decay Mechanism

Most studies recognize that spillover effects are characterized by spatial decay. It is because the cost of logistics and information exchange will rise with increased geographical distance and the restriction of administrative boundaries. As a result, the spatial spillover effect of *ILU* may show a specific attenuation pattern and boundary effect as the possibility of factor spillover decreases.

First, we consider the role of geographic distance. Studies have confirmed that information dissemination shows the law of attenuation with increased geographical distance. As the distance between neighboring cities and knowledge-center cities increases, the efficiency of information dissemination decreases [45]. Furthermore, increased geographic distance leads to interfirm transaction costs, transportation costs, and risk control. This former is a function of the geographic distance between the cooperating parties [46,47], and the latter is reflected in the increased risk of breach of contract and the reduced level of trust, among others. These are not conducive to inter-city technology, exchange, and cooperation, leading to spatial limitations in the spillover effects of *ILU* on low-carbon transitions in neighboring cities.

Second, we consider the role of administrative boundaries. Local governments in China can deeply participate in the urban economic growth model, which provides many administrative and financial resources needed to promote urban development. However, this model may cause local governments to ignore the big picture of policies and hinder the joint promotion of inter-city policies. Specifically, local governments may impose explicit or implicit administrative restrictions on factor mobility to protect local markets. Without an effective inter-regional coordination mechanism, it is not easy to realize the effective interconnection of cross-regional infrastructure. It will not be conducive to the optimal allocation of resources and the dissemination of advanced technologies, hindering interregional cooperation in production links and industrial synergistic development and causing the spatial spillover of *ILU* to form a specific border effect.

Hypothesis (H3). *Influenced by geographic distance and administrative boundaries, there is spatial attenuation in the spillover effect of intensive land use on low-carbon transition.*

3. Materials and Methods

3.1. Data Sources

This paper takes Chinese prefecture-level cities as the research object. After deleting the samples with missing severe data, we selected 283 prefecture-level cities in China (except Tibet, Hong Kong, Macao, and Taiwan) from 2006 to 2019 as the research sample. Variable data were obtained from *China Urban Statistical Yearbook* (2007–2020), *China Statistical Yearbook* (2007–2020), and statistics published by the National Bureau of Statistics. Green patent authorizations were obtained from the China Research Data Service Platform (CNRDS). In addition, we verified all data obtained to ensure accuracy, and some missing values were supplemented by interpolation. Some key variables were logarithmized to ensure consistent statistical caliber.

3.2. Research Method

3.2.1. Spatial Autocorrelation Test

We refer to the study of Elhorst (2014) [48] for the first step of testing the applicability of the spatial measurement model to determine the spatial correlation of the variables. We choose *Moran's I* index, which is more common in the existing literature, to determine whether the spatial autocorrelation of low-carbon transition exists in each city. In addition, we also chose to use *Geary's C* index to conduct the spatial autocorrelation test from a global perspective. It is more sensitive than *Moran's I* index in the localized test (a value less than 1 indicates a positive correlation).

$$Moran's\ I = \sum_{i=1}^{n}\sum_{j=1}^{n} W_{ij}(X_i - \overline{X})(X_j - \overline{X}) / S^2 \sum_{i=1}^{n}\sum_{j=1}^{n} W_{ij} \tag{1}$$

$$\begin{gathered} Geary's\ C = (n-1)\sum_{i=1}^{n}\sum_{j=1}^{n} W_{ij}(X_i - X_j)^2 / 2\left(\sum_{i=1}^{n}\sum_{j=1}^{n} W_{ij}\right)\left[\sum_{i=1}^{n}(X_i - \overline{X})^2\right] \\ S^2 = \sum_{i=1}^{n}(X_i - \overline{X})^2/n;\quad \overline{X} = \frac{1}{n}\sum_{i=1}^{n} X_i \end{gathered} \tag{2}$$

where X_i and X_j are the actual observed values of regions "i" and "j", respectively. W_{ij} is the spatial weight matrix. n represents the total number of geography units that refers to the sample cities.

3.2.2. Spatial Econometric Models

First, we construct neighborhood spatial weights (W_{adj}), inverse distance spatial weight matrix (W_{dis}), economic spatial weight matrix (W_{econ}), and economic distance spatial weight matrix ($W_{econdis}$) from two perspectives: geographic distance and economic distance, respectively. For the geographic distance spatial weight matrix, the closer the distance, the greater the influence of the neighborhood. For the economic distance spatial weight matrix, cities with higher economic levels significantly influence the neighborhood more than cities with lower economies.

$$w_{adj} = \begin{cases} 0 & (\text{urban area } i,\ j \text{ are not adjacent}) \\ 1 & (\text{urban area } i,\ j \text{ are adjacent}) \end{cases} \tag{3}$$

$$w_{dis} = \begin{cases} 0 & (i \neq j) \\ \frac{1}{d_{ij}} & (i = j), \end{cases} d_{ij} \text{ is the distance between two urban centers} \tag{4}$$

$$w_{econ} = \begin{cases} 0 & (i \neq j) \\ \frac{1}{|X_i - X_j|} & (i = j), \end{cases} X \text{ is the economic aggregate, measured using GDP} \tag{5}$$

$$w_{econdis} = aW_{dis} + bW_{econ},\quad a = b = 0.5 \tag{6}$$

Spatial econometric model was used to identify and verify the spatial spillover effects of *ILU* on low-carbon transition. The spatial econometric model is constructed as follows:

$$\begin{aligned} LCT_{it} &= \alpha + \rho \sum_{j=1,j\neq i}^{N} \omega_{ij} LCT_{jt} + \beta X_{it} + \theta \sum_{j=1}^{N} \omega_{ij} X_{ijt} + \varphi_i + v_t + \varepsilon_{it} \\ \varepsilon_{it} &= \psi \sum_{j=1,j\neq i}^{N} \omega_{ij} \varepsilon_{jt} + \mu_{it}; \mu_{it} \sim N(0, \sigma^2 I) \end{aligned} \tag{7}$$

where *LCT* denotes the urban low-carbon transition variable; *X* is each explanatory variable, including *ILU*; *w* is the spatial weight matrix; ρ is the coefficient of the effect of the local explanatory variable on the explanatory variables in other urban areas. When $\rho > 0$, it indicates the spatial spillover effect between adjacent regions; when $\rho < 0$, it indicates the spatial negative effect between adjacent regions. β, θ are the parameter estimates of *X* explanatory variables; φ and v denote the area effect and time effect, respectively; ψ denotes the spatial correlation between residuals; ε is the random error term; *i*, *t* are the area individual and time dimensions.

In Equation (7), if $\rho = 0, \theta = 0, \psi = 0$, then Equation (7) is the spatial lag model (SLM), if $\rho = 0, \theta = 0, \psi \neq 0$, then it is the spatial error model (SEM), if $\rho \neq 0, \theta \neq 0, \psi = 0$, then it is the spatial Durbin model (SDM). In this paper, the Wald test and LR test are used to screen SLM, SEM, and SDM. If the tests both reject the original hypothesis of setting H_0: $\theta = 0$ and H_0: $\theta + \rho\beta = 0$, SDM is selected, and one of the original hypotheses is accepted, then the choice is made between SLM and SEM.

When the spatial Durbin model (SDM) was selected for the model, the direct, indirect, and total effects proposed by LeSage & Pace (2010) were used to further examine the spatial effects of the impact of *ILU* on urban low-carbon transition [49]. Using Y_t to represent the *LCT* vector, the SDM model is rewritten into the following vector form.

$$Y_t = (1-\rho W)^{-1}(\beta X_t + \theta W X_t) + (1-\rho W)^{-1}\mu_t \tag{8}$$

In Equation (8), we derive the partial differential matrix by taking the *k* explanatory variable as the independent variable. The mean of the diagonal elements represents the average effect of the change in the explanatory variables on the explanatory variables of local areas, which is the direct effect. The mean of the non-diagonal elements represents the average effect of the change in the explanatory variables on the variables of the explanatory variables in other urban areas, which is the indirect effect.

$$\left[\frac{\partial Y}{\partial X_{1k}} \cdots \frac{\partial Y}{\partial X_{Nk}}\right]_t = (1-\rho W)^{-1} \begin{bmatrix} \beta_k & \omega_{12} v_k & \cdots & \omega_{1N} v_k \\ \omega_{21} v_k & \beta_k & \cdots & \omega_{2N} v_k \\ \vdots & \vdots & \ddots & \vdots \\ \omega_{N1} v_k & \omega_{N2} v_k & \cdots & \beta_k \end{bmatrix} \tag{9}$$

3.2.3. Spatial Difference-in-Differences Model

According to the guideline of the Ministry of Land and Resources of China, the economic and intensive use of land is a strategic choice for new urbanization. Land-use intensification is an important policy tool in promoting the implementation of new urbanization. Therefore, this paper considers the new urbanization contention policy a proxy variable for *ILU*. In this study, we combine the difference-in-differences model (DID) with the spatial econometric model and relax the original assumption that the experimental group will not affect individuals in the control group. As a result, we construct the spatial difference-in-differences model (SDID) as follows:

$$CEE_{it} = \gamma_0 + \rho_1 \sum_{j=1,j\neq i}^{N} \omega_{ij} CEE_{jt} + \gamma_1 policy_{it} + \theta_1 \sum_{j=1}^{N} \omega_{ij} policy_{ijt} + \gamma_2 Control_{it} + \theta_2 \sum_{j=1}^{N} \omega_{ij} Control_{ijt} + \varphi_i + v_t + \varepsilon_{it} \tag{10}$$

where *policy* denotes the exogenous policy shock of "new urbanization"; γ_1 denotes the impact coefficient of policy on local low-carbon economic transition; θ_1 denotes the estimated coefficient of policy on low-carbon economic transition in neighboring areas; the rest of variables are explained in the same way as Equation (7).

3.3. Research Method

3.3.1. Explained Variable

Low-carbon transition (*LCT*) is the target explanatory variable. Considering the low-carbon transition (*LCT*) should balance carbon dioxide emission reduction and economic development [50], which can obtain the maximum economic output with the least factor input and the lowest carbon emission. Therefore, this paper uses the efficiency tool to represent the urban low-carbon transition. Stochastic frontier analysis utilizing the "input-output" paradigm can effectively deal with efficiency issues. In this study, specific indicators are selected with reference to existing studies [51]. Among them, input indicators include energy, labor, and capital, using the urban electricity consumption to measure the electricity input, the total urban employment to measure the labor input [52], and the urban capital stock based on the perpetual inventory method to measure the capital input [53]. Output indicators include desired output and non-desired output, and the urban GDP is used to measure the desired output of the city. The indicator system is shown in Table 1.

Table 1. Indicator system of low-carbon transition.

Indicator	Variable	Description
Input	Electricity	Urban total social electricity consumption (Unit: 10,000 kw-h)
	Labor	Total employment in the city (Unit: 10,000 persons)
	Capital	Urban capital deposit (Unit: 10,000 yuan)
Desired output	Economic efficiency output	Urban GDP (Unit: 10,000 yuan)
Non-desired output	Carbon dioxide emissions	Urban CO_2 emissions (Unit: 10,000 tons)

We also refer to the IPCC 2006 methodology to measure the undesired urban output using the carbon emissions generated during the consumption of natural gas, liquefied petroleum gas, electricity, and thermal energy for the whole city community [54]. Combined with the methodology of Tone and Tsutsui (2010) [55], we measure *CCE* using an excess efficiency model (EBM).

$$r^* = \min\left(\theta - \varepsilon^- \sum_{i=1}^{m} \frac{\omega_i^- s_i^-}{x_{i0}}\right) / \left[\varphi + \varepsilon^+ \left(\sum_{r=1}^{s} \frac{\omega_r^+ s_r^+}{y_{r0}} + \sum_{p=1}^{q} \frac{\omega_p^{u-} s_p^{u-}}{u_{p0}}\right)\right] \tag{11}$$

$$s.t.\begin{cases} \sum_{j=1}^{n} x_{ij}\lambda_j + s_i^- = \theta x_{i0} & (i = 1, 2 \ldots, m) \\ \sum_{j=1}^{n} y_{rj}\lambda_j - s_r^+ = \theta y_{r0} & (r = 1, 2 \ldots, s) \\ \sum_{j=1}^{n} u_{pj}\lambda_j + s_p^- = \theta u_{p0} & (p = 1, 2 \ldots, q) \\ \lambda_j \geq 0; s_i^-, s_r^+, s_p^- \geq 0 \end{cases} \tag{12}$$

In Equation (11), r^* represents the optimal efficiency value of the *LCT* measured by the super total factor productivity model. There are $m + 1$ parameters in this model. x, y and u represent the inputs, expected outputs, and unexpected outputs of DMU0, respectively, θ is the radial efficiency value; s represents the input slack vector. ε is a core parameter that determines the importance of the non-radial part of the computation of the efficiency value of r^*, and it takes the value in the range of [0, 1]. When $\varepsilon = 0$, it is equivalent to the radial model, and when taking the value of 1, it is equivalent to the SBM model. In Equation (12), λ denotes the weight coefficient.

3.3.2. Core Explanatory Variables

According to the different construction bases and standards of the evaluation index system, the evaluation index system mainly includes the "input-output", the "economic-social-ecological", and the "intensive-efficient-coordinated" index system. In selecting indicators, many scholars believe that the number of indicators is lower than possible but should be reasonably screened. Evaluation methods mainly include multiple single-indicator methods, factor synthesis evaluation methods, fuzzy synthesis evaluation methods, ideal value correction models, etc. Considering that intensive land use is a dynamic process, the degree of intensification can be effectively enhanced by improving the land use intensity and optimizing the land output efficiency and land use structure within a certain period. Based on the principle of dominant factors and local conditions, this paper selects indicators according to the evaluation system of "economy-society-ecology." Specific indicators and their measurement methods are shown in Table 2.

Table 2. Indicator system of intensive land use.

Primary Indicators	Description (Unit)	Indicator Attributes
Land use density	Built-up area/total urban area (%)	Negative
	Capital stock/built-up area (yuan/KM^2)	Positive
	Road area/built-up area (m^2/KM^2)	Positive
	House area/built-up area (m^2/KM^2)	Positive
	Urban population/built-up area (10,000 people/KM^2)	Positive
	Employment/built-up area (person/KM^2)	Positive
Economic and social efficiency	GDP/built-up area (yuan/KM^2)	Positive
	Non-agricultural industry output value/built-up area (yuan/KM^2)	Positive
	General income of fiscal budget/built-up area (yuan/KM^2)	Positive
	Disposable income of urban residents/built-up area (yuan/KM^2)	Positive
	Built-up area/urban resident population (m^2/person)	Negative
Ecological benefits	Per capita green area (m^2/person)	Positive
	Green coverage rate of urban built-up areas (%)	Positive
	Industrial sewage discharge per capita (ton/person)	Negative

In this paper, the TOPSIS model is chosen to measure the *ILU* composite indicators. TOPSIS model is an effective method in multi-objective decision analysis. It is a ranking method close to the ideal solution, which ranks the indicators by detecting the distance between the evaluation object and the optimal solution and the worst solution. In the calculation process, it is necessary to normalize the positive and negative indicators separately.

$$C_i = \sqrt{\sum_{j=1}^{m} (Z_{\min j} - Z_{ij})^2} / (\sqrt{\sum_{j=1}^{m} (Z_{\max j} - Z_{ij})^2} + \sqrt{\sum_{j=1}^{m} (Z_{\min j} - Z_{ij})^2}) \tag{13}$$

$$\begin{aligned} &Z_{ij} = X_{ij} / \sqrt{\sum_{i=1}^{n} X_{ij}^2} \quad or \quad Z_{ij} = \tfrac{1}{X_{ij}} / \sqrt{\sum_{i=1}^{n} \left(\tfrac{1}{X_{ij}}\right)^2} \\ &Z^+ = (Z_{\max 1} Z_{\max 2} Z_{\max 3} \ldots Z_{\max m}) \\ &Z^- = (Z_{\min 1} Z_{\min 2} Z_{\min 3} \ldots Z_{\min m}) \end{aligned} \tag{14}$$

where n is the number of cities participating in the evaluation; m is the number of evaluation indicators. C_i denotes the proximity of evaluation object i to the optimal solution, and finally, the comprehensive evaluation results of *ILU* in Chinese cities are obtained by sorting them according to the size of C_i.

3.3.3. Control Variables

In this study, other factors that may affect the low-carbon urban transition are included in the empirical model in order to mitigate omitted variable bias as much as possible. The main ones include. For environmental regulation (*ER*), we choose three indicators: sulfur dioxide removal rate (industrial sulfur dioxide removal/industrial sulfur dioxide generation), industrial soot removal rate (industrial soot removal/industrial soot generation), and comprehensive industrial solid waste utilization rate (comprehensive industrial solid waste utilization/(comprehensive industrial solid waste generation + comprehensive utilization of previous years' storage)), and use the entropy value method to calculate the intensity of environmental regulation. For the annual average temperature (*TEM*), we used the cumulative daily temperature average to represent this variable. Openness to foreign investment (*OPEN*), we use the annual real foreign investment (converted to RMB based on the average RMB exchange rate) as a share of GDP. Government intervention (*GOV*), we use the share of fiscal expenditure net of science and education in total fiscal expenditure. In industrial agglomeration (*AGG*), we use the Location Quotient method to calculate the manufacturing agglomeration status of each city. Marketization (MARK), which we measure using the share of self-employment and private employment in total employment. Financial development (*FIAN*), which we measure using the year-end loan balance as a share of GDP.

3.3.4. Other Variables

1. Instrumental variable

In this study, we will further examine the implementation effect of the new urbanization pilot policy (*Policy*) in the robustness test. In 2013, China established a new "people-oriented" urbanization policy. In 2015 and 2016, China's National Development and Reform Commission (NDRC) announced three batches of comprehensive national pilot projects for new urbanization. Since then, Chinese government departments have continued to improve the program and expand the pilot project scope into developing replicable and replicable experiences. This study assigns a value of 1 to the approved pilot cities (experimental group) and 0 to the unapproved non-pilot cities (control group), denoted as $Treated_i$. Among the pilot cities, this study assigns a value of 1 to the year in which the pilot cities are approved and subsequent years and 0 to the remaining years. All year's corresponding to the non-pilot cities are assigned a value of 0 and denoted as $Time_t$. In this case, $Policy_{it} = Treated_i * Time_t$.

2. Channel variables

Based on the theoretical analysis in the previous subsections, the two key mechanism variables for channel analysis are industrial structure transformation (*IS*) and technology spillover (*TS*).

Industrial structure transformation (*IS*). The upgrading of industrial structures towards cleanliness is the key to realizing the goal of green development. The current indicators for industrial structure upgrading mainly use internal structure change, energy consumption per unit GDP of industry, and product sales of pollution-intensive industries. We utilize the entropy value method to determine the degree of cleaner transformation of industrial structure. In this paper, we refer to Zhang et al. (2023) to construct the indicator system from two aspects of clean energy consumption and clean production [31]. Clean energy consumption is measured by the ratio of total industrial energy consumption to industrial added value; clean production is expressed by the ratio of regional industrial added value to carbon emissions. Through the dimensionless quantization of the indicators, the entropy value method is then used to identify the degree of cleaner transformation of the industrial structure.

Technology spillover (*TS*). Generally speaking, due to China's imperfect patent guarantee mechanism and relatively backward R&D capability, it is difficult for enterprises to convert R&D inputs into green innovation outputs. In contrast, the number of green patents can reflect the actual innovation outputs more objectively. We use the number of green patent acquisitions obtained in one year as a proxy variable for technology spillovers.

The results of descriptive statistics for each variable are shown in Table 3.

Table 3. Descriptive statistics of variables.

Variable	Mean	N	SD	Min	p25	p50	p75	Max
LCT	0.500	3962	0.150	0.130	0.400	0.480	0.580	1.160
ILU	0.0800	3962	0.0700	0.0100	0.0500	0.0600	0.100	0.630
ER	0.610	3962	0.200	0.0600	0.460	0.660	0.760	0.990
TEM	14.60	3962	5.100	−1.090	10.91	15.54	17.90	25.68
OPEN	1.900	3962	1.980	0	0.460	1.280	2.690	15.32
GOV	0.800	3962	0.0400	0.610	0.780	0.800	0.830	0.980
AGG	0.860	3962	0.480	0.0200	0.520	0.770	1.140	3.050
MAK	0.480	3962	0.140	0	0.380	0.480	0.580	0.940
FIAN	0.880	3962	0.560	0.0800	0.540	0.710	1.010	9.620

4. Results

4.1. Baseline Regression Analysis

4.1.1. Spatial Autocorrelation Test and Spatio-Temporal Distribution

We calculated the spatial correlation of *LTC* in China using ArcGis 10.2 software. As shown in Appendix A Table A2, the spatial correlation indices were significant for all years. Among them, the global *Moran's I* value are all greater than zero, and the *Geary's C* values are all within the interval [0, 1]. It indicates a significant positive spatial correlation of *LCT* at the four-city level in China. In the time dimension, the global *Moran's I* index increases yearly, revealing that the spatial correlation of low carbon transition among cities has been strengthened year by year in recent years.

4.1.2. Baseline Result

We first performed the Wald and LR tests, and the results showed that both passed the 1% significance test. This result rejects the original hypothesis of using the SLM or SEM model, indicating that the spatial error and lag terms exist simultaneously. Therefore, we use the spatial Durbin model for the empirical analysis. The Hausman test results pass the 1% significance test, indicating that the selection of the fixed-effects model is consistent with the model set. Table 4 reports the regression results of the spatial Durbin model for the four spatial weight matrices (regression results of Equation (7)). First, the coefficient of the effect of *ILU* on urban low-carbon transition is significantly positive under all four weights. It indicates a significant positive relationship between *ILU* and urban low-carbon efficiency.

Second, the results from the decomposition coefficients are shown. The results in column (2) of Table 4 show that the estimated coefficient of the indirect effect of *ILU* is 1.995 under the neighboring weights, which is significantly positive at the 1% statistical level. It indicates that *ILU* contributes to the local low-carbon transition and has a significant positive spatial spillover effect on the low-carbon transition of cities adjacent to the local one. The results from column (5) show that the indirect effect of *ILU* under geographical weight is 8.195 and passes the significance level test at the 1% level. The results in columns (8) and (11) show that the indirect effect of *ILU* under the economic weight is 3.710, and the indirect effect of *ILU* under the economic distance weight is 8.224. Both coefficients are significant at the 1% level. Specifically, *ILU* has a significant positive spillover effect on urban low-carbon transition regardless of the spatial weights used.

Finally, in terms of coefficient magnitude, the indirect effect of *ILU* on urban low-carbon transition shows a consistent feature across all four spatial weights, i.e., the indirect effect coefficient is higher than the direct effect coefficient. It indicates that we need to test the relationship between *ILU* and urban low-carbon transition based on the perspective of spatial spillover. This facilitative spillover effect has a stronger explanatory power in the total effect.

Table 4. Results of spatial spillover effects.

Panel A: Results of Adjacent Space Weights and Distance Spatial Weights.						
	W_{adj} **(Model 1)**			W_{dis} **(Model 2)**		
Variable	**(1)**	**(2)**	**(3)**	**(4)**	**(5)**	**(6)**
	LR_Direct	**LR_Indirect**	**LR_Total**	**LR_Direct**	**LR_Indirect**	**LR_Total**
ILU	0.6155 ***	1.9447 ***	2.5602 ***	0.5692 ***	8.1948 ***	8.7640 ***
	(−4.54)	(−8.983)	(−11.114)	(−4.192)	(−4.726)	(−5.102)
ER	−0.0363 **	0.0779 **	0.0416	−0.0395 **	0.7691 ***	0.7296 ***
	(−2.226)	(−2.439)	(−1.215)	(−2.438)	(−2.833)	(−2.683)
TEM	−0.0108	−0.0155	−0.0263 ***	−0.0086	−0.0402	−0.0489
	(−0.971)	(−1.311)	(−4.250)	(−0.923)	(−1.110)	(−1.627)
OPEN	−0.0032 *	−0.0016	−0.0048*	−0.0043 **	0.0317 *	0.0274 *
	(−1.726)	(−0.526)	(−1.753)	(−2.449)	(−1.932)	(−1.718)
GOV	−0.4000 ***	0.3100 **	−0.09	−0.3665 ***	−0.011	−0.3775
	(−4.669)	(−2.311)	(−0.668)	(−4.290)	(−0.013)	(−0.461)
AGG	−0.0260 ***	0.0162	−0.0098	−0.0326 ***	0.1029	0.0703
	(−3.490)	(−1.215)	(−0.741)	(−4.406)	(−1.28)	(−0.887)
MAK	0.0301	0.0093	0.0394	0.0159	0.3887	0.4046
	(−1.206)	(−0.201)	(−0.79)	(−0.64)	(−1.061)	(−1.102)
FIAN	−0.0383 ***	−0.0549 ***	−0.0932 ***	−0.0324 ***	−0.4038 ***	−0.4362 ***
	(−5.824)	(−4.227)	(−7.246)	(−4.928)	(−4.248)	(−4.628)
City FE	YES	YES	YES	YES	YES	YES
Year FE	YES	YES	YES	YES	YES	YES
Observations	3962	3962	3962	3962	3962	3962
R-squared	0.542	0.572	0.572	0.445	0.673	0.491
Panel B: Results of Economic Spatial Weights and Economic Distance Spatial Weights.						
	W_{econ} **(Model 3)**			$W_{econdis}$ **(Model 4)**		
Variable	**(7)**	**(8)**	**(9)**	**(10)**	**(11)**	**(12)**
	LR_Direct	**LR_Indirect**	**LR_Total**	**LR_Direct**	**LR_Indirect**	**LR_Total**
ILU	0.4631 ***	3.7100 ***	4.1732 ***	0.5688 ***	8.2240 ***	8.7928 ***
	(3.532)	(8.331)	(8.967)	(4.189)	(4.730)	(5.105)
ER	−0.0452 ***	0.1454 **	0.1002	−0.0395 **	0.7717 ***	0.7323 ***
	(−2.791)	(2.473)	(1.562)	(−2.435)	(2.835)	(2.686)
TEM	−0.0246 ***	0.0446 **	0.0200	−0.0087	−0.0399	−0.0485
	(−4.495)	(2.326)	(0.946)	(−0.931)	(−1.096)	(−1.612)
OPEN	−0.0025	−0.0189 ***	−0.0214 ***	−0.0043 **	0.0320 *	0.0277 *
	(−1.538)	(−3.008)	(−3.189)	(−2.445)	(1.942)	(1.728)
GOV	−0.2572 ***	−0.9358 ***	−1.1930 ***	−0.3656 ***	−0.0158	−0.3814
	(−3.249)	(−3.884)	(−4.579)	(−4.280)	(−0.019)	(−0.465)
AGG	−0.0230 ***	0.0713 ***	0.0483 *	−0.0327 ***	0.1036	0.0709
	(−3.244)	(2.692)	(1.750)	(−4.411)	(1.285)	(0.893)
MAK	0.0255	−0.0119	0.0136	0.0159	0.3878	0.4037
	(1.031)	(−0.150)	(0.155)	(0.639)	(1.057)	(1.098)
FIAN	−0.0493 ***	0.0067	−0.0426 **	−0.0324 ***	−0.4051 ***	−0.4376 ***
	(−7.795)	(0.341)	(−2.043)	(−4.928)	(−4.247)	(−4.625)
City FE	YES	YES	YES	YES	YES	YES
Year FE	YES	YES	YES	YES	YES	YES
Observations	3962	3962	3962	3962	3962	3962
R-squared	0.470	0.309	0.498	0.527	0.559	0.558

Note: The numbers in parentheses are robust t-statistics. ***, ** and * represent significance levels of 1%, 5% and 10%, respectively.

Regarding the effects of other control variables: Environmental regulation (*ER*) shows consistent characteristics across the four spatial weights. Its direct effect coefficient (effect on local low-carbon transition) is significantly negative yet. However, the indirect effect coefficient (effect on the low-carbon transition of neighboring and associated cities) is significantly positive. The possible reason for this is that environmental regulation policies emphasize the role of regulation, which reduces the total local pollution index while also impacting the development of local industries, thus showing a negative effect on low carbon efficiency. Further, under a more stringent environmental regulation policy, polluting industries in cities are at risk of closing down. It can force the development of local clean technologies. With the spillover of knowledge and technology, neighboring or associated cities can promote low-carbon transition development through secondary innovation and technology imitation. The indirect effect of government intervention (*GOV*) is significantly positive (0.310) with the neighboring weights and significantly negative (0.936) with the economic weights. The possible reason is that geographical proximity increases the "demonstration effect" of urban areas. The prosperity of local economic development can lead to technological progress and knowledge accumulation in neighboring cities, thus promoting the low-carbon transition of neighboring cities. Regarding economic distance, the development of local cities tends to eliminate pollution-intensive industries and make them move to cities with higher economic connectivity. It causes an increase in carbon emissions in the receiving region, which harms its green transformation. Financial development (*FIAN*) shows a significant indirect effect under the neighborhood, geographic, and economic geography weight. Moreover, it has a negative sign of the coefficient. Further deepening of financial development may have a siphoning effect on the neighboring areas' financial resources and human capital. This effect is not conducive to the expansion of production of the plot industries in the neighboring and associated cities nor to the updating and R&D of clean technologies. It will eventually lead to the inhibition of their low-carbon transition.

4.2. Robustness Tests

We first conduct a parallel trend test to analyze whether policy evaluation can be conducted using the double difference approach. After the results showed that this important test was passed, we proceeded to model estimation. Table 5 reports the regression results of the spatial difference-in-differences model for the four spatial weight matrices (regression results of Equation (10)). As we can see from Table 5, the indirect effects of the *ILU* ("New Urbanization" pilot policy dummy variables) are significant at different spatial weight matrices, and the coefficients of the variables are positive. It implies that the *ILU* construction will significantly promote the low-carbon transformation of the surrounding and associated cities. It also indicates a non-negligible spatial correlation in the error term of the model, and if not taken into account, the regression results will produce biased estimates. Using spatial econometric models in this paper is necessary.

In particular, there are some characteristics of Models 1–4 results based on spatial weights of different geographical elements. As can be seen, the coefficient of distance spatial weights (W_{dis}, 0.342) and the coefficient of e economic distance spatial weights ($W_{econdis}$, 0.343) are higher than the coefficient of an indirect effect of economic, spatial weights (W_{econ}, 0.112) and higher than the coefficient of neighboring spatial weights (W_{adj}, 0.051). It may be because the pilot cities have made more efforts to promote low-carbon, green, inclusive, and intelligent cities. The demonstration effect and the economic correlation effect on the neighboring cities and associated cities exemplify the positive effect on the low-carbon transition of the cities. Furthermore, the size of the coefficient indicates that this role-modeling effect is more likely to be constrained by geographical distance.

4.3. Mechanism Verification

4.3.1. Channel Mechanism of Industrial Structure Transformation

Model 1 in Table 6 shows the impact of intensive land use on industrial structure upgrading, and it can be seen that the impact is positive and significant, indicating that *ILU* promotes China's low-carbon transformation through the *IS* influence mechanism.

Specifically, *ILU*, as a long-term national policy, will constrain the disorderly expansion of enterprise land use and promote the economization of enterprise production and operation in the coming period. In addition, the objective constraints of *ILU* on urban space will limit the entry of highly polluting and low-value-added industries. Therefore, in cities with intensive land utilization, industries can obtain green development and regulate the industrial layout by regulating the proportion of clean industries, thus promoting low-carbon development.

Table 5. Robustness test results of exogenous shock.

Panel A: Results of Adjacent Space Weights and Distance Spatial Weights.						
	W_{adj} (Model 1)			**W_{dis} (Model 2)**		
Variable	**(1)**	**(2)**	**(3)**	**(4)**	**(5)**	**(6)**
	LR_Direct	**LR_Indirect**	**LR_Total**	**LR_Direct**	**LR_Indirect**	**LR_Total**
ILU	0.0051	0.0510 ***	0.0561 ***	−0.0073	0.3422 ***	0.3349 ***
	(0.507)	(3.500)	(4.082)	(−0.731)	(3.702)	(3.723)
ER	−0.0500 ***	0.0425	−0.0075	−0.0427 ***	0.6401 **	0.5975 *
	(−3.046)	(1.263)	(−0.206)	(−2.614)	(2.062)	(1.919)
TEM	−0.0117	−0.0120	−0.0237 ***	−0.0098	−0.0382	−0.0481
	(−1.048)	(−0.999)	(−3.663)	(−1.048)	(−0.938)	(−1.389)
OPEN	−0.0030	−0.0083 ***	−0.0113 ***	−0.0053 ***	−0.0002	−0.0055
	(−1.628)	(−2.637)	(−4.000)	(−2.975)	(−0.013)	(−0.321)
GOV	−0.4750 ***	0.0111	−0.4639 ***	−0.4279 ***	−0.7720	−1.1999
	(−5.595)	(0.083)	(−3.450)	(−5.117)	(−0.828)	(−1.310)
AGG	−0.0255 ***	0.0113	−0.0141	−0.0309 ***	0.1323	0.1014
	(−3.389)	(0.820)	(−1.015)	(−4.162)	(1.416)	(1.095)
MAK	0.0294	−0.0105	0.0190	0.0100	0.0321	0.0420
	(1.165)	(−0.218)	(0.362)	(0.399)	(0.076)	(0.099)
FIAN	−0.0388 ***	−0.0567 ***	−0.0955 ***	−0.0311 ***	−0.4983 ***	−0.5294 ***
	(−5.884)	(−4.224)	(−7.054)	(−4.715)	(−4.233)	(−4.516)
City FE	YES	YES	YES	YES	YES	YES
Year FE	YES	YES	YES	YES	YES	YES
Observations	3962	3962	3962	3962	3962	3962
R-squared	0.475	0.440	0.433	0.587	0.502	0.588
Panel B: Results of Economic Spatial Weights and Economic Distance Spatial Weights.						
	W_{econ} (Model 3)			**$W_{econdis}$ (Model 4)**		
Variable	**(7)**	**(8)**	**(9)**	**(10)**	**(11)**	**(12)**
	LR_Direct	**LR_Indirect**	**LR_Total**	**LR_Direct**	**LR_Indirect**	**LR_Total**
ILU	0.0179 **	0.1121 ***	0.1300 ***	−0.0073	0.3438 ***	0.3365 ***
	(2.008)	(3.555)	(3.826)	(−0.735)	(3.707)	(3.727)
ER	−0.0534 ***	0.1415 **	0.0881	−0.0426 ***	0.6422 **	0.5996 *
	(−3.257)	(2.267)	(1.292)	(−2.613)	(2.063)	(1.920)
TEM	−0.0207 ***	0.0664 ***	0.0457 **	−0.0099	−0.0377	−0.0476
	(−3.747)	(3.300)	(2.059)	(−1.058)	(−0.922)	(−1.371)
OPEN	−0.0045 ***	−0.0317 ***	−0.0362 ***	−0.0052 ***	−0.0001	−0.0054
	(−2.804)	(−4.866)	(−5.227)	(−2.970)	(−0.006)	(−0.313)
GOV	−0.3906 ***	−1.5333 ***	−1.9239 ***	−0.4269 ***	−0.7834	−1.2102
	(−5.038)	(−6.378)	(−7.540)	(−5.105)	(−0.838)	(−1.319)
AGG	−0.0226 ***	0.0578 **	0.0352	−0.0310 ***	0.1334	0.1024
	(−3.145)	(2.068)	(1.203)	(−4.167)	(1.423)	(1.103)
MAK	0.0101	−0.0701	−0.0601	0.0099	0.0286	0.0386
	(0.404)	(−0.839)	(−0.654)	(0.397)	(0.067)	(0.090)
FIAN	−0.0458 ***	0.0243	−0.0215	−0.0311 ***	−0.4998 ***	−0.5309 ***
	(−7.197)	(1.168)	(−0.974)	(−4.715)	(−4.227)	(−4.508)
City FE	YES	YES	YES	YES	YES	YES
Year FE	YES	YES	YES	YES	YES	YES
Observations	3962	3962	3962	3962	3962	3962
R-squared	0.533	0.552	0.522	0.455	0.662	0.431

Note: The numbers in parentheses are robust t-statistics. ***, ** and * represent significance levels of 1%, 5% and 10%, respectively.

Table 6. Channel mechanism verification.

Variable	IS	TS
	Model 1	Model 2
ILU (Main)	0.022 **	0.071 ***
	(2.11)	(5.18)
ILU·W	0.156 ***	4.966 ***
	(4.94)	(2.99)
Control	YES	YES
City FE	YES	YES
Year FE	YES	YES
Observations	3962	3962

Note: The numbers in parentheses are robust t-statistics. *** and ** represent significance levels of 1% and 5%, respectively.

4.3.2. Channel Mechanism of Technology Spillover

As shown in Model 2 in Table 6, the effect of intensive land use on technology spillovers is significant, indicating that *ILU* promotes China's low-carbon transition through the influence mechanism of *TS*. Specifically, *ILU*'s land use restrictions on enterprises can first force enterprises to increase investment and research in green products and new materials. Second, compared with the standardized and large-scale production of the secondary industry, the knowledge, and technology-intensive tertiary sector tends to have higher value-added, lower energy consumption, and is more in line with the need for intensive land use. It means cities with intensive land utilization have built a good platform for technology R&D and dissemination. Third, the positive psychological effect of *ILU* on the low-carbon development of industries should not be ignored. Positive public opinion encourages regional industrial enterprises and regional enterprises to imitate each other and technological innovation, thus promoting the low-carbon transformation of the region.

4.4. *Test of Spatial Attenuation Boundary*

In order to examine the regional boundaries of *ILU* on urban low-carbon transition, this section uses threshold inverse distance spatial weights for multiple spatial Durbin model estimation analysis, and the weights are specifically set as follows:

$$w_{ij} = \begin{cases} 0 & (i = j \ or \ d_\mu > d_{ij} \ or \ d_{ij} > d_l \\ \frac{1}{d_{ij}} & (i \neq j \ and \ d_\mu < d_{ij} < d_l) \end{cases} \quad (15)$$

In Equation (15), d_{ij} represents the distance between city i and city j regions, d_μ is the lower limit of spatial threshold distance, and d_l is the upper limit of spatial threshold distance, whose values are set autonomously. When the distance between two cities is within the spatial threshold range, the spatial relationship between the two cities is considered to exist, and the weight value is the inverse of the distance between the two. Below or above this range, it is considered that there is no spatial relationship between the two cities, and its weight value is 0. Thus, we derive the spatial weight matrix of distances in different distance ranges and re-estimate the spatial Durbin model to measure the spatial spillover effects of *ILU* affecting urban low-carbon transition in different distance ranges (as shown in Table 7 and Figure 3).

The relationship between the local effect, spillover effect, and the corresponding distance threshold of *ILU* on urban low-carbon transition is shown in Figure 3. The effect of *ILU* on local low-carbon transition is relatively stable. However, the spillover effect exhibits an inverted "U" shape concerning the increase of distance threshold. It indicates that although *ILU* can promote the low-carbon transformation of neighboring and associated cities across space with their demonstration effects, their spillover effects have geographical peaks and decay boundaries due to the limitations of industrial development, infrastructure coverage, and administrative division. Moreover, the coefficient of the spillover effect peaks at 250 km and then decays, while the indirect effect at 450 km is not significant. It implies that the *ILU* has the highest facilitation effect at 250 km. This radiation range is close to the radius of the regional scope of Chinese provinces, indicating that strengthening the *ILU* cooperation among regional urban clusters will help promote regional low-carbon transition.

Table 7. Spatial attenuation coefficients.

Distance (KM)	LR_Direct	LR_Indirect	LR_Total
50	0.7916 ***	0.1158 **	0.9073 ***
	(6.033)	(2.239)	(6.583)
100	0.6747 ***	0.4863 ***	1.1611 ***
	(5.066)	(4.274)	(7.150)
150	0.5934 ***	1.3236 ***	1.9170 ***
	(4.439)	(7.235)	(9.276)
200	0.6166 ***	1.7716 ***	2.3882 ***
	(4.633)	(7.057)	(8.862)
250	0.6151 ***	2.0986 ***	2.7137 ***
	(4.622)	(7.318)	(8.965)
300	0.6511 ***	0.7324 ***	1.3835 ***
	(4.671)	(3.002)	(5.638)
350	0.6601 ***	0.7008 ***	1.3609 ***
	(4.740)	(2.784)	(5.417)
400	0.6847 ***	0.6299 **	1.3146 ***
	(4.927)	(2.427)	(5.060)
450	0.7149 ***	0.5342 **	1.2491 ***
	(5.155)	(1.970)	(4.588)
500	0.7441 ***	0.3775	1.1216 ***
	(5.351)	(1.355)	(4.054)
550	0.7544 ***	0.3050	1.0594 ***
	(5.426)	(1.059)	(3.704)
600	0.7519 ***	0.3038	1.0557 ***
	(5.412)	(1.044)	(3.656)
650	0.7449 ***	0.3561	1.1010 ***
	(5.370)	(1.178)	(3.658)
700	0.7509 ***	0.3165	1.0674 ***
	(5.415)	(1.044)	(3.538)
750	0.7632 ***	0.2695	1.0327 ***
	(5.510)	(0.888)	(3.423)

Note: The numbers in parentheses are robust t-statistics. *** and ** represent significance levels of 1% and 5%, respectively.

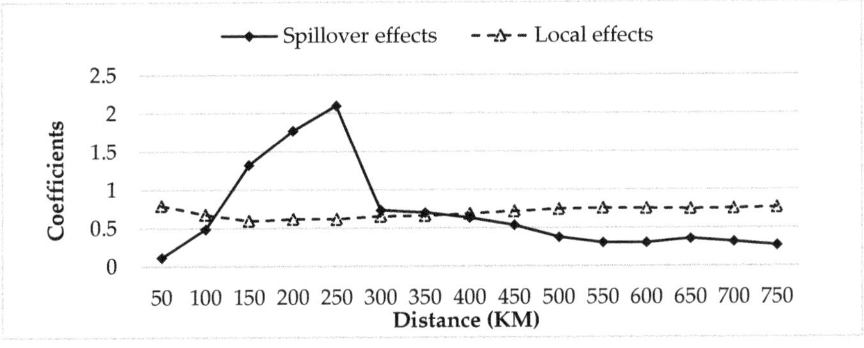

Figure 3. Regional boundary of spatial spillover effects.

4.5. Heterogeneity Tests

4.5.1. Heterogeneity at the *ILU* Dimension

In the previous section, we designed a comprehensive indicator to represent *ILU* based on the connotation of "intensive land use". The indicator contains three dimensions: density of land use, economic and social efficiency, and ecology benefits. In order to further analyze the influence mechanism of *ILU* on neighboring and associated towns, we conducted empirical tests on each of the three dimensions of *ILU*. The regression results are shown in Table 8.

Table 8. Heterogeneity test results of the *ILU* dimension.

Variable	*UCR*	*UCR*	*UCR*
	(1)	**(2)**	**(3)**
	LR_Direct	**LR_Indirect**	**LR_Total**
density of land use	0.3838	23.2338 ***	23.6175 ***
	(1.096)	(4.245)	(4.299)
economic and social efficiency	0.8405 ***	20.7071 ***	21.5477 ***
	(2.894)	(3.656)	(3.797)
ecology benefits	4.1700 ***	75.0155 ***	73.8306 ***
	(10.986)	(3.301)	(3.208)
Control	YES	YES	YES
City FE	YES	YES	YES
Year FE	YES	YES	YES
Observations	3962	3962	3962

Note: The numbers in parentheses are robust t-statistics. *** represent significance levels of 1%, respectively.

The density of land use provides a good impetus to the low-carbon transition of cities (direct effect of 0.384 and indirect effect of 23.234). However, the direct effect is not significant. This suggests that the density of land has a better demonstration effect on neighboring and associated cities, but attention needs to be paid to the local low-carbon transition. *ILU* in the land, social and ecological dimensions have significant positive spatial spillover. It shows that the rational and efficient use of land, increased social inclusion, and strengthened ecological constraints can form positive spillover to neighboring and associated cities. Among them, the coefficient of ecological dimension is the highest, at 75.016.

4.5.2. Spatial Heterogeneity

We further examine the spatial heterogeneity of the spillover effects of *ILU* on the low-carbon transformation of cities. We specifically focus on three perspectives: geographic location, city level, and city circle (economic location). Since this part of the test mainly considers the effect of spatial heterogeneity, we use geographic weights for estimation.

Cities with different geographic locations. As for the natural location of the province, different regions have distinct economic development goals, land use regulations, and contaminant pressure. Accordingly, the impact of IER may be affected by the geographical location, so we have divided the sample into three sample subgroups, including eastern, western, and central regions. The results are shown in Panel A of Table 9. In the eastern region, the spillover effect of *ILU* on urban low-carbon transition is 5.132, and this coefficient is positive and significant. The spillover effect of *ILU* on low-carbon transition for cities in the central region is also greater than 0 (coefficient of 3.824) but not significant. The results are different in the western region, where the spillover effect coefficient is negative and insignificant (−0.961). This result indicates that the *ILU* affects urban low-carbon transition differently depending on the geographical location. It plays an important driving role only in the eastern region.

Table 9. Results of spatial heterogeneity test.

Panel A: Geographic Location.			
Variable	*UCR*	*UCR*	*UCR*
	LR_Direct	**LR_Indirect**	**LR_Total**
East	0.3455 **	5.1323 ***	5.4778 ***
	(−1.962)	(−7.386)	(8.276)
Central	1.6698 ***	3.8238	5.4937
	(−4.304)	(−1.093)	(1.534)
West	0.5520 **	−0.9605	−0.4085
	(−2.066)	(−0.435)	(−0.191)
Panel B: City Level.			
Variable	*UCR*	*UCR*	*UCR*
	LR_Direct	**LR_Indirect**	**LR_Total**
Extra Large	2.4806 ***	7.9541 ***	7.9253 ***
	(−3.913)	(−3.021)	(3.065)
Large	0.1135	5.8359 ***	5.9494 ***
	(−0.634)	(−4.801)	(4.901)
Moderate	0.367	−0.8509	1.6297
	(−1.240)	(−0.657)	(1.395)
Small	−0.0288	−7.4345 **	−7.0676 **
	(−0.078)	(−2.167)	(−2.026)
Panel C: City Circle.			
Variable	*UCR*	*UCR*	*UCR*
	LR_Direct	**LR_Indirect**	**LR_Total**
Yangtze River Delta	0.9410 *	0.2061	1.1471
	(−1.753)	(−0.108)	(0.625)
Beijing–Tianjin–Hebei City Circle	5.3022 ***	2.8549	8.1571 ***
	(−7.842)	(−0.960)	(2.619)
Middle Yangtze River Economic Belt	2.8349 ***	−2.1027	0.7322
	(−3.767)	(−0.754)	(0.251)
Pearl River Delta	−1.1533 ***	2.7095	1.5562
	(−3.996)	(−1.546)	(0.825)
Chengdu–Chongqing City Circle	8.1728 ***	−4.4565	3.7163
	(−5.553)	(−0.533)	(0.417)

Note: The numbers in parentheses are robust t-statistics. ***, ** and * represent significance levels of 1%, 5% and 10%, respectively.

The cities are of different sizes. We use the year-end population of the city district as a proxy variable for city size. Given the frequent changes in the administrative divisions of city districts in many cities, we use the city size division criteria published by China in 2014 to select cities. Based on the total year-end population of city municipal districts, the 283 city samples can be categorized into four groups: extra-large cities (more than 5 million people), large cities (1 million to 5 million people), moderate cities (half a million to 1 million people), and small cities (less than half a million people). Among all the cities in the sample, there are 13 extra-large cities, 127 large cities, 98 moderate cities, and 45 small cities. As shown in Panel B of Table 9: The indirect effects for medium-sized and large cities are 7.954 and 5.836, respectively, and are significant at the 1% level. It indicates that the *ILU* of extra-large and large cities has a positive spillover effect on the low-carbon transition of neighboring cities. The indirect effect coefficient of moderate and small cities is negative, and small cities are significant at the 5% level. It implies that moderate and small cities in China have caused some degree of pollution to the development of neighboring cities in promoting *ILU*. In addition, extra-large cities significantly promote local green transformation, with a coefficient value of 2.486.

We also test the spillover effects of *ILU* in different economic circles on urban low-carbon transformation separately according to the current policies of building urban circles implemented in China. Panel C of Table 9 shows that the *ILU* of the Yangtze River Delta, Beijing–Tianjin–Hebei urban circle, and Pearl River Delta cities form positive spillover to the low-carbon transition of neighboring cities. The middle reaches of the Yangtze River Economic Belt and Chengdu–Chongqing City Circle form a negative spillover. The coefficients of the indirect effects of the above five urban areas do not pass the significance test. It indicates that the *ILU* in China's urban areas did not significantly affect the low-carbon transition of neighboring cities in the promotion process. However, the *ILU* of the five urban areas all play a significant role locally. Only the *ILU* in the Pearl River Delta inhibits the local low-carbon transition, while the other four urban areas effectively promote the local low-carbon transition.

5. Discussion

In order to comply with the global low-carbon development trend, the Chinese government has proposed two significant goals of achieving "carbon peak" by 2030 and "carbon neutrality" by 2060, as well as the goals of promoting pollution reduction and carbon neutrality. Comprehensively promoting green and low-carbon transformation is an important strategic direction for economic and social development in the coming period. The baseline regression results in this paper verify that *ILU* has a significant positive spillover effect on low-carbon transformation, providing ideas for promoting *ILU* development to realize low-carbon urban transformation and carbon neutrality. This result is consistent with the findings of Shang et al. (2022) [56], but we go a step further by considering the influence of spatial factors and drawing conclusions about spillover effects and spatial boundaries. In addition, our findings further confirm the greenness and sustainability of China's *ILU* policy and urban spatial optimization [57]. China has implemented and is implementing integrated land use policies (e.g., Provisions on Saving and Intensive Land Use (2014) [16].As the spatial mainstay of economic development, land use intensification, while emphasizing the principles of "structural optimization" and "prioritizing conservation," has green, livable, and ecologically friendly connotations that contribute to the sustainable development of China's economy and the construction of an ecological civilization.

The existence of spatial boundaries provides some empirical evidence for local governments in China to break down inter-provincial administrative barriers and promote the articulation of urban spatial planning and ecological governance mechanisms between provinces. According to the division of China's administrative boundaries, 450 km is the average distance between China's provincial capital cities, and 250 km is the average radius of China's provinces. While the spillover effect of *ILU* on low-carbon city transformation can reach as far as 450 km, the peak of the spillover effect's coefficient is around 250 km. The driving force of *ILU* can reach as far as across China's provincial boundaries to affect the low-carbon urban transition in other provinces. However, local governments also need to be aware that the benefits are more significant in neighboring areas of the province.

The results on the heterogeneity of *ILU* dimensions and different spatial weights reflect the parts of the *ILU* policy promotion process that need attention and improvement. One of them is the need to pay attention to the goals of *ILU* policy promotion. According to the results of the four spatial weights, it can be found that *ILU* creates a driving effect on the low-carbon transition of neighboring and economically linked cities. It suggests cities which committed to promoting *ILU* can exert targeting, demonstration, and economic linkage effects. Considering that the driving effect of *ILU* is stronger in large cities, local governments can use large cities as their hinterland to play a "point-surface" driving role and thus promote the realization of the overall low-carbon transformation goals of

Chinese cities. Second, it is essential to emphasize the means of *ILU* policies. Compared to ecological carbon sink improvement and land use efficiency, the increase in land use density cannot significantly promote the local low-carbon transition. This result is similar to that of Baur et al. (2015) [58] based on data from European cities. Considering that land use density is related to the city's actual built-up area and the population's carrying capacity. This situation may arise because the current land intensification in China is still in the stage of capital intensification [59]. The most important vehicle for urban development is the construction land, and capital investment is concentrated in construction land. In this case, an increase in land use density will increase infrastructure and energy investment. Therefore, decisions about urban folding, spatial planning, and urbanization development need to be implemented prudently.

To promote China's low-carbon transition, paying attention to the spatial differences and regional cooperation in the environmental benefits of *ILU* policies is essential. This is consistent with the findings of current research in other countries [60,61]. First, it is necessary to pay attention to geographic location differences. The results of the heterogeneity regression show that only the cities in the eastern region can implement *ILU* policies while generating positive spillover effects on the low-carbon transition of neighboring cities, and the results in the central and western regions are not significant. In fact, the land carbon emissions of the eastern region, including Beijing, Tianjin, Shanghai, Jiangsu Province, Zhejiang Province and other provinces, account for a high proportion of the national emissions (see Figure 1); however, eco-efficiency and energy use efficiency are also higher in the eastern coastal region than in the central and western regions [62,63]. Taken together, although the eastern region faces more substantial pressure to reduce emissions, it has developed a more inclusive green land use system over the years and has experience in urban spatial planning and low-carbon environmental management, thus creating a demonstration effect on neighboring cities. These achievements may be related to the greater concentration of talent, technology, and innovation in the eastern region, which could receive further attention. On the contrary, the central and western regions, such as Sichuan, Hubei, Henan, and other provinces, are all in rapid economic development, with large populations and high pressure on land use carbon emissions. Promoting *ILU* in cities in the central and western regions is more challenging and requires more advanced experience from the eastern regions. Secondly, it is necessary to pay attention to urban-level differences. The positive spillover effect of *ILU* on urban low-carbon transition is not evident in small and medium-sized cities and urban circles. These findings reflect two aspects: small and medium-sized cities have weaker governance capacity and may face more difficult ecological governance and spatial layout adjustment [57]. The second is that administrative barriers within the city region have not yet been broken down [64], and there is less willingness to cooperate between city clusters [65], which limits the direct and spillover effects of *ILU* policies in the city region on the low-carbon transition. The above conclusions provide ideas for future synergistic promotion of low-carbon transformation in Chinese cities.

Although this study complements the shortcomings of the studies related to *ILU* and urban low-carbon transition, it also provides a theoretical reference for the study of urban low-carbon transition at the spatial level. However, there are still certain shortcomings in this study that need to be improved. First, the low-carbon transition of cities mainly includes two components of low-carbon and economic development. In the future, we will continue exploring the low-carbon land use in the low-carbon transition of cities and further enrich the urban environment-related research. Second, limited by data availability, the measurement of *ILU* still needs to be improved. Meanwhile, the connotation of *ILU* is constantly being improved and has likewise changed regarding green and blue infrastructure construction. We expect to expand on these two aspects in the subsequent study to improve the measure of *ILU*. Third, the research questions and scope can be extended to other developing countries.

6. Conclusions and Policy Implications

6.1. Conclusions

This study incorporates China's *ILU* policy and low-carbon transition into a unified analytical framework from the perspective of spatial spillover. The spatial autocorrelation, evolution, spillover effects, spatial decay, and spatial heterogeneity of the intensive land use construction in China's low-carbon transition are investigated based on theoretical and empirical analyses. Many valuable conclusions are drawn as follows:

(1) intensive land use has significant spatial spillover effects on the low-carbon transition of cities. It not only plays a positive role for neighboring cities but also promotes the low-carbon transformation of economically related cities. The exogenous shock test of the pilot cities of new urbanization also verifies this result. Furthermore, the spillover effect of new urbanization exists in the range of 0–450 km and peaks at about 250 km.

(2) The results of mechanism validation indicate that industry transfer and technology spillover are dual mechanism channels for intensive land use for low-carbon transition in China.

(3) intensive land use plays a catalytic role in the low-carbon transition of surrounding cities through three dimensions: the density of land use, economic and social efficiency, and ecology benefits, among which the ecological dimension has the most potent effect. In contrast, land use density does not contribute to the local low-carbon transition.

(4) As an essentially urban development strategy, the intensive land use in the eastern region and large-sized cities can significantly contribute to the low-carbon transition of neighboring cities. However, intensive land use in urban economic zones cannot perform the radiation effect to drive the low-carbon transformation of neighboring cities.

On average, the obtained results are broadly consistent with the existing literature. At the same time, it provides more novel and convincing evidence on the effectiveness of intensive land use and the importance of clean technology development and production transformation for carbon neutrality in emerging countries such as China. It will help promote Chinese government departments to promote and implement *ILU* policies and action plans for dual carbon goals in a targeted manner and also help provide theoretical and empirical references for developing countries to promote environmentally friendly and coordinated development.

6.2. Policy Implications

We propose the following policy recommendations mainly from the perspective of developing countries choosing governments as facilitators.

First, the government should increase *ILU* through specific measures to ensure that it plays an active role in the low-carbon transition of the city. The government needs to focus on formulating economic, social, and ecological policies. Quality improvement should not be neglected in favor of economic growth, and excessive "accumulation" of factors of production on land will ultimately lead to lower efficiency and higher emissions. At the same time, the government should control the increase of land and explore the land stock, coordinate the safeguarding of development and the protection of resources, optimize the allocation of factors through industrial upgrading, strengthen the control of urban growth boundaries, rationally allocate land resources, and enhance the intensive level of urban land use.

Second, the government needs to use the technical means of regional integration to optimize the intensive use of urban land in the city circle. Integrating national economic and social development planning, land use master planning, land improvement planning, area development planning, ecological environment planning, and enterprise access catalogs are important designs worth trying by local governments. This top-level design can strengthen the long-term mechanism of land use optimization under the control and integration of multiple regulations and cultivate the concept of regional integration. As a result, it will enhance the synergistic and consistent mechanism for optimizing the allocation of land resources.

Thirdly, the local government should use land-space constraints on economic activities to promote a cleaner transformation of the industrial structure and the process of promoting green technologies. Transforming the existing unreasonable and unclean industrial structure is essential for developing an intensive and efficient economic growth mode. On the one hand, local governments need to protect patented technologies and promote enterprises' green technology research and development process. On the other hand, the government needs to promote the cleanliness of the energy structure and strengthen the rational allocation of industrial development and energy structure. In addition, the government should play an organizing and guiding role in promoting green and healthy production methods such as recycling, high efficiency, and emission reduction and improve the city's scientific and technological innovation capacity and technology spillover effect. In this way, a virtuous green circular economy can be formed within and among cities.

Finally, local governments should choose and implement *ILU* policy tools according to local conditions. The government needs to form a differentiated idea of intensive land use. Specifically, the government can build a differentiated control system for intensive land use based on the zoning of land intensive use evaluation results, zoning of resource and environmental carrying status, and zoning of dominant industries, and from the perspective of economic development priority and resource and environmental utilization and protection priority.

Author Contributions: Conceptualization, X.L.; methodology, Y.G.; software, G.W.; validation, X.L.; formal analysis, X.L.; investigation, X.L.; resources, Y.G.; data curation, X.L.; writing—original draft preparation, X.L.; writing—review and editing, G.W.; visualization, Y.G.; supervision, G.W.; project administration, Y.G.; funding acquisition, Y.G. All authors have read and agreed to the published version of the manuscript.

Funding: This research was funded by the National Social Science Foundation of China, grant number: 20BJL142, Postdoctoral Science Foundation of China, grant number: 2022M720131, Key Research and Special Popularization (Soft Science Research) Projects of Henan Province in 2022, grant number: 222400410135, and Carbon Emissions Trading Provincial and Ministerial Joint Construction Collaborative Innovation Project Center of Hubei Province, grant number 22CICETS-YB011.

Data Availability Statement: The data are available on request.

Acknowledgments: Thanks for the support from Hubei University and Guizhou University.

Conflicts of Interest: The authors declare no conflict of interest.

Appendix A

Table A1. Summary table of acronyms.

Acronyms	Description
LCT	Low-carbon transition
ILU	Intensive land use
ER	Environmental regulation
TEM	annual average temperature
OPEN	Openness to foreign investment
GOV	Government Intervention
AGG	Industrial agglomeration
MAK	Marketization
FIAN	Financial development
IS	Industrial structure transformation
TS	Technology spillover

Table A2. Results of spatial autocorrelation test.

Year	*Moran's I*		*Geary's C*	
2006	0.076 ***	(3.137)	0.919 ***	(−2.697)
2007	0.082 ***	(3.359)	0.912 ***	(−2.942)
2008	0.117 ***	(4.765)	0.880 ***	(−4.013)
2009	0.124 ***	(5.035)	0.879 ***	(−4.074)
2010	0.125 ***	(5.049)	0.874 ***	(−4.241)
2011	0.138 ***	(5.558)	0.855 ***	(−4.890)
2012	0.127 ***	(5.160)	0.860 ***	(−4.661)
2013	0.133 ***	(5.358)	0.863 ***	(−4.674)
2014	0.099 ***	(4.040)	0.904 ***	(−3.258)
2015	0.134 ***	(5.399)	0.863 ***	(−4.696)
2016	0.164 ***	(6.596)	0.843 ***	(−5.500)
2017	0.061 **	(2.535)	0.962	(−1.353)
2018	0.158 ***	(6.351)	0.850 ***	(−5.250)
2019	0.202 ***	(8.097)	0.816 ***	(−6.303)

Notes: *** $p < 0.01$, ** $p < 0.05$. Z-values in parentheses.

References

1. Zhou, A.; Xin, L.; Li, J. Assessing the Impact of the Carbon Market on the Improvement of China's Energy and Carbon Emission Performance. *Energy* **2022**, *258*, 124789. [CrossRef]
2. Amri, F.; Zaied, Y.B.; Lahouel, B.B. ICT, Total Factor Productivity, and Carbon Dioxide Emissions in Tunisia. *Technol. Forecast. Soc. Chang.* **2019**, *146*, 212–217. [CrossRef]
3. Ahmed, R.; Liu, G.; Yousaf, B.; Abbas, Q.; Ullah, H.; Ali, M.U. Recent Advances in Carbon-Based Renewable Adsorbent for Selective Carbon Dioxide Capture and Separation—A Review. *J. Clean. Prod.* **2020**, *242*, 118409. [CrossRef]
4. Lv, Q.; Liu, H.; Yang, D.; Liu, H. Effects of Urbanization on Freight Transport Carbon Emissions in China: Common Characteristics and Regional Disparity. *J. Clean. Prod.* **2019**, *211*, 481–489. [CrossRef]
5. CO_2 Emissions in 2022. Available online: https://www.iea.org/reports/co2-emissions-in-2022 (accessed on 13 July 2023).
6. Scafetta, N. Detection of Non-climatic Biases in Land Surface Temperature Records by Comparing Climatic Data and Their Model Simulations. *Clim. Dyn.* **2021**, *56*, 2959–2982. [CrossRef]
7. Ribes, A.; Qasmi, S.; Gillett, N.P. Making Climate Projections Conditional on Historical Observations. *Sci. Adv.* **2021**, *7*, eabc0671. [CrossRef] [PubMed]
8. Bp Statistical Review of World Energy Report in 2021. Available online: https://www.bp.com.cn/content/dam/bp/country-sites/zh_cn/china/home/reports/statistical-review-of-world-energy/2021/BP_Stats_2021.pdf (accessed on 14 July 2023).
9. Liu, C.; Sun, W.; Li, P.; Zhang, L.; Li, M. Differential Characteristics of Carbon Emission Efficiency and Coordinated Emission Reduction Pathways under Different Stages of Economic Development: Evidence from the Yangtze River Delta, China. *J. Environ. Manag.* **2023**, *330*, 117018. [CrossRef]
10. Campbell, C.A.; Zentner, R.P.; Liang, B.C.; Roloff, G.; Gregorich, E.C.; Blomert, B. Organic C Accumulation in Soil over 30 Years in Semiarid Southwestern Saskatchewan—Effect of Crop Rotations and Fertilizers. *Can. J. Soil Sci.* **2000**, *80*, 179–192. [CrossRef]
11. Houghton, R.A.; House, J.I.; Pongratz, J.; van der Werf, G.R.; DeFries, R.S.; Hansen, M.C.; Le Quéré, C.; Ramankutty, N. Carbon Emissions from Land Use and Land-Cover Change. *Biogeosciences* **2012**, *9*, 5125–5142. [CrossRef]
12. IPCC. Climate change 2013: The physical science basis. In *Contribution of Working Group I to the Fifth Assessment Report of the Intergovernmental Panel on Climate Change*; Stocker, T., Qin, D., Eds.; Cambridge University Press: Cambridge, UK; New York, NY, USA, 2013; pp. 11–12.
13. Ali, G.; Nitivattananon, V. Exercising Multidisciplinary Approach to Assess Interrelationship between Energy Use, Carbon Emission and Land Use Change in a Metropolitan City of Pakistan. *Renew. Sustain. Energy Rev.* **2012**, *16*, 775–786. [CrossRef]
14. Fu, B.; Wu, M.; Che, Y.; Yang, K. Effects of Land-Use Changes on City-Level Net Carbon Emissions Based on a Coupled Model. *Carbon Manag.* **2017**, *8*, 245–262. [CrossRef]
15. Martínez-Zarzoso, I.; Maruotti, A. The Impact of Urbanization on CO_2 Emissions: Evidence from Developing Countries. *Ecol. Econ.* **2011**, *70*, 1344–1353. [CrossRef]
16. Provisions for the Conservation and Intensive Utilization of Land. Available online: https://www.gov.cn/gongbao/content/2014/content_2736111.htm (accessed on 14 July 2023).
17. Land Use Optimization Based on Carbon Neutrality. Available online: https://www.163.com/dy/article/GMMQ1R8505346KFL.html (accessed on 14 July 2023).
18. Xie, H.; Chen, Q.; Lu, F.; Wu, Q.; Wang, W. Spatial-Temporal Disparities, Saving Potential and Influential Factors of Industrial Land Use Efficiency: A Case Study in Urban Agglomeration in the Middle Reaches of the Yangtze River. *Land Use Policy* **2018**, *75*, 518–529. [CrossRef]

19. Brännlund, R.; Lundgren, T.; Marklund, P.-O. Carbon Intensity in Production and the Effects of Climate Policy—Evidence from Swedish Industry. *Energy Policy* **2014**, *67*, 844–857. [CrossRef]
20. Meyfroidt, P. Environmental Cognitions, Land Change, and Social–Ecological Feedbacks: An Overview. *J. Land Use Sci.* **2013**, *8*, 341–367. [CrossRef]
21. Ke, Y.; Yi, X. Path Selection for Low-Carbon Economic Land Use Pattern in China. *Energy Procedia* **2011**, *5*, 452–456. [CrossRef]
22. Liu, H.; Lei, J. The Impacts of Urbanization on Chinese Households' Energy Consumption: An Energy Input-Output Analysis. *J. Renew. Sustain. Energy* **2018**, *10*, 015903. [CrossRef]
23. Zhang, W.; Xu, H. Effects of Land Urbanization and Land Finance on Carbon Emissions: A Panel Data Analysis for Chinese Provinces. *Land Use Policy* **2017**, *63*, 493–500. [CrossRef]
24. Wang, S.; Yang, C.; Li, Z. Spatio-Temporal Evolution Characteristics and Spatial Interaction Spillover Effects of New-Urbanization and Green Land Utilization Efficiency. *Land* **2021**, *10*, 1105. [CrossRef]
25. Zhang, L.; Jin, G.; Wan, Q.; Liu, Y.; Wei, X. Measurement of Ecological Land Use/Cover Change and Its Varying Spatiotemporal Driving Forces by Statistical and Survival Analysis: A Case Study of Yingkou City, China. *Sustainability* **2018**, *10*, 4567. [CrossRef]
26. Deng, X.; Gibson, J. Improving Eco-Efficiency for the Sustainable Agricultural Production: A Case Study in Shandong, China. *Technol. Forecast. Soc. Chang.* **2019**, *144*, 394–400. [CrossRef]
27. Jin, G.; Chen, K.; Wang, P.; Guo, B.; Dong, Y.; Yang, J. Trade-Offs in Land-Use Competition and Sustainable Land Development in the North China Plain. *Technol. Forecast. Soc. Chang.* **2019**, *141*, 36–46. [CrossRef]
28. Xie, H.; Zhai, Q.; Wang, W.; Yu, J.; Lu, F.; Chen, Q. Does Intensive Land Use Promote a Reduction in Carbon Emissions? Evidence from the Chinese Industrial Sector. *Resour. Conserv. Recycl.* **2018**, *137*, 167–176. [CrossRef]
29. Goh, C.S.; Junginger, M.; Potter, L.; Faaij, A.; Wicke, B. Identifying Key Factors for Mobilising Under-Utilised Low Carbon Land Resources: A Case Study on Kalimantan. *Land Use Policy* **2018**, *70*, 198–211. [CrossRef]
30. Peng, J.; Zheng, Y.; Liu, C. The Impact of Urban Construction Land Use Change on Carbon Emissions: Evidence from the China Land Market in 2000–2019. *Land* **2022**, *11*, 1440. [CrossRef]
31. Zhang, M.; Liu, X.; Peng, S. Effects of Urban Land Intensive Use on Carbon Emissions in China: Spatial Interaction and Multi-Mediating Effect Perspective. *Environ. Sci. Pollut. Res.* **2023**, *30*, 7270–7287. [CrossRef]
32. Zhu, L.; Xing, H.; Hou, D. Analysis of Carbon Emissions from Land Cover Change during 2000 to 2020 in Shandong Province, China. *Sci. Rep.* **2022**, *12*, 8021. [CrossRef]
33. Lee, C.-C.; Feng, Y.; Peng, D. A Green Path towards Sustainable Development: The Impact of Low-Carbon City Pilot on Energy Transition. *Energy Econ.* **2022**, *115*, 106343. [CrossRef]
34. Feng, H.; Wang, S.; Zou, B.; Yang, Z.; Wang, S.; Wang, W. Contribution of Land Use and Cover Change (LUCC) to the Global Terrestrial Carbon Uptake. *Sci. Total Environ.* **2023**, *901*, 165932. [CrossRef]
35. Chuai, X.; Xia, M.; Ye, X.; Zeng, Q.; Lu, J.; Zhang, F.; Miao, L.; Zhou, Y. Carbon Neutrality Check in Spatial and the Response to Land Use Analysis in China. *Environ. Impact Assess. Rev.* **2022**, *97*, 106893. [CrossRef]
36. Nichols, D.A. Land and economic growth. *Am. Econ. Rev.* **1970**, *60*, 332–340.
37. Popp, A.; Humpenöder, F.; Weindl, I.; Bodirsky, B.L.; Bonsch, M.; Lotze-Campen, H.; Müller, C.; Biewald, A.; Rolinski, S.; Stevanovic, M.; et al. Land-use protection for climate change mitigation. *Nat. Clim. Chang.* **2014**, *4*, 1095–1098. [CrossRef]
38. Shao, Q.; Zhang, W.; Cao, X.; Yang, J.; Yin, J. Threshold and Moderating Effects of Land Use on Metro Ridership in Shenzhen: Implications for TOD Planning. *J. Transp. Geogr.* **2020**, *89*, 102878. [CrossRef]
39. Yao, Y.; Pan, H.; Cui, X.; Wang, Z. Do Compact Cities Have Higher Efficiencies of Agglomeration Economies? A Dynamic Panel Model with Compactness Indicators. *Land Use Policy* **2022**, *115*, 106005. [CrossRef]
40. Dong, Y.; Jin, G.; Deng, X. Dynamic Interactive Effects of Urban Land-Use Efficiency, Industrial Transformation, and Carbon Emissions. *J. Clean. Prod.* **2020**, *270*, 122547. [CrossRef]
41. Chen, W.; Shen, Y.; Wang, Y.; Wu, Q. The Effect of Industrial Relocation on Industrial Land Use Efficiency in China: A Spatial Econometrics Approach. *J. Clean. Prod.* **2018**, *205*, 525–535. [CrossRef]
42. Peng, C.; Song, M.; Han, F. Urban Economic Structure, Technological Externalities, and Intensive Land Use in China. *J. Clean. Prod.* **2017**, *152*, 47–62. [CrossRef]
43. Wang, F.; Yang, F. A Review of Research on China's Carbon Emission Peak and Its Forcing Mechanism. *Chin. J. Popul. Resour. Environ.* **2018**, *16*, 49–58. [CrossRef]
44. Liu, X.; Zhang, W.; Cheng, J.; Zhao, S.; Zhang, X. Green Credit, Environmentally Induced R&D and Low Carbon Transition: Evidence from China. *Environ. Sci. Pollut. Res.* **2022**, *29*, 89132–89155. [CrossRef]
45. Eriksson, R.H. Localized Spillovers and Knowledge Flows: How Does Proximity Influence the Performance of Plants? *Econ. Geogr.* **2011**, *87*, 127–152. [CrossRef]
46. Klagge, B.; Martin, R. Decentralized versus Centralized Financial Systems: Is There a Case for Local Capital Markets? *J. Econ. Geogr.* **2005**, *5*, 387–421. [CrossRef]
47. Desmet, K.; Rossi-Hansberg, E. Spatial Development. *Am. Econ. Rev.* **2014**, *104*, 1211–1243. [CrossRef]
48. Elhorst, J.P. Spatial Panel Data Models. In *Spatial Econometrics: From Cross-Sectional Data to Spatial Panels*; Elhorst, J.P., Ed.; SpringerBriefs in Regional Science; Springer: Berlin/Heidelberg, Germany, 2014; pp. 37–93.
49. LeSage, J.P.; Pace, R.K. Spatial Econometric Models. In *Handbook of Applied Spatial Analysis: Software Tools, Methods and Applications*; Fischer, M.M., Getis, A., Eds.; Springer: Berlin/Heidelberg, Germany, 2010; pp. 355–376.

50. Feng, R.; Shen, C.; Huang, L.; Tang, X. Does Trade in Services Improve Carbon Efficiency?—Analysis Based on International Panel Data. *Technol. Forecast. Soc. Chang.* **2022**, *174*, 121298. [CrossRef]
51. Lin, B.; Jia, H. Does the Development of China's High-Speed Rail Improve the Total-Factor Carbon Productivity of Cities? *Transp. Res. Part D Transp. Environ.* **2022**, *105*, 103230. [CrossRef]
52. Zhang, R.; Tai, H.; Cheng, K.; Zhu, Y.; Hou, J. Carbon Emission Efficiency Network Formation Mechanism and Spatial Correlation Complexity Analysis: Taking the Yangtze River Economic Belt as an Example. *Sci. Total Environ.* **2022**, *841*, 156719. [CrossRef]
53. Fang, G.; Gao, Z.; Tian, L.; Fu, M. What Drives Urban Carbon Emission Efficiency?—Spatial Analysis Based on Nighttime Light Data. *Appl. Energy* **2022**, *312*, 118772. [CrossRef]
54. Zhang, W.; Liu, X.; Wang, D.; Zhou, J. Digital Economy and Carbon Emission Performance: Evidence at China's City Level. *Energy Policy* **2022**, *165*, 112927. [CrossRef]
55. Tone, K.; Tsutsui, M. An Epsilon-Based Measure of Efficiency in DEA—A Third Pole of Technical Efficiency. *Eur. J. Oper. Res.* **2010**, *207*, 1554–1563. [CrossRef]
56. Shang, Y.; Xu, J.; Zhao, X. Urban Intensive Land Use and Enterprise Emission Reduction: New Micro-Evidence from China towards COP26 Targets. *Resour. Policy* **2022**, *79*, 103158. [CrossRef]
57. Li, X.; Wang, L. Does Administrative Division Adjustment Promote Low-Carbon City Development? Empirical Evidence from the "Revoke County to Urban District" in China. *Environ. Sci. Pollut. Res.* **2023**, *30*, 11542–11561. [CrossRef] [PubMed]
58. Baur, A.H.; Förster, M.; Kleinschmit, B. The Spatial Dimension of Urban Greenhouse Gas Emissions: Analyzing the Influence of Spatial Structures and LULC Patterns in European Cities. *Landsc. Ecol.* **2015**, *30*, 1195–1205. [CrossRef]
59. Chen, J.; Wu, F. Housing and Land Financialization under the State Ownership of Land in China. *Land Use Policy.* **2022**, *112*, 104844. [CrossRef]
60. Bridge, G.; Bouzarovski, S.; Bradshaw, M.; Eyre, N. Geographies of Energy Transition: Space, Place and the Low-Carbon Economy. *Energy Policy* **2013**, *53*, 331–340. [CrossRef]
61. Hill, D. Regional Cooperation and Asia's Low Carbon Economy Transition: The Case of New Zealand. In *Investing on Low-Carbon Energy Systems: Implications for Regional Economic Cooperation*; Anbumozhi, V., Kalirajan, K., Kimura, F., Yao, X., Eds.; Springer: Singapore, 2016; pp. 309–326.
62. Zhang, Y.; Wang, L.; Tang, Z.; Zhang, K.; Wang, T. Spatial Effects of Urban Expansion on Air Pollution and Eco-Efficiency: Evidence from Multisource Remote Sensing and Statistical Data in China. *J. Clean. Prod.* **2022**, *367*, 132973. [CrossRef]
63. Li, J.; Ma, J.; Wei, W. Study of regional differences in energy carbon emission efficiency among eight comprehensive economic zones in China. *J. Quant. Tech. Econ.* **2020**, *37*, 109–129. [CrossRef]
64. Xin, L.; Sun, H.; Xia, X. Spatial–Temporal Differentiation and Dynamic Spatial Convergence of Inclusive Low-Carbon Development: Evidence from China. *Environ. Sci. Pollut. Res.* **2023**, *30*, 5197–5215. [CrossRef]
65. Zhou, W.; Cheng, F. Study on the willingness of inter-governmental cooperation of city clusters under the orientation of regional coordinated development—A textual analysis based on the government work reports of cities in three major city clusters. *Urban Probl.* **2022**, *7*, 12–23. [CrossRef]

land

Article

Accounting for and Comparison of Greenhouse Gas (GHG) Emissions between Crop and Livestock Sectors in China

Jinyu Han [1], Jiansheng Qu [1,2,*], Dai Wang [2] and Tek Narayan Maraseni [3]

1 National Science Library (Chengdu), Chinese Academy of Sciences, Chengdu 610299, China; hanjy@clas.ac.cn
2 College of Earth and Environmental Sciences, Lanzhou University, Lanzhou 730000, China; 120220907491@lzu.cn
3 Institute for Agriculture and the Environment, University of Southern Queensland, Toowoomba, QLD 4350, Australia; tek.maraseni@usq.edu.au
* Correspondence: qujs@clas.ac.cn

Abstract: The synergistic greenhouse gas (GHG) emission reduction of the crop production (CP) and livestock farming (LF) sectors is of great significance for food security and low-carbon development, especially for China, the world leader in agricultural production. In this paper, the GHG emissions from the CP and LF sectors are accounted for and compared, and the spatial econometric model is adopted for comparative study based on the panel data from 1997 to 2021. The results show that: (1) The total amount and intensity of GHG emissions from both sectors showed obvious spatial heterogeneity and spatial dependence, and the spatial distribution pattern was relatively stable. (2) The influence of each factor on the GHG intensity and spatial characteristics of CP and LF varies widely. For the CP sector, economic development (local effect −0.29, adjacent effect +1.13), increased urbanization rate (−0.24, +0.16), agricultural structure (−0.29, +0.05), and urban-rural disparity (−0.03, +0.17) all reduce the GHG intensity of local region, while increasing the GHG intensity of its adjacent areas, signifying leakage. The economic structure (+0.06, +0.16), agricultural finance support (+0.02, +0.26), mechanization level (+0.05, +0.03), and land occupancy rate (+0.54, +0.44) all play a role in increasing the GHG intensity of CP in the local region and its adjacent areas. The disaster degree (−0.03, −0.03) also reduced the GHG intensity of CP. For the LF sector, economic structure (+0.08, +0.11), urban-rural disparity (+0.11, +0.21), agricultural development level (+0.03, +0.50), and increased land occupancy rate (+0.05, +0.01) can improve the GHG intensity of the one region and adjacent areas. Economic development (+0.03, −0.15), urbanization rate (+0.04, −0.30), agricultural structure (+0.09, −0.03), and disaster degree (+0.02, −0.06) can increase the GHG intensity of the local region while reducing the GHG intensity of adjacent areas. Based on the results, under the background of carbon peaking and carbon neutralization(dual-carbon) goals, this study first puts forward collaborative emission reduction measures for CP and LF, respectively, then further rises to sector synergy and regional synergy, and constructs the countermeasure system framework of collaborative emission reduction from three levels, to provide guidance and reference for the realization of dual goals of agricultural GHG reduction and food security.

Keywords: crop production GHG emission; livestock farming GHG emission; spatial dependence; influencing factors; spatial Durbin model; synergetic measures

1. Introduction

The rapid development of agriculture is inevitably accompanied by the deterioration of the environment and the emergence of a series of ecological problems, especially greenhouse gas (GHG) emissions. This issue has been widely concerning due to the increasing climate change [1]. Agriculture has become one of the major emitters of GHG, producing about 14% of global GHG emissions and 58% of global non-carbon (CH_4, N_2O) GHG emissions [2,3]. Moreover, if effective measures are not taken as soon as possible, the agricultural GHG is expected to increase by 30% by 2050 [4], making it hard to realize the emissions reduction target of the Paris Agreement. As a world leader in agriculture production, China feeds 20% of the world's population with only 8% of global cropland [5,6]. Since the launch of the reform and opening up policy, China has made remarkable and rapid achievements in agriculture; the output of grain, meat, and aquatic products in 2017 has reached 1/5, 1/4, and 1/3 of world supplies after a 2-fold, 10-fold, and 14-fold increase since 1978, respectively [7]. However, great achievements in agriculture have come at the cost of large amounts of GHG emissions. China's agricultural GHG have accounted for 17% of the national emissions [8], of which agricultural CH_4 and N_2O emissions are much higher than other industries; agricultural CH_4 emissions accounted for 50.15% of the total emissions, and N_2O accounted for 92.43% of total emissions [9,10].

Unlike other sectors, agriculture is more dependent on region-specific factors, such as topography, soil, and climate [11], as well as on socioeconomic factors, including mechanization, irrigation, and the supply-demand situation of agricultural products. Therefore, the differences among region-specific factors have led to heterogeneity in agriculture, which may have caused spatial variations in agricultural GHG emissions. China has a vast territory and a wide distribution of agriculture. Due to significant differences in agricultural production conditions and resource endowments among provinces, there are large disparities in the agricultural development level and its structure. Extensive research has been conducted on these regional disparities using various indicators, such as total agricultural GHG emissions [12–14], agricultural GHG intensity [15,16], net agricultural GHG [17], and agricultural GHG efficiency(productivity) [16,18,19]. These indicators all show obvious regional heterogeneity in agricultural GHG emissions. However, most of the related studies used the concept of "agriculture" to account for GHG emissions, treating crop production (CP) and livestock farming (LF) as one whole subject. A small number of studies separately examine GHG emissions from LF and find significant spatial significance in both the total amount and intensity of GHG emissions. When it comes to the factors influencing GHG, researchers have found that the level of economic development [20,21], urbanization [20], technological development [22], agricultural economic level [23], agricultural structure [20,24], level of agricultural mechanization [22], agricultural human resources [25], and agricultural disaster severity [20] are the main factors influencing agricultural GHG emissions and their spatial heterogeneity. Additionally, there is a certain degree of spatial spillover effect, meaning that the agricultural GHG emissions of one province are not independent but are influenced by its surrounding provinces [23,26,27].

Although there have been studies on the spatial heterogeneity of agricultural or livestock GHG emissions, the majority of these studies included LF GHG emissions in agriculture. However, the distribution of the CP and LF sectors varies across provinces in China, resulting in spatial distribution heterogeneity for CP and LF GHG emissions. Treating them as a whole in research would hide or weaken the spatial heterogeneity at a more micro level. It would also mask the specific mechanisms of certain influencing factors, leading to a significant discount in the targeted formulation of GHG reduction policies. Moreover, the CP and LF sectors have strong complementarity, as CP provides feed for LF, and LF provides organic fertilizers for CP sector. The synergetic action between the two can theoretically achieve win-win benefits and GHG emissions reduction. Furthermore, there is currently limited literature on the spatial spillover effects of agricultural GHG emissions, and the existing studies mainly focus on the existence of spillover effects, paying less attention to the magnitude and direction of these effects.

Therefore, this research aims to fill the gap by following aspects. First of all, the paper innovatively divides agriculture into CP and LF, investigates the spatial distribution of GHG emissions for each sector, respectively, and explores the mechanisms of their respective interactions with relevant factors. Then, the spatial heterogeneity and influencing mechanisms of certain factors can be presented more specifically at a more micro level. Secondly, the research further examines spatial spillover mechanisms of both sectors. The spillover effect of agricultural GHG among provinces is widespread [23,26,27], while limited studies have explored it in depth. Last but not least, a strategic system for coordinated emission reduction in both sectors is designed based on the empirical results, which is a breakthrough in the research of agricultural GHGs. In the context of China's dual-carbon goals, this study has both theoretical value and practical significance.

The rest of this paper is organized as follows (Figure 1). In Section 2, we present the accounting process of GHG emissions from CP and LF and the theoretical aspects of the spatial Durbin model (SDM). In Section 3, the spatial variation of GHG emissions from CP and LF is firstly demonstrated from the scale and intensity, followed by the results and discussion of SDM results. Synergic measures were put forward in Section 4, and we concluded in Section 5.

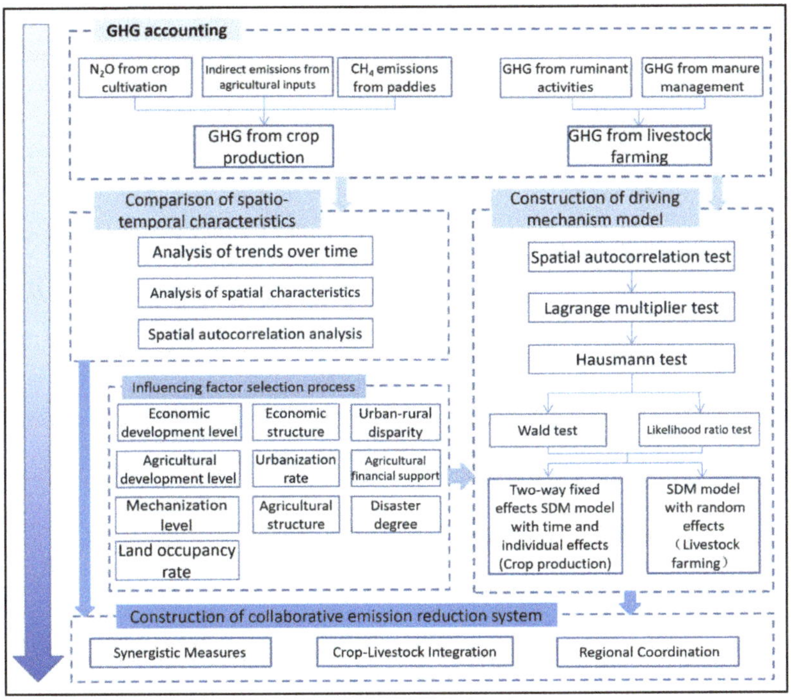

Figure 1. Methodology and process of this study.

2. Materials and Methods

2.1. GHG Accounting

The agricultural GHG accounting system based on the life cycle assessment (LCA) method has been well-developed. The current system mainly includes the CO_2, N_2O, and CH_4 emissions generated throughout the entire production process, including soil emissions, energy input, and material input [28–30] (Table 1). Specifically, the agricultural GHG encompasses four main parts: (a) N_2O emissions from crop production. This mainly refers to N_2O emissions during soil tillage, and the emission coefficients (Table A1) per unit area of different crops vary [20]. (b) Indirect emissions from agricultural inputs: This mainly includes the indirect GHG generated by the use of pesticides, plastic films, electricity, fertilizers, diesel, and other agricultural inputs during the production process (Table A2). (c) CH_4 emissions from paddy fields. This refers to the direct CH_4 emissions generated by paddy fields. The emission coefficients (Table A3) per unit area vary due to the hydrological, climatic, and soil conditions of different provinces, as well as the rice planting season (early-, middle-, or late-season rice) [1]. (d) GHG emissions from livestock. This includes the CH_4 and N_2O emissions generated by manure management and ruminant activities of herbivorous animals (Table A4). The sum of emissions from a, b, and c represents the GHG emissions from crop cultivation, while d represents the emissions from livestock breeding. The accounted N_2O and CH_4 emissions are converted into CO_2 equivalents using the conversion factors for greenhouse gases provided by the IPCC. The CO_2 equivalent values are divided by the output value of crop cultivation and livestock breeding, respectively, to obtain the GHG intensities of crop cultivation and livestock breeding for each province in different years.

Table 1. GHG accounting process and data sources.

GHG Types	GHG Sources	Accounting Process and Data Sources
GHG from crop production	a. N_2O from crop cultivation	The planting area of different crops such as rice, wheat (spring and winter wheat), soybean, maize, vegetables, sorghum, millet, potato, and peanut are multiplied by their respective N_2O emission coefficients and then converted into the CO_2 equivalent. The planting area of various crops comes from the China Statistical Yearbook and the China Agricultural Yearbook.
	b. Indirect emissions from agricultural inputs	The quantity of different inputs such as chemical fertilizer, diesel, pesticide, agricultural film, machinery power, and irrigation area is multiplied by the emission coefficients to obtain the quantity of CO_2 emission. The data on various types of agricultural inputs come from the China Agricultural Yearbook and New China Agriculture 60 Years Statistics.
	c. CH_4 emissions from paddies	CH_4 emissions from early, late, and mid-season rice (single-cropping late rice, winter paddy field, and wheat stubble rice) in different provinces were obtained by multiplying the planting areas with respective emission coefficients and then converted into CO_2 equivalent. The area data of various types of paddy fields come from the China Agricultural Yearbook.
GHG from livestock farming	d. CH_4 and NO_2 from ruminant activities and manure management	After converting the sales quantity and stock quantity of pigs, cattle, sheep, horses, donkeys, and mules into the annual average feeding quantity, the CH_4 and N_2O emissions obtained by multiplying the annual average feeding quantity of different animals with the emission coefficients are converted into CO_2 equivalent. Data on the number of animals sold out and the number of animals in stock are from the China Agricultural Yearbook.

2.2. Model Setting

Given the spatial correlation and spatial heterogeneity of GHG emission intensities in CP and LF, this study adopts a spatial econometric model to explore the spatial heterogeneity effects and its influencing factors. To validate the rational selection of the model, the spatial autocorrelation of GHG intensities in both sectors needs to be tested before entering the spatial econometric model. Spatial autocorrelation can be divided into global autocorrelation and local autocorrelation [31], which respectively investigate whether there is a spatial correlation among all spatial units as a whole and the specific form of correlation between individual spatial units and their surrounding units. In this study, only the global spatial autocorrelation of CP and LF GHG intensity is verified to demonstrate the scientific and rational application of the spatial econometric model. The commonly used indicator for testing global autocorrelation is *Moran's I*, and the formula for calculation is as follows [21,32]:

$$Moran's\ I = \frac{n\sum_{i=1}^{n}\sum_{j\neq 1}^{n} W_{ij}(x_i - \overline{x})(x_j - \overline{x})}{\sum_{i=1}^{n}(x_i - \overline{x})\sum_{i=1}^{n}\sum_{j=1}^{n} W_{ij}} \quad (1)$$

where n represents each province, x_i represents the GHG intensity of province i, $\overline{x}$ represents the average GHG intensity of all provinces and W_{ij} represents the spatial matrix between provinces i and j. Considering the model test results and the province-level agricultural situation, after systematic comparison and reference to similar literature, this study uses a simple and classical binary adjacency matrix, where two regions with a common boundary are considered adjacent [31,33]. The values on the main diagonal are set to 0, and W_{ij} for adjacent provinces is set to 1; otherwise, it is set to 0 (Hainan is considered adjacent to Guangdong). *Moran's I* ranges between −1 and 1, where a value greater than 0 indicates positive spatial correlation and clustering of GHG intensity among provinces, a value less than 0 indicates discrete distribution, and *Moran's I* = 0 indicates random distribution. The larger the absolute value of *Moran's I*, the greater the spatial correlation of GHG intensity among provinces.

Spatial econometric models effectively address the limitations of traditional regression models that assume spatial homogeneity, making them more reliable when applied to research subjects involving spatial autocorrelation [31]. Spatial econometric models can be divided into the spatial error model (SEM), spatial lag model (SLM), and SDM [33]. Among them, the SEM focuses on analyzing the differences in the form of interactions between different regions, and the SLM is commonly used to study the spillover effects of variables on regions outside the focal region. The SDM can be seen as a synthesis of the SLM and the SEM, which can be simplified to a SEM or a SLM under certain conditions [33]. The theoretical form of the SDM is as follows:

$$Y_t = \delta * W * Y_t + \beta_1 * X_t + \beta_2 * W * X_t + v_t \tag{2}$$

In the Equation (2), Y_t represents a 31 × 1 vector of GHG intensity in each province at time t (number of provinces), X_t represents a 31 × K matrix of exogenous explanatory variables, where K is the number of selected explanatory variables, W represents a 31 × 31 spatial weight matrix, which is also based on geographical adjacency, $W * X_t$ represents the interaction term between the spatial weight matrix and the exogenous explanatory variables, and δ, β represents the corresponding coefficients to be estimated. If β_2 is zero, the SDM can be simplified to a SLM, and if $\beta_2 + \delta * \beta_1 = 0$, the SDM can be simplified to a SEM. The theoretical form of the SDM, further simplified by removing the subscript t, is as follows:

$$Y = (I - \delta W)^{-1} * (\beta_1 W + \beta_2 WX) + (I - \delta W)^{-1} \tag{3}$$

Taking the partial derivative of Y with respect to the k-th explanatory variable of the i-th province yields:

$$\begin{aligned}\left[\frac{\partial Y}{\partial x_{1k}} \cdots \frac{\partial Y}{\partial x_{Nk}}\right] &= \begin{bmatrix} \frac{\partial y_1}{\partial x_{1k}} & \cdots & \frac{\partial y_1}{\partial x_{Nk}} \\ \vdots & \ddots & \vdots \\ \frac{\partial y_N}{\partial x_{1k}} & \cdots & \frac{\partial y_N}{\partial x_{Nk}} \end{bmatrix} = (I_N - \delta W)^{-1} \begin{bmatrix} \beta_{1k} & w_{12}\beta_{2k} & \cdots & w_{1N}\beta_{2k} \\ w_{21}\beta_{2k} & \beta_{1k} & \cdots & w_{2N}\beta_{2k} \\ \vdots & \vdots & \ddots & \vdots \\ w_{N1}\beta_{2k} & w_{N2}\beta_{2k} & \cdots & \beta_{1k} \end{bmatrix} \\ &= (I - \delta W)^{-1}(\beta_{1k} I_N + \beta_{2k} W)\end{aligned} \tag{4}$$

The direct effects of the SDM represent the average change in the dependent variable (GHG emission intensity) in a province caused by the explanatory variable of that province, which is the average of the diagonal elements of Equation (4) ($\overline{d}$ denotes the average of the diagonal elements of the matrix):

$$\text{direct effects} = \left[(I_N - \delta W)^{-1}(\beta_{1k} I_N + \beta_{2k} W)\right]^{\overline{d}} \tag{5}$$

The indirect effects of the SDM refer to the average change in the dependent variable (GHG emission intensity) in neighboring provinces caused by the explanatory variable of a province, which is the average of the off-diagonal elements of Equation (4) ($\overline{rsum}$ denotes the average of the off-diagonal elements of the matrix):

$$\text{indirect effects} = \left[(I_N - \delta W)^{-1}(\beta_{1k} I_N + \beta_{2k} W)\right]^{\overline{rsum}} \quad (6)$$

The total effect is the sum of the direct effects and indirect effects [33]. As for whether the SDM in this study can be simplified to a SEM or a SLM, as well as the choice of fixed effects or random effects models, they can be determined through the Wald test, LR test, and Hausman test to select the most suitable model form. The model testing and empirical analysis in this study were conducted using the Stata 15.0.

2.3. Variable Definition

In this study, the dependent variables are the GHG intensity of the CP and LF sectors. When selecting the independent variables, we try to choose variables that could potentially affect both the CP and LF sectors in order to compare the different mechanisms of the same variable on the GHG intensity of both sectors. Based on relevant studies on the factors influencing agricultural production and GHG emissions [34–40], two categories of 10 indicators are chosen as explanatory variables (Table 2). The first category represents the macro development of each province, including economic development level, economic structure, urbanization rate, and urban-rural disparity, totaling four indicators. The second category represents the agricultural development situation of each province, including agricultural structure, agricultural financial support, disaster degree, agricultural development level, mechanization level, and arable land occupancy rate, totaling six specific indicators. It is worth noting that in the process of calculating these indicators, data such as output value and value-added have been adjusted to constant prices in 1997. For some provinces and years, rural population data were missing, and the annual changes were minimal. Therefore, the moving average method was used to fill in the missing data. Before entering the empirical model, all indicators were standardized. Furthermore, before the regression analysis, we first tested the multicollinearity. The variance inflation factor (VIF) index of all the selected variables was less than 10, indicating that there was no significant collinearity between them. The meanings and descriptive statistics of each indicator are shown in Table 2.

Table 2. Model variables.

Variable Type	Variable Name	Description	Max	Min.	Mean	SD
Independent variable	CP GHG intensity	GHG emissions/crop production value	0.3827	0.0172	0.0966	0.453
	LF GHG intensity	GHG emissions/livestock production value	1.9294	0.0080	0.2045	2.952
Explanatory variable	Economic development level	Per capita GDP	33.04	2.21	8.26	0.546
	Economic structure	Proportion of added value of primary industry	37.840	0.360	13.426	7.448
	Urbanization rate	Urban population/total population	0.896	0.149	0.481	0.163
	Urban-rural disparity	Urban/rural consumption level	8.900	1.500	3.036	0.829
	Agricultural structure	Output value of crop production/output value of livestock farming	5.224	0.803	2.124	0.775
	Agricultural financial support	The proportion of financial support for agriculture in total financial expenditure	0.190	0.021	0.092	0.033
	Disaster degree	Disaster-affected area/crop planting area	0.936	0.000	0.257	0.163
	Agriculture development level	Agricultural added value/rural population	1.354	0.133	0.510	0.268
	Mechanization level	Agricultural machinery power/rural population	10.845	0.026	1.196	0.810
	Land occupancy rate	Arable land area/rural population	10.301	0.634	2.228	1.678

3. Results

3.1. Spatial Distribution of GHG Emissions for CP and LF Sectors

In terms of the national total, the agricultural GHG emissions in 1997, 2009, and 2021 reached 256.24 million tons, 282.74 million tons, and 293.19 million tons, respectively (Figure 2). The total agricultural carbon emissions show an increasing trend at the beginning, but the growth rate has slowed down sharply and is even showing signs of a peak point. Actually, some studies have shown a peak in China's agricultural GHG emissions in recent years [21,41]. Comparing the GHG emissions from the CP and LF sectors of each province (Figure 2), it can be seen that provinces with a strong tradition of agriculture have higher total GHG emissions, and in most provinces, GHG emissions from the CP sector are higher than those from the LF sector. Provinces in the northwest such as Qinghai, Tibet, Ningxia, and Inner Mongolia, which are mainly focused on the LF sector, have relatively low total GHG emissions, but the GHG emissions from the LF sector are significantly higher than those from the CP sector. Yunnan, Sichuan, and other provinces also have slightly higher GHG emissions from the LF sector compared to the CP sector. The comparison of the CP and LF GHG emissions reflects the spatial pattern of CP and LF production in each province. Furthermore, the total agricultural GHG emissions and the proportion of GHG emissions from the CP and LF sectors have remained relatively stable over the years, indicating that the structure of the CP and LF industries in each province is relatively stable.

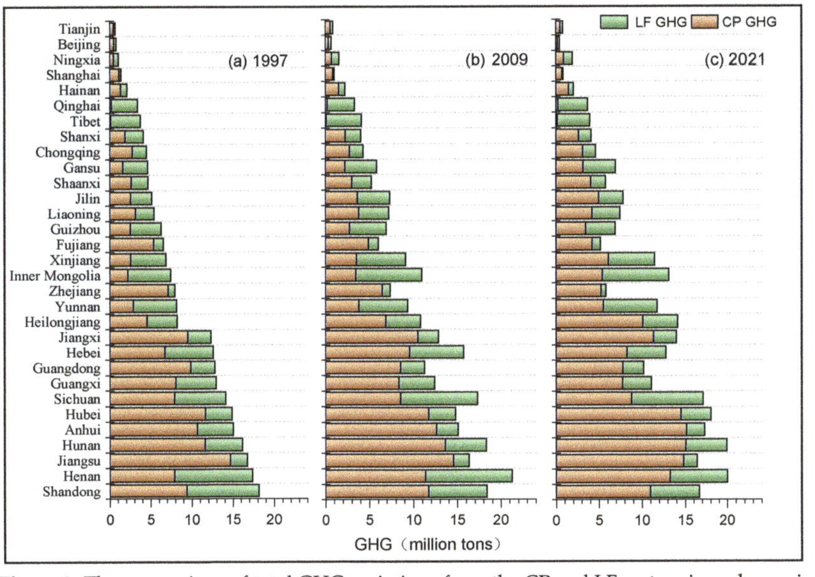

Figure 2. The comparison of total GHG emissions from the CP and LF sectors in each province in (**a**) 1997, (**b**) 2009, and (**c**) 2021.

Individually, looking at the GHG intensity of the CP sector (Figure 3), traditional grain-producing provinces such as Hubei, Hunan, Jiangxi, and Guangxi generally have higher GHG intensity. Among the above provinces, southern ones have higher GHG intensity than those in the northern part. This is mainly because of the higher proportion of rice in the crop structure of southern provinces. The GHG footprint of rice in China is 3.3 times that of maize and 2.1 times that of wheat [20]. In terms of GHG intensity in the LF sector, provinces such as Tibet, Qinghai, Inner Mongolia, and Xinjiang are significantly higher than other provinces, and they also have higher GHG emissions from livestock compared to provinces with high livestock GHG emissions, such as Henan and Sichuan. This is because these provinces have a higher proportion of ruminant animals, such as cattle and sheep, in their LF sector, and these animals have much higher GHG intensity due to the CH_4 emissions from rumination.

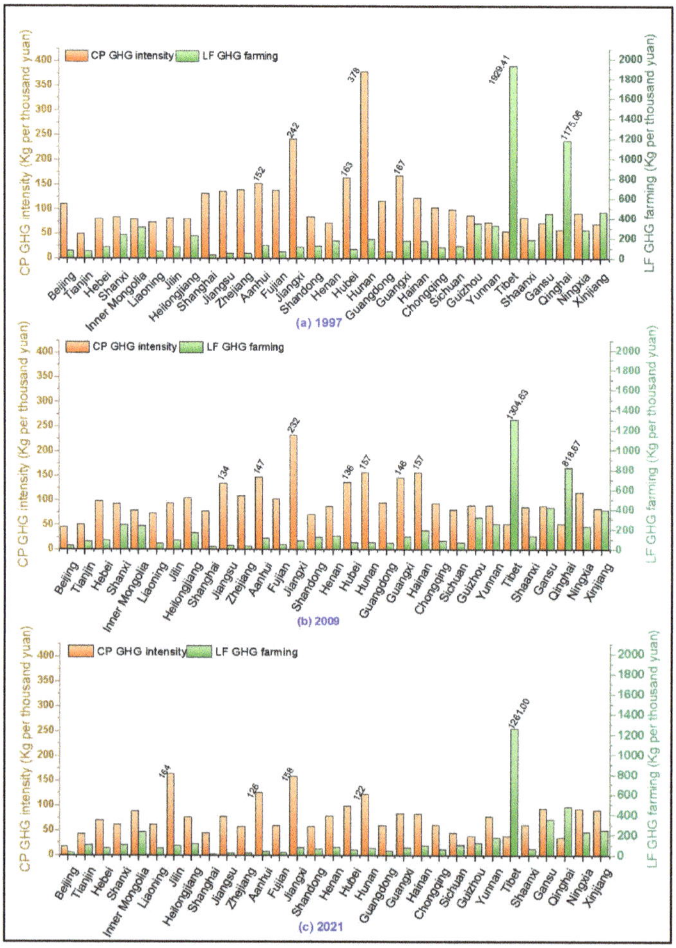

Figure 3. The comparison of GHG intensity from the CP and LF sectors in each province in (**a**) 1997, (**b**) 2009, and (**c**) 2021.

Comparing the carbon intensity of the CP and LF sectors, it can be seen that the GHG intensity in the LF sector is much higher than that in the CP sector in all provinces, further confirming the viewpoint that the GHG footprint of livestock is higher than that of crop production [42]. Looking at the trends over the years (Figure 3), the GHG intensity of CP and LF sectors in each province has decreased significantly, owing to the substantial improvement in agricultural production efficiency in China in recent years [5]. However, the distribution pattern of GHG intensity in the CP and LF sectors remains relatively stable. The LF GHG intensity is still high in provinces such as Qinghai, Tibet, Ningxia, and Inner Mongolia, which focus on LF, while the CP GHG intensity in provinces such as Hunan and Jiangxi has also been consistently higher than that in other provinces.

3.2. Spatial Autocorrelation Test

A spatial autocorrelation test on the GHG emission intensity of the CP and LF sectors is conducted to explore whether provinces with similar GHG emission intensity show spatial clustering and some degree of spatial heterogeneity.

During the entire study period, the p-values and z-values of *Moran's I* for LF GHG intensity passed the test, and *Moran's I* for every year were greater than 0, indicating significant spatial autocorrelation and strong spatial clustering of LF GHG intensity. For CP GHG intensity, most years also showed spatial autocorrelation, but a few years (2012–2017) did not pass the test (Figure 4). Nevertheless, this still suggests the presence of spatial autocorrelation and spatial clustering in the CP GHG intensity. The main reason is that the calculated *Moran's I* is based on a simple binary geographic adjacency matrix, which assumes that if spatial units are not adjacent, they do not influence each other, and even if they are adjacent, it assumes equal influence, which cannot fully explain the spatial clustering of GHG intensity. For example, in 2017, Sichuan Province had a crop-to-livestock output ratio of 2.4, while the ratios of Chongqing and Guizhou in the eastern neighborhood were 1.9 and 3.2, respectively, and the ratios of Qinghai and Tibet in the western neighborhood are only 1.1 and 1.6.

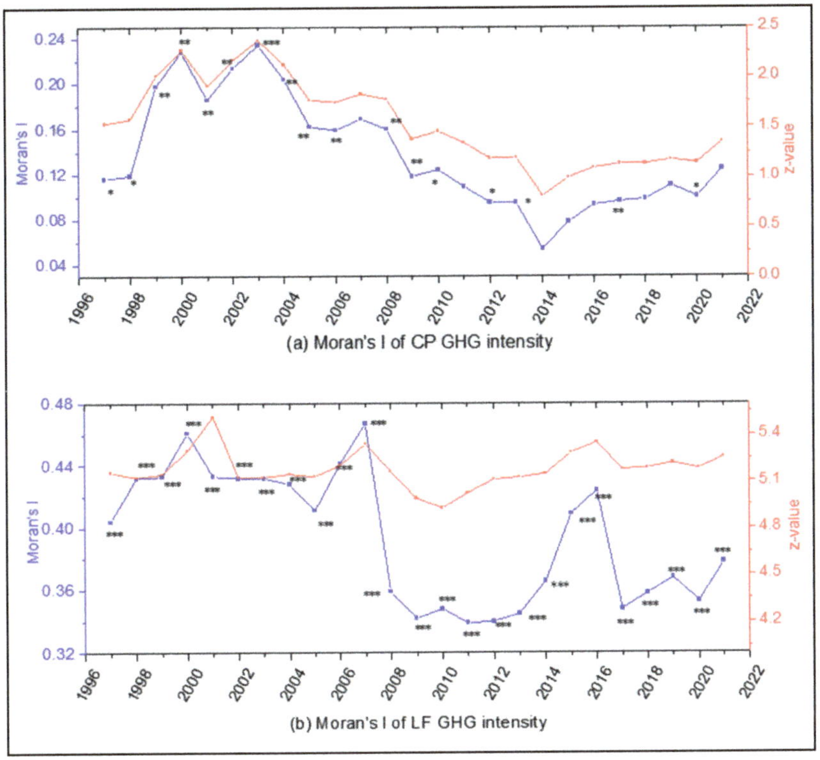

Figure 4. The *Moran's I* of GHG intensity of the (**a**) CP and (**b**) LF sectors. *** means $p < 0.01$, and accordingly, ** means $p < 0.05$, * means $p < 0.1$.

The spatial autocorrelation is relatively stable in the historical trends, especially for *Moran's I* of LF GHG emission intensity. The *Moran's I* of LF GHG intensity shows a downward trend over time (Figure 4), which indicates that in the context of significantly improved overall agricultural production efficiency in China [5], the provinces that originally had high GHG emission intensity experienced a gradual decrease in production efficiency improvement. The gap between them and provinces with low emission intensity is gradually narrowing. It should be noted that *Moran's I* aims to prove the existence of spatial spillover effect and is only the first step to verifying the rationality of SDM [43–45]. The following steps, such as the Hausmann test, LM test, and LR test, will be conducted to show the existence of spatial effect and prove the suitability of spatial econometric models.

3.3. Model Test

A Lagrange multiplier (LM) test was conducted to further examine the suitability of the spatial econometric models. According to the criteria proposed by Anselin (1991) and the LM test results (Table 3), it is found that the CP GHG intensity is better suited for a SEM, while the LF GHG intensity is better suited for a SDM. Subsequently, the Hausman test results (Table 4) indicate that the CP GHG intensity should use a fixed effects model, while the LF sector is better suited for a random effects model. It can be seen from the likelihood ratio (LR) test that both the time fixed effects and individual fixed effects are significant. Therefore, the appropriate model for the CP sector is the time-individual fixed effects model. The Wald test and LR test results (Table 5) reject the hypothesis that the SDM can degenerate into the SEM and the spatial autoregressive model at a 1% significance. To summarize, the CP GHG emission intensity is best analyzed using an individual-time fixed effects spatial Durbin model, while the LF sector is more suitable for a random effects spatial Durbin model.

Table 3. LM test statistics and significance.

LM Test		CP Sector	LF Sector
Spatial error model	Lagrange multiplier	213.494 ***	257.791 ***
	Robust Lagrange multiplier	128.813 ***	14.577 ***
Spatial lag model	Lagrange multiplier	85.530 ***	321.804 ***
	Robust Lagrange multiplier	0.849	78.590 ***

Note: *** $p < 0.01$.

Table 4. Hausmann test results.

Variable Classification	Statistic	*p*-Value
CP GHG intensity	10.59	0.5646
LF GHG intensity	486.05	0.0000

Table 5. Results of Wald test and LR test.

Test Types	Variables	Can SDM Be Simplified to SAR?	Can SDM Be Simplified to SEM?
LR test	CP GHG intensity	41.70 ***	40.07 ***
	LF GHG intensity	86.61 ***	157.55 ***
Wald test	CP GHG intensity	33.26 ***	40.88 ***
	LF GHG intensity	87.90 ***	150.10 ***

Note: *** $p < 0.01$.

3.4. Results of SDM

The regression results of the SDM (Table 6) show that the autoregressive coefficients of the CP and LF emission intensity pass the test at the 10% and 1% confidence levels, respectively.

From the regression coefficients and their significance, it can be observed that for the CP sector, factors such as economic development level, urbanization level, agricultural structure, and agricultural development level can locally reduce GHG intensity. In particular, the inhibitory effect of agricultural development level is the most significant. On the other hand, the mechanization level and land occupancy rate increase CP GHG emission intensity, with the latter having a larger impact. In terms of spatial effects, factors such as economic development level, rural-urban disparity, agricultural financial support, and land occupancy rate can increase the GHG intensity of neighboring areas, with economic development level having the most significant influence.

Table 6. SDM estimation results.

Explanatory Variables	CP GHG Intensity		LF GHG Intensity	
	Main Effects (Main)	**Spatial Effects (Wx)**	**Main Effects (Main)**	**Spatial Effects (Wx)**
Economic development level	−0.315 *** (0.113)	1.053 *** (0.304)	0.0422 (0.0504)	−0.101 (0.107)
Economic structure	0.0638 (0.0544)	0.117 (0.135)	0.0767 *** (0.0242)	0.0285 (0.0361)
Urbanization rate	−0.244 *** (0.0552)	0.169 (0.124)	0.0545 ** (0.0264)	−0.202 *** (0.0560)
Urban-rural disparity	−0.0360 ** (0.0297)	0.158 ** (0.0685)	0.0915 *** (0.0137)	0.0808 *** (0.0230)
Agricultural structure	−0.296 *** (0.0436)	0.0719 (0.0871)	0.0859 *** (0.0196)	−0.0516 (0.0323)
Agricultural financial support	0.0141 (0.0400)	0.234 *** (0.0898)	−0.0728 *** (0.0182)	−0.158 *** (0.0272)
Disaster degree	−0.0255 ** (0.0189)	−0.0260 ** (0.0382)	0.0214 ** (0.00939)	−0.0426 ** (0.0179)
Agriculture development level	−0.636 *** (0.0643)	−0.148 ** (0.141)	−0.00245 (0.0294)	0.294 *** (0.0519)
Mechanization level	0.0464 ** (0.0235)	0.0210 (0.0504)	−0.0346 *** (0.0114)	−0.0148 (0.0216)
Land occupancy rate	0.533 *** (0.0814)	0.345 * (0.181)	0.0530 ** (0.0360)	−0.0165 (0.0678)
Constant			0.00997 (0.146)	
ρ	0.113 * (0.0594)		0.440 *** (0.0458)	
R^2	0.6158		0.6578	
Log-likelihood	−927.1855		−927.1855	

Note: *** $p < 0.01$, ** $p < 0.05$, * $p < 0.1$; Values in parentheses are the standard deviations.

For the LF sector, the mechanization level and financial support for agriculture play a certain inhibitory role in the region, although their effects are relatively weak. On the other hand, economic structure, urbanization rate, rural-urban disparity, agricultural structure, and disaster degree contribute to increased LF GHG emissions. Rural-urban disparity and agricultural development level can promote GHG intensity in neighboring provinces, while urbanization rate, agricultural financial support, and disaster degree can inhibit GHG intensity in adjacent areas.

After determining whether various factors have an impact on the GHG intensity of the CP and LF sectors in the local and neighboring areas, the direct effects, indirect effects, and total effects of these factors are discussed to distinguish the effects of each factor more accurately (Figure 5). It can be observed that there are significant differences in the effects of various factors on the GHG intensity of the CP and LF sectors, whether in the local or adjacent areas. Although the regression coefficients of some factors' direct effects or indirect effects are not significant, the magnitude and direction of these effects on the GHG intensity of the CP and LF sectors can still be observed to some extent.

Firstly, for the CP sector, factors such as economic development level (direct effect −0.29; indirect effect +1.13), urbanization level (−0.24; +0.16), rural-urban disparity (−0.03; +0.17), and agricultural structure (−0.29; +0.05) can reduce the GHG intensity in the local area while increasing the GHG intensity in neighboring areas, with the increasing effect of economic development level being particularly significant. Economic structure (+0.06; +0.16), agricultural financial support (+0.02; +0.26), mechanization level (+0.05; +0.03), and land occupancy rate (+0.54; +0.44) can increase the GHG intensity of the CP sector in both the local and adjacent areas, while the disaster degree (−0.03; −0.03) can reduce the GHG intensity in both areas.

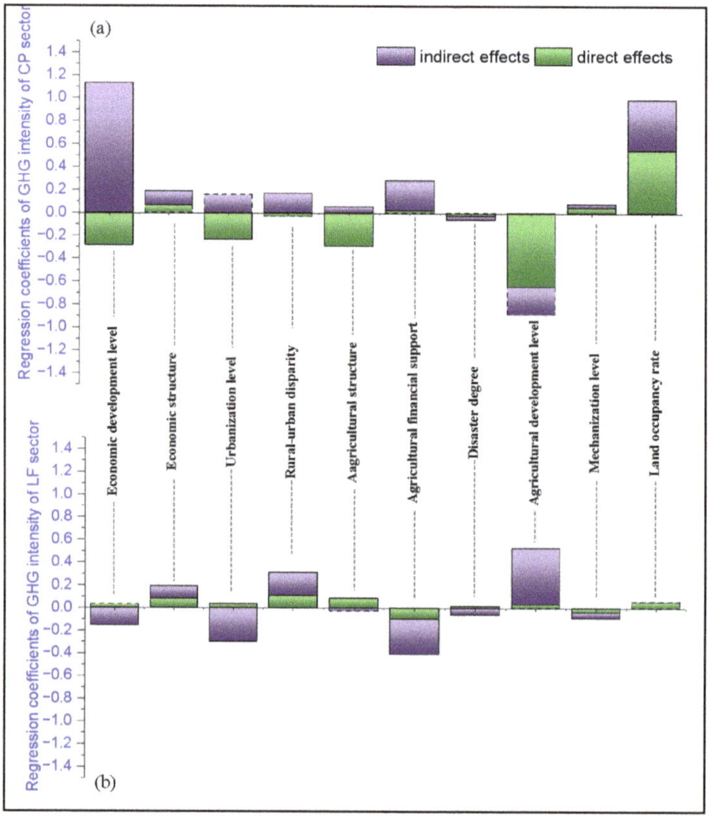

Figure 5. Comparison of regression coefficients between GHG intensity and variables in the (**a**) CP and (**b**) LF sectors. The dotted line in the figure indicates the variable with an insignificant regression coefficient. The total effect is the sum of direct and indirect effects.

For the LF sector, factors such as economic structure (+0.08; +0.11), rural-urban disparity (+0.11; +0.21), agricultural development level (+0.03; +0.50), and land occupancy rate (+0.05; +0.01) can increase the GHG intensity in both the local and adjacent areas, although the increasing effect of land occupancy rate is small. Economic development level (+0.03; −0.15), urbanization level (+0.04; −0.30), agricultural structure (+0.09; −0.03), and disaster degree (+0.02; −0.06) have an increasing effect on GHG intensity in the local area but reduce the GHG intensity in adjacent areas, with urbanization level having the most significant effect on reducing the LF GHG intensity in adjacent areas. Both agricultural financial support (−0.10; −0.31) and mechanization level (−0.04; −0.05) factors can reduce the carbon emission intensity of the livestock sector in both the local and adjacent areas, but the regression coefficient former factor is much larger than that of mechanization level.

In summary, the mechanisms of various factors on the GHG intensity of the CP and LF sectors are significantly different. In terms of the magnitude of their effects on both sectors, factors such as economic development level, agricultural development level, and land occupancy rate have a greater impact on the GHG intensity of the CP sector while having a smaller impact on the GHG intensity of the LF sector. In terms of the direction of their effects on both sectors, factors such as economic development level, urbanization level, agricultural structure, agricultural financial support, agricultural development level, and mechanization level show completely opposite effects, i.e., while increasing the GHG intensity of the CP sector, they can reduce the GHG intensity of the LF sector, and vice versa. It is generally believed that an increase in per capita arable land will improve production efficiency due to the scale effect of agricultural production, thereby reducing agricultural GHG emissions. However, this study found that per capita arable land has a certain increasing effect on the GHG intensity of the CP sector after separating the CP and LF sectors. This may be because provinces with a higher per capita arable land are mainly grain-producing areas, such as the northeast provinces, compared to other provinces producing cash crops, which have relatively lower value-added products, resulting in relatively higher GHG intensity (GHG emissions per unit of value-added).

4. Construction of a Synergistic GHG Reduction System

As agriculture plays a fundamental role in food supply, emission reduction measures in the CP and LF sectors should ensure a coordinated and comprehensive approach. It is necessary to guarantee food security while reducing GHG emissions in the production process, contributing to the achievement of the carbon reduction goals. In the context of dual-carbon goals, based on the differences in spatial distribution and influencing mechanisms of the CP and LF sectors, this study first proposes specific emission reduction measures tailored to each sector. Then, based on the synergy of multiple measures, it further progresses to the coordination of crop–livestock integration and regional coordination, proposing a strategic framework for coordinated emission reduction at three levels (Figure 6). This framework aims to provide guidance and reference for achieving dual goals of agricultural GHG reduction and food security.

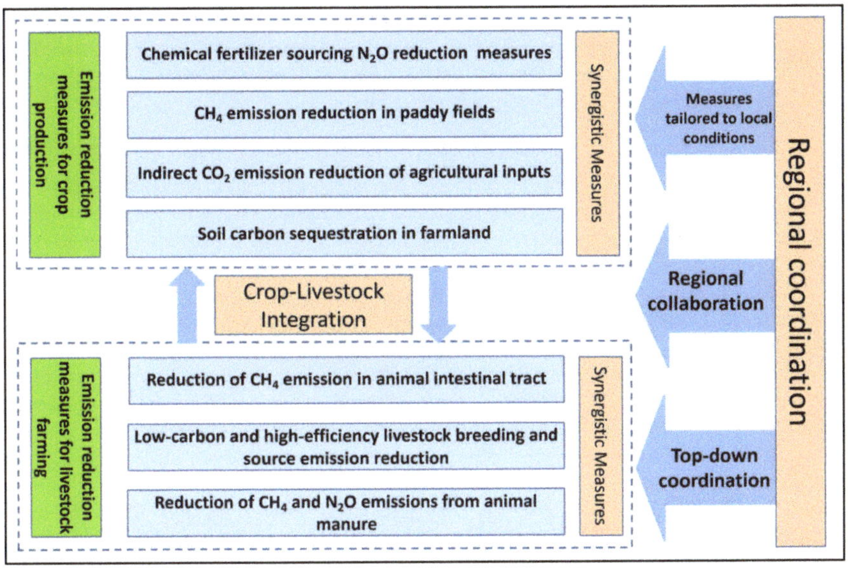

Figure 6. Framework for coordinated emission reduction measures at three levels.

4.1. Synergistic Measures

Because of the significant differences in the spatial distribution and influencing mechanisms of the CP and LF sectors, specific emission reduction measures need to be formulated for each sector. For the CP sector, while ensuring food supply, measures such as improving nitrogen fertilizer efficiency, optimizing irrigation patterns, developing nitrification inhibitors, and exploring new nitrogen fertilizer technologies can reduce emissions of N_2O from fertilizer sources [46]. Measures such as precise fertilizer regulation, optimizing cultivation practices, implementing organic matter return, and optimal water management can help reduce CH_4 emissions from paddy fields. Implementing plans to reduce inputs and increase the efficiency of agrochemicals such as fertilizers, pesticides, and agricultural films can indirectly achieve GHG reduction by improving agricultural eco-efficiency and reducing input quantities. In addition to emissions reduction, the carbon sequestration capacity of soil should not be overlooked. By promoting the development of technologies related to soil organic carbon, emission reduction, and carbon sequestration can be achieved in coordination. For the LF sector, the main sources of GHG emissions are enteric fermentation and manure management. Measures such as precision feeding, rapid CH_4 monitoring, and optimizing feeding structures can be implemented to reduce CH_4 emissions from enteric fermentation. Regarding manure management, measures such as manure return to fields, biogas utilization, and inhibition of GHG synthesis can be adopted to reduce emissions. Furthermore, promoting standardized and ecological farming practices and optimizing the structure of LF can be effective means of reducing GHG emissions.

4.2. Crop-Livestock Integration

Continued efforts should be made to promote the transformation of agricultural production towards a circular "resources—products—renewable resources—products" mode and accelerate the development of integrated crop-livestock circular agriculture. This approach will achieve overall economic, ecological, and social benefits greater than the sum of its parts. Promoting the recycling of crop straws is an important step. Establishing a sound system for straw collection, storage, and transportation, promoting the utilization of straw as feed, and popularizing technologies such as straw silage, baling, ammonization, and pellet production can serve as a linkage for driving integrated crop-livestock systems. Additionally, the utilization of livestock manure for biogas production, through the construction of biogas digesters, can tightly connect the livestock and crop sectors, achieving the integrated development of crop-livestock systems and biogas industries. This can effectively reduce agricultural GHG emissions and achieve energy substitution for energy savings and emission reductions in other sectors.

4.3. Regional Coordination

Given the spatial heterogeneity of the CP and LF sectors and their different mechanisms of factors, regional synergy in agricultural GHG reduction should be implemented from three aspects: (a) Measures need to be tailored to local conditions. Considering varying economic, social, and agricultural conditions, each province or region should formulate GHG reduction policies in the CP and LF sectors that are tailored to their specific circumstances. For example, agricultural financial support policies can increase CP GHG emissions intensity for both local provinces and neighboring provinces, but for the LF sector, it can significantly reduce GHG intensity in the local provinces and surrounding provinces. Therefore, using agricultural financial support policies to achieve GHG emissions reduction goals is only applicable to major livestock-producing provinces, while traditional major CP provinces may need to rely on other measures. (b) Regional collaboration is crucial. Both the CP and LF sectors have evident spatial spillover effects on GHG intensity. One single factor that reduces local GHG intensity may also affect or even increase GHG intensity in neighboring areas. This "domino effect" necessitates increased cooperation among provinces when formulating relevant GHG reduction measures. Joint exploration of GHG reduction technologies and improved agricultural resource utilization efficiency should be pursued. (c) Top-down coordination is necessary. At the national level, a unified strategy should be employed, considering overall grain supply and food security. This involves coordinating and optimizing the production layout of the CP and LF sectors.

5. Conclusions

Broadly defined, GHG emissions in agriculture include both the CP and LF sectors. However, studying them as a whole may obscure or weaken the micro-level spatial characteristics and specific mechanisms of factors. This study separates the CP and LF sectors from the macro "agriculture" and conducts separate research on their GHG emissions characteristics. Spatial econometric models are used to explore and compare the spatial characteristics and mechanisms of factors of both sectors. A system of coordinated measures, integrated crop-livestock production, and regional collaboration for emissions reduction is then proposed. The main conclusions and policy implications are as follows.

The spatial distribution of GHG emissions in the CP and LF sectors is consistent with the spatial patterns in each province. Because the GHG emission intensity of LF is much higher than that of the CP sector, and the GHG intensity of rice planting is higher than that of other planting, the GHG emission intensity of all provinces shows obvious spatial heterogeneity. The growth rate of total agricultural GHG has slowed down sharply and is even showing signs of an inflection point in recent years due to the significant drop in intensity for both sectors caused by the increase in agricultural productivity. At a critical time when agricultural GHG is approaching the peak point and with the need for carbon neutrality, further improvement in agricultural productivity is necessary. However, agriculture is a prerequisite for human survival and development, and the GHG reduction in this field must be made only if food supplies are met.

The magnitude of the impact of different factors on GHG intensity in the CP and LF sectors also varies dramatically. Traditionally, the increase of agricultural financial support and mechanization level are all important policy tools to boost agricultural productivity. However, our more specific empirical research showed that the increase in agricultural financial support and mechanization level can increase the GHG intensity of the CP sector while decreasing the GHG intensity in the LF sector. Other factors also affect both CP and LF sectors at different magnitudes and directions, indicating that agricultural GHG reduction policies need to be tailored to specific sectors. The spatial spillover effects of both sectors also have important policy implications. When formulating a certain policy tool to reduce local GHG emissions, its increasing effect on the GHG of neighboring areas must be considered comprehensively, which requires the coordination of higher-level authorities. Provinces with higher CP GHG intensity are often the main food-supplying regions that play a strategic role in the whole country or even worldwide, so when considering GHG reduction, their contribution to food security should be emphasized.

This study has certain inadequacies, which require further research in the future. Although a more detailed study than previous research was conducted, the classification of the sectors still needs to be more specific. Research on specific crop species or animal types is necessary. Furthermore, our investigation focuses on a provincial perspective. In the future, the following research should deepen the research to a more microscopic level. The research at the county level is of greater significance to the micro-level GHG emission mechanism and GHG reduction policies.

Author Contributions: Conceptualization, J.Q. and J.H.; methodology, J.H.; software, J.H.; validation, J.H. and D.W.; formal analysis, J.H.; data curation, D.W.; writing—original draft preparation, J.H.; writing—review and editing, D.W. and T.N.M.; visualization, J.H. and D.W.; supervision, J.Q.; project administration, J.Q.; funding acquisition, J.Q. All authors have read and agreed to the published version of the manuscript.

Funding: This research was funded by the General Program of the National Natural Science Foundation of China (No. 42171300), the Consulting Program of Academic Divisions of the Chinese Academy of Sciences (No. E3C00008) and Innovation Fund of National Science Library (Chengdu), Chinese Academy of Sciences (No. E3Z0000808).

Data Availability Statement: Not applicable.

Conflicts of Interest: The authors declare no conflict of interest.

Appendix A

Table A1. Coefficients for N_2O from crop cultivation.

Sources	Emission Coefficients ($kg{\cdot}hm^{-2}$)
Rice	0.24
Spring-season wheat	0.4
Winter-season wheat	1.75
Soybean	2.29
Maize	2.532
Vegetables	4.944
Other dryland crops	0.95

Table A2. Coefficients for indirect emissions from agricultural inputs.

GHG Sources	Emission Coefficients
Pesticide	4.9341 $kg{\cdot}kg^{-1}$
Chemical fertilizer	0.8956 $kg{\cdot}kg^{-1}$
Agricultural film	5.18 $kg{\cdot}kg^{-1}$
Agricultural irrigation	266.48 $kg{\cdot}hm^{-2}$
Agricultural machinery	0.18 $kg{\cdot}kW^{-1}$
Agricultural energy(diesel)	0.5927 $kg{\cdot}kg^{-1}$

Table A3. Coefficients for CH_4 emissions from paddies.

Provinces	Early-Season Rice	Mid-Season Rice	Late-Season Rice
Beijing	0	13.23	0
Tianjin	0	11.34	0
Hebei	0	15.33	0
Shanxi	0	6.62	0
Inner Mongolia	0	8.93	0
Liaoning	0	9.24	0
Jilin	0	5.57	0
Heilongjiang	0	8.31	0
Shanghai	12.41	53.87	27.5
Jiangsu	16.07	53.55	27.6
Zhejiang	14.37	57.96	34.5
Anhui	16.75	51.24	27.6
Fujian	7.74	43.47	52.6
Jiangxi	15.47	65.42	45.8
Shandong	0	21	0
Henan	0	17.85	0
Hubei	17.51	58.17	39
Hunan	14.71	56.28	34.1
Guangdong	15.05	57.02	51.6
Guangxi	12.41	47.78	49.1
Hainan	13.43	52.29	49.4
Chongqing	6.55	25.73	18.5
Sichuan	6.55	25.73	18.5
Guizhou	5.1	22.05	21
Yunnan	2.38	7.25	7.6
Tibet	0	6.83	0
Shaanxi	0	12.51	0
Gansu	0	6.83	0
Qinghai	0	0	0
Ningxia	0	7.35	0
Xinjiang	0	10.5	0

Table A4. Coefficients for GHGs from ruminant activities and manure management.

Sources	CH_4 from Ruminant Activities (kg per Year)	CH_4 from Manure Management (kg per Year)	NO_2 from Manure Management (kg per Year)
Non-dairy cattle	51.4	1.5	1.37
Dairy cattle	68	16	1
Horses	18	1.64	1.39
Donkeys	10	0.9	1.39
Mules	10	0.9	1.39
Pigs	1	3.5	0.53
Sheep	5	0.16	0.33

References

1. Qu, J.S.; Han, J.Y.; Liu, L.N.; Xu, L.; Li, H.J.; Fan, Y.J. Inter-provincial correlations of agricultural GHG emissions in China based on social network analysis methods. *China Agric. Econ. Rev.* **2020**, *13*, 229–246. [CrossRef]
2. IPCC. *Climate Change 2007: Mitigation of Climate Change. Contribution of Working Group III to the Fourth Assessment Report of the Intergovernmental Panel on Climate Change*; Cambridge University Press: Cambridge, UK, 2007.
3. Beach, R.H.; Deangelo, B.J.; Rose, S.; Li, C.; Salas, W.; DelGrosso, S.J. Mitigation potential and costs for global agricultural greenhouse gas emissions. *Agric. Econ.* **2008**, *38*, 109–115. [CrossRef]
4. Tubiello, F.N.; Salvatore, M.; Golec, R.D.C.; Ferrara, A.; Rossi, S.; Biancalani, R.; Federici, S.; Jacobs, H.; Flammini, A. Agriculture, forestry and other land use emissions by sources and removals by sinks: 1990–2011 analysis. In *FAO Statistics Division*; FAO: Rome, Italy, 2014; Volume 4, pp. 375–376.
5. Zeng, J.J.; Han, J.Y.; Qu, J.S.; Maraseni, T.N.; Xu, L.; Li, H.J.; Liu, L.N. Ecoefficiency of China's agricultural sector: What are the spatiotemporal characteristics and how are they determined? *J. Clean. Prod.* **2021**, *325*, 129346. [CrossRef]
6. Zhang, L.; Pang, J.X.; Chen, X.P.; Lu, Z.M.N. Carbon emissions, energy consumption and economic growth: Evidence from the agricultural sector of China's main grain-producing areas. *Sci. Total Environ.* **2019**, *665*, 1017–1025. [CrossRef] [PubMed]
7. Liu, Y.S.; Zou, L.L.; Wang, Y.S. Spatial-temporal characteristics and influencing factors of agricultural eco-efficiency in China in recent 40 years. *Land Use Policy* **2020**, *97*, 104794. [CrossRef]
8. He, Y.Q.; Cheng, X.Y.; Wang, F.; Cheng, Y. Spatial correlation of China's agricultural greenhouse gas emissions: A technology spillover perspective. *Nat. Hazards* **2020**, *104*, 2561–2590. [CrossRef]
9. Han, M.; Zhang, B.; Zhang, Y.; Guan, C.H. Agricultural CH_4 and N_2O emissions of major economies: Consumption-vs. production-based perspectives. *J. Clean. Prod.* **2019**, *210*, 276–286. [CrossRef]
10. Zhang, H.Y.; Xu, Y.; Lahr, M. The greenhouse gas footprints of China's food production and consumption (1987–2017). *J. Environ. Manag.* **2021**, *301*, 113934. [CrossRef]
11. Lepadatu, C. Effects of market reform on agricultural policy community and rural areas. *Sci. Pap.* **2012**, *12*, 39–42.
12. Cui, Y.; Khan, S.U.; Deng, Y.; Zhao, M. Regional difference decomposition and its spatiotemporal dynamic evolution of Chinese agricultural carbon emission: Considering carbon sink effect. *Environ. Sci. Pollut. Res.* **2021**, *28*, 38909–38928. [CrossRef]
13. Xu, X.C.; Zhang, N.; Zhao, D.X.; Liu, C.J. The effect of trade openness on the relationship between agricultural technology inputs and carbon emissions: Evidence from a panel threshold model. *Environ. Sci. Pollut. Res.* **2021**, *28*, 9991–10004. [CrossRef] [PubMed]
14. He, Y.Q.; Wang, H.C.; Chen, R.; Hou, S.Q.; Xu, D.D. The Forms, Channels and Conditions of Regional Agricultural Carbon Emission Reduction Interaction: A Provincial Perspective in China. *Int. J. Environ. Res. Public Health* **2022**, *19*, 10905. [CrossRef] [PubMed]
15. Gu, R.L.; Duo, L.H.; Guo, X.F.; Zou, Z.L.; Zhao, D.X. Spatiotemporal heterogeneity between agricultural carbon emission efficiency and food security in Henan, China. *Environ. Sci. Pollut. Res.* **2023**, *30*, 49470–49486. [CrossRef] [PubMed]
16. Zhu, Y.; Huo, C.Y. The impact of agricultural production efficiency on agricultural carbon emissions in China. *Energies* **2022**, *15*, 4464. [CrossRef]
17. Wu, H.; Huang, H.; Tang, J.; Chen, W.; He, Y. Net greenhouse gas emissions from agriculture in China: Estimation, spatial correlation and convergence. *Sustainability* **2019**, *11*, 4817. [CrossRef]
18. Wu, H.J.; Yuan, Z.W.; Geng, Y.; Ren, J.Z.; Jiang, S.Y.; Sheng, H.; Gao, L.M. Temporal trends and spatial patterns of energy use efficiency and greenhouse gas emissions in crop production of Anhui province, China. *Energy* **2017**, *133*, 955–968. [CrossRef]
19. Huang, X.Q.; Feng, C.; Qin, J.H.; Wang, X.; Zhang, T. Measuring China's agricultural green total factor productivity and its drivers during 1998–2019. *Sci. Total Environ.* **2022**, *829*, 154477. [CrossRef]
20. Han, J.Y.; Qu, J.S.; Maraseni, T.N.; Xu, L.; Zeng, J.J.; Li, H.J. A critical assessment of provincial-level variation in agricultural GHG emissions in China. *J. Environ. Manag.* **2021**, *296*, 113190. [CrossRef]
21. Su, L.J.; Wang, Y.T.; Yu, F.F. Analysis of regional differences and spatial spillover effects of agricultural carbon emissions in China. *Heliyon* **2023**, *9*, E16752. [CrossRef]
22. West, T.O.; Marland, G. A synthesis of carbon sequestration, carbon emissions, and net carbon flux in agriculture: Comparing tillage practices in the United States. *Agric. Ecosyst. Environ.* **2022**, *91*, 217–232. [CrossRef]

23. Wu, Y.G.; Feng, K.W. Spatial-temporal differentiation features and correlation effects of provincial agricultural carbon emissions in China. *Environ. Sci. Technol.* **2019**, *42*, 180–190.
24. Xiong, C.H.; Yang, D.G.; Huo, J.W.; Wang, G.L. Agricultural net carbon effect and agricultural carbon sink compensation mechanism in Hotan Prefecture, China. *Pol. J. Environ. Stud.* **2017**, *26*, 365–373. [CrossRef] [PubMed]
25. Adams, S.; Acheampong, A.O. Reducing carbon emissions: The role of renewable energy and democracy. *J. Clean. Prod.* **2019**, *240*, 118245. [CrossRef]
26. Warner, J.; Zawahri, N. Hegemony and asymmetry: Multiple-chessboard games on transboundary rivers. *Int. Environ. Agreem. Politics Law Econ.* **2012**, *2*, 215–229. [CrossRef]
27. Zhou, W.F.; He, J.; Liu, S.Q.; Xu, D.D. How does trust influence farmers' low-carbon agricultural technology adoption? Evidence from rural Southwest, China. *Land* **2023**, *12*, 466. [CrossRef]
28. Johnson, J.M.F.; Franzluebbers, A.J.; Weyers, S.L.; Reicosky, D.C. Agricultural opportunities to mitigate greenhouse gas emissions. *Environ. Pollut.* **2007**, *150*, 107–124. [CrossRef]
29. Li, B.; Zhang, J.B.; Li, H.P. Research on spatial-temporal characteristics and affecting factors decomposition of agricultural carbon emission in China. *China Popul. Resour. Environ.* **2011**, *21*, 80–86.
30. Wu, G.Y.; Liu, J.D.; Yang, L.S. Dynamic evolution of China's agricultural carbon emission intensity and carbon offset potential. *China Popul. Resour. Environ.* **2021**, *31*, 69–78.
31. Anselin, L.; Syabri, I.; Kho, Y. GeoDa: An Introduction to Spatial Data Analysis. *Geogr. Anal.* **2006**, *38*, 5–22. [CrossRef]
32. Sridharan, S.; Tunstall, H.; Lawder, R.; Mitchell, R. An exploratory spatial data analysis approach to understanding the relationship between deprivation and mortality in Scotland. *Soc. Sci. Med.* **2007**, *65*, 1942–1952. [CrossRef]
33. Lesage, J.; Pace, R. *Introduction to Spatial Econometrics*; CRC Press/Taylor & Francis: Boca Raton, FL, USA, 2009.
34. Guo, H.P.; Fan, B.J.; Pan, C.L. Study on mechanisms underlying changes in agricultural carbon emissions: A case in Jilin Province, China, 1998–2018. *Int. J. Environ. Res. Publ. Health.* **2021**, *18*, 919. [CrossRef] [PubMed]
35. Xiong, C.H.; Chen, S.; Xu, L.T. Driving factors analysis of agricultural carbon emissions based on extended STIRPAT model of Jiangsu Province, China. *Growth Chang.* **2020**, *51*, 1401–1416. [CrossRef]
36. Chen, Y.H.; Li, M.J.; Su, K.; Li, X.Y. Spatial-temporal characteristics of the driving factors of agricultural carbon emissions: Empirical evidence from Fujian, China. *Energies* **2019**, *12*, 3102. [CrossRef]
37. Ali, B.; Ullah, A.; Khan, D. Does the prevailing Indian agricultural ecosystem cause carbon dioxide emission? A consent towards risk reduction. *Environ. Sci. Pollut. Res.* **2021**, *28*, 4691–4703. [CrossRef]
38. Smith, W.N.; Grant, B.B.; Desjardins, R.L.; Kroebel, R.; Li, C.; Qian, B.; Worth, D.E.; McConkey, B.G.; Drury, C.F. Assessing the effects of climate change on crop production and GHG emissions in Canada. *Agric. Ecosyst. Environ.* **2013**, *179*, 139–150. [CrossRef]
39. Xu, Q.H.; Zhang, G.S. Spatial spillover effect of agricultural mechanization on agricultural carbon emission intensity: An empirical analysis of panel data from 282 cities. *China Popul. Resour. Environ.* **2022**, *32*, 23–33.
40. Zhang, D.; Shen, J.B.; Zhang, F.S.; Li, Y.E.; Zhang, W.F. Carbon footprint of grain production in China. *Sci. Rep.* **2017**, *7*, 4126. [CrossRef]
41. Yang, F. Impact of agricultural modernization on agricultural carbon emissions in China: A study based on the spatial spillover effect. *Environ. Sci. Pollut. Res.* **2023**, *30*, 91300–91314. [CrossRef]
42. Anselin, L.; Rey, S. Properties of Tests for Spatial Dependence in Linear Regression Models. *Geogr. Anal.* **1991**, *23*, 112–131. [CrossRef]
43. Zhao, P.J.; Zeng, L.E.; Lu, H.Y.; Zhou, Y.; Hu, H.Y.; Wei, X.Y. Green economic efficiency and its influencing factors in China from 2008 to 2017: Based on the super-SBM model with undesirable outputs and spatial Dubin model. *Sci. Total Environ.* **2020**, *741*, 140026. [CrossRef]
44. Yang, W.Y.; Wang, W.L.; Ouyang, S.S. The influencing factors and spatial spillover effects of CO_2 emissions from transportation in China. *Sci. Total Environ.* **2019**, *696*, 133900. [CrossRef] [PubMed]
45. Wang, G.F.; Deng, X.Z.; Wang, J.Y.; Zhang, F.; Liang, S.Q. Carbon emission efficiency in China: A spatial panel data analysis. *China Econ. Rev.* **2019**, *56*, 101313. [CrossRef]
46. Maraseni, T.N.; Mushtaq, S.; Reardon-Smith, K. Climate change, water security and the need for integrated policy development: The case of on-farm infrastructure investment in the Australian Irrigation Sector. *Environ. Res. Lett.* **2012**, *7*, 034006. [CrossRef]

land

MDPI

Article

Decoupling CO_2 Emissions from Economic Growth in China's Cities from 2000 to 2020: A Case Study of the Pearl River Delta Agglomeration

Zhixiong Wang [1,†], Fuhan Li [2,†], Zihan Xie [3,4,*], Qingyin Li [5], Yongli Zhang [5] and Meilin Dai [5]

1 Guangdong Urban & Rural Planning and Design Institute Co., Ltd., Guangzhou 510290, China; wangzhixiong@gdupi.com
2 Faculty of Hospitality and Tourism Management, Macau University of Science and Technology, Macau 999078, China; 2009853gb111040@student.must.edu.mo
3 Guangzhou Institute of China Academy of Information and Communications Technology, Guangdong Provincial Key Laboratory of Urbanization and Geo-Simulation, Guangzhou 510699, China
4 School of Geography and Planning, Sun Yat-sen University, Guangzhou 510275, China
5 School of Architecture and Urban Planning, Guangdong University of Technology, Guangzhou 510062, China; 2112110025@mail2.gdut.edu.cn (Q.L.); 2112110048@mail2.gdut.edu.cn (Y.Z.); 3219008409@mail2.gdut.edu.cn (M.D.)
* Correspondence: xiezh35@mail2.sysu.edu.cn
† These authors contributed equally to this work.

Abstract: As one of the most densely populated, economically developed, and outwardly open urban agglomerations in China, the Pearl River Delta (PRD) urban agglomeration is a key player in achieving China's carbon peak and carbon neutrality targets. This study analyzes low-emission development by examining the evolutionary patterns of carbon dioxide (CO_2) emissions and the decoupling relationship between economic growth and CO_2 emissions, using the latest available data from 2000 to 2020. Here are the main findings: (1) We found a significant fluctuation in the decoupling statuses between economic advancements and CO_2 emissions within the PRD domain. Predominantly, a weak decoupling scenario was observed, where economic proliferations were paralleled by nearly equivalent increments in CO_2 emissions. (2) The growth rate of carbon emissions increased significantly relative to economic expansion during 2015–2020, especially pronounced in cities such as Guangdong, Zhuhai, Foshan, and Dongguan. This delineates the persistent challenges in steering towards a pathway of energy conservation and emission abatement in the region. (3) Furthermore, a differential role of elasticity factors was noted across cities: Guangzhou and Shenzhen witnessed a significant influence of energy-saving elasticity in fostering a decoupling between economic surge and CO_2 emissions, whereas in other cities, the emphasis shifted towards emission-reduction elasticity as a more vital determinant. The results of this study are of great significance for guiding policy makers and stakeholders in urban clusters across China and in similar regions globally to achieve low carbon development goals.

Keywords: CO_2 emissions; decoupling; tapio indicator; urban agglomeration; China

1. Introduction

The flourishing development of the global economy and acceleration of urbanization and industrialization have led to a significant increase in irreversible fossil energy consumption and an unprecedented rise in carbon dioxide (CO_2) emissions [1–3]. Global warming is now a serious worldwide public environmental concern [4,5]. China, being the largest CO_2 emitter globally, has set emission-reduction targets to reach an emission peak before 2030 [6]. Achieving this ambitious goal will require region-level efforts, particularly in urban carbon reduction. Urban agglomeration has emerged as a key issue in reducing emissions, and China has implemented policy measures to promote the development of urban agglomeration and use its effects to reduce emissions.

The formation, development, and expansion of urban clusters are clearly the next major trend in urban spatial organization, in the dual processes of global urbanization and economic globalization. According to the 2015 Sustainable Competitiveness of Cities Worldwide Report by the United Nations Habitat Assembly, the world's giant cities have begun to merge into super-giants or super-urban clusters [7]. Because urban economic activity accounts for approximately 80% of global gross domestic product (GDP) values, 66% of energy consumption, and a staggering 70% of carbon emissions [8], this trend clearly indicates that urban clusters will become one of the most important geographical units for reducing emissions. Therefore, it is essential to focus on sustainable development strategies in urban clusters to reduce their impact on the environment and promote a low-carbon future. China has undergone industrialization and urbanization at a pace and scale that surpasses any other country in the world, with many new cities emerging and growing in the process [9–11]. Among them, the Pearl River Delta (PRD) Urban Agglomeration is a world-class urban agglomeration in China [12]. The PRD Urban Agglomeration, located in the central-southern part of Guangdong Province. It covers approximately 54,000 km^2 and comprising nine cities, this region encapsulates a mere 20% of Guangdong's land area but contributes over 80% to the provincial GDP [13,14]. This dynamic urban cluster, which contributed 8.65% to the nation's GDP in 2022, not only supports the economic spine of the province but also plays a pivotal role in China's One Belt, One Road (Belt and Road) Initiative.

Despite its economic prowess, the PRD's growth model has leaned heavily on resource consumption, manifesting a high reliance on fossil fuels and a notable environmental toll [15]. The region's transformation of nonurban land into urban spaces to facilitate its growth has been at the expense of natural ecosystems, farmland, and water bodies. The current socio-economic structure, centered on resource consumption, poses a significant obstacle to the transition towards a green and low-carbon urban development, with limited synergistic benefits observed in terms of resource environment, economic development, and social well-being at this stage.

As China navigates the era of low-carbon economies, the PRD stands at the forefront of efforts aiming for sustainable development and emission reduction, notably under the Greater Bay Area project. Addressing greenhouse gas emissions here is not just imperative but holds the promise of showcasing a model of sustainable development in China. However, to truly materialize this vision, understanding and addressing the CO_2 emissions intricacies within the PRD is vital. In this context, previous studies that concentrated primarily on regional CO_2 emission differences appear to have missed the nuanced developmental disparities across cities within the PRD. This leaves a significant gap in our understanding of regional CO_2 emissions and their underlying complexities [16]. Furthermore, overlooking developmental differences within the city cluster can hinder the targeted control of CO_2 emissions, leading to unbalanced regional environmental development. Considering the imbalance in regional development, it is beneficial to formulate emission-reduction policies suitable for the development of different cities rather than using a one-size-fits-all approach. Therefore, this study focuses on analyzing CO_2 emissions in the PRD from the perspective of regional development imbalances and proposes targeted emission-reduction measures for the region.

The complex nexus between regional economic development, energy consumption, and carbon emissions has held a prominent place in scholarly and policy dialogues globally, serving as a critical determinant in shaping sustainable developmental strategies [17–19]. This relationship, often convoluted, brings forward a multi-faceted interaction where the aggressive pursuit of economic growth sometimes translates into escalated levels of energy consumption and heightened carbon footprints, affecting climate change patterns and environmental stability.

Globally, a myriad of studies has ventured deep into understanding these interlinkages, unfolding varying dynamics across different nations and regions. For instance, Chontanawat et al. (2008) embarked on a systematic analysis encompassing over a hundred countries, thereby presenting an elaborate canvas that allows for international comparison and policy crafting [20]. Complementing this, Bella et al. (2014) further analyzed the relationship within OECD countries, examining the interdependencies between CO_2 emissions, electrical power consumption, and GDP, which offers insightful inputs into understanding the dynamics in developed nations [21]. Furthermore, the collective work of Ntanos et al. (2015) charted the global trends in energy consumption and CO_2 emissions, presenting an encompassing view that helps to dissect the global patterns and implications [22].

In this ever-expanding urban landscape marked by relentless industrialization, cities globally are grappling with escalating carbon emissions, which in turn place an enormous strain on the environment. This forms a critical backdrop for the intensified scholarly and policy focus on the decoupling of economic growth from resource consumption and environmental degradation. As cities witness rapid expansion and the continuation of industrialization, there has been an increasing trend in carbon emissions, exerting a substantial burden on the environment. In this context, the decoupling of economic growth from resource consumption and environmental pollution emerges as a pivotal and pressing topic of exploration. In recent years, the theory of decoupling has served as a vital tool in assessing and scrutinizing the relationship between economic growth and CO_2 emissions. Initially rooted in the field of physics to measure disruptions or disconnections between physical quantities, it has evolved to encompass the intricate interplay between the economy and the environment. To mitigate the challenges brought forth by urban expansion, the Organization for Economic Co-operation and Development (OECD) introduced the concept of decoupling, aiming to sever the links between economic growth, resource consumption, and environmental degradation. Using the OECD decoupling theory, de Freitas and Kaneko [23] investigated the complete decoupling of Brazil's energy consumption, CO_2 emissions, and economic growth rate from 2004 to 2009. Tapio's decoupling elasticity method [24] used the European transportation industry to examine the relationship between economic growth and CO_2 emissions, indicating a weak decoupling between CO_2 emissions in the transportation industry and GDP. Since then, Tapio's elasticity decoupling theory has gained widespread application in various fields [8,25,26]. For example, Raza and Lin [27] used Tapio's decoupling method to estimate the decoupling status and mitigation potential of CO_2 emissions from the transport sector in Pakistan for 1984 to 2018. Lin et al. [28] employed the OECD and Tapio decoupling analysis to evaluate the relationship between CO_2 emissions and GDP in South Africa from 1990 to 2012. The findings indicated that South Africa experienced expansive negative decoupling between CO_2 emissions and GDP during 1990 to 1994, followed by weak decoupling between 1994 and 2010, and achieved strong decoupling from 2010 to 2012.

The examination of the decoupling relationships between energy consumption, economic growth, and CO_2 emissions has been extensively studied through various research models. Notwithstanding, the majority of these investigations have been conducted at national or interprovincial scale [29–31], leaving a noticeable gap in literature concerning the urban scale. Moreover, these research models predominantly focus on total CO_2 emissions, offering an overly macroscopic view and neglecting the pivotal aspects of energy conservation and emission reduction elasticity.

In light of the existing literature and methodologies employed, our study ventures to fill this research void by introducing a nuanced approach to the decoupling analysis. We innovatively decompose decoupling elasticity into two distinct facets: emission reduction elasticity and energy-saving elasticity. This methodological innovation permits an in-depth exploration of the primary factors influencing the decoupling of CO_2 emissions from economic growth, extending the understanding of decoupling dynamics at the urban agglomeration level.

This study presents several noteworthy contributions to the existing body of literature. Firstly, it pioneers in spotlighting the disparities in CO_2 emissions within the PRD urban agglomerations, offering a fresh and critical lens to scrutinize emission reduction strategies. This nuanced approach potentially sets a precedent for similar analyses in other urban agglomerations, fostering a more comprehensive understanding of urban-scale decoupling phenomena. Secondly, it enhances the existing decoupling theory which has been primarily applied at broader scales like national, regional, or sectoral levels, by zooming in on the intricacies of urban agglomerations. By meticulously dissecting the decoupling index into emission-reduction and energy-saving elasticity components, this study not only sheds light on the multifaceted nature of decoupling processes but also paves the way for crafting more targeted and effective strategies for sustainable urban development.

2. Data and Methods

2.1. *Tapio Decoupling Method*

The OECD defined decoupling as breaking the link between economic growth and environmental degradation, and developed a method for calculating decoupling that has been widely used since then [23]. However, the OECD decoupling index is too sensitive to the choice of the base period, and thus the stability of its calculation results is poor [32]. Based on the OECD decoupling model, Tapio introduced the concept of decoupling elasticity to construct the decoupling index, which addressed the difficulties in choosing the base period in the OECD decoupling model [33]. Tapio's decoupling theory provides reasonable decoupling positions for eight possible combinations of environmental pollution variables and economic variables, and is currently the most widely used method to study decoupling relationships [8].

$$E_{(C,\ GDP)} = \%\Delta C \Big/ \%\ GDP = \frac{(C^t - C^0)/C^0}{\left(GDP^t - GDP^0\right)/GDP^0} \tag{1}$$

2.2. *Decoupling Decomposition Method*

In expanding upon the Tapio decoupling method, this study conceptualizes a framework for analyzing the relationships among economic activities, energy consumption, and carbon emissions. The approach integrates insights from the Tapio decoupling model with the Economy–Energy–Environment (3E) system theory to foster a deeper understanding of the interactions between these elements.

In this analysis, energy consumption is introduced as a mediating decomposition variable in the decoupling model between economic growth and CO_2 emissions. Thus, a causal chain for the decoupling model is formulated as shown in Equation (2):

$$E_{(C,\ GDP)} = \frac{\%\Delta C}{\%\Delta EC} \times \frac{\%\Delta EC}{\%\Delta GDP} = \frac{(C^t - C^0)/C^0}{\left(EC^t - EC^0\right)/EC^0} \times \frac{\left(EC^t - EC^0\right)/C^0}{\left(GDP^t - GDP^0\right)/GDP^0} \tag{2}$$

where $\frac{\%\Delta C}{\%\Delta GDP}$ represents the decoupling elasticity between economic growth and CO_2 emissions; $\frac{\%\Delta C}{\%\Delta EC}$ indicates the emission-reduction elasticity between energy consumption and CO_2 emissions; and $\frac{\%\Delta EC}{\%\Delta GDP}$ is the energy-saving elasticity between economic growth and energy consumption.

2.3. *Data Sources*

The CO_2 emission data used in this paper are based on those provided by the China Carbon Accounts and Datasets (CEADs) (https://www.ceads.net/data/county/, accessed on 10 June 2023), which are derived from the inversion of Defense Meteorological Satellite Program/the Operational Linescan System (DMSP/OLS) and National Polar-orbiting Partnership/Visible Infrared Imager Radiometer Suite (NPP/VIIRS) nighttime lighting data provided by the National Geomatics Center of China (NGCC) from 1997–2017. The two sets of DMSP/OLS and NPP/VIIRS sensors can finely capture low-intensity nighttime light sources produced by urban centers or even by small-scale residential land or traffic, which provides an ideal data source for estimating energy consumption from human activities [34]. In the calibration of nighttime lighting data, Chen et al. [35] adopted the particle swarm optimization-back propagation (PSO-BP) algorithm to unify DMSP/OLS and NPP/VIIRS satellite images, which resulted in high-quality, stable nighttime light data with long time span, wide coverage, and uniform aperture. This data is not only useful for remote sensing, but also for population distribution, urban expansion, GDP prediction, and pollutant estimation [36]. This data is publicly available in the China Carbon Accounting Database, which has been widely used in academic CO_2 emission-related studies [15].

3. Results and Discussion

3.1. Spatiotemporal Characteristics of Energy Consumption and CO_2 Emissions

Figure 1 shows the CO_2 emission structure of the PRD urban agglomeration from 2000 to 2020. The share of carbon emissions from non-PRD cities in the province increased by 10.92% in 2020 relative to the base period of the study as a result of accelerated economic development and urbanization. However, the PRD is still the main area affecting changes in the carbon emission levels in Guangdong Province. Among the cities in the PRD, Guangzhou's total carbon emissions are significantly higher due to the influences of population size, economic growth, industrial manufacturing, and industrial structure. The total carbon emissions of Foshan and Dongguan are just behind those of Guangzhou. This is mainly because these two cities have been influenced by industrial transfers in Hong Kong, Macao, Guangzhou, and Shenzhen in recent years. Shenzhen has knowledge-, technology-, and capital-intensive industries as its pillar industries, among which high-tech, cultural and creative, and financial industries have large shares. Therefore, Shenzhen is less dependent on energy, and even though it has the largest GDP, its carbon emission level is not the highest among the PRD cities.

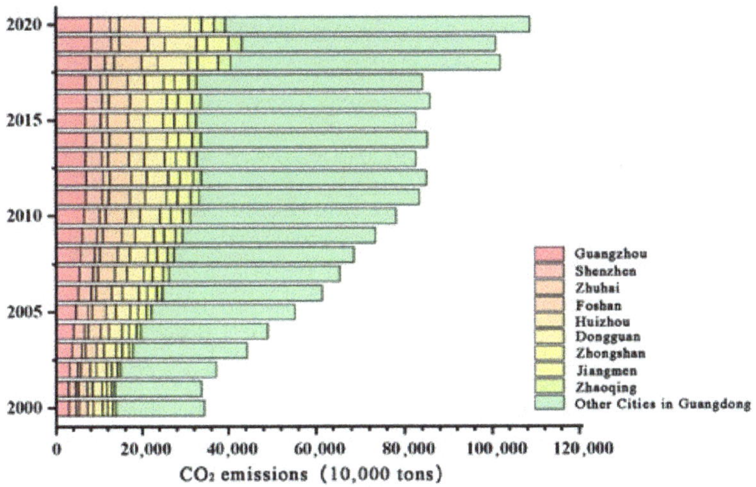

Figure 1. CO_2 Emissions of cities in the Pearl River Delta (PRD) urban agglomeration, 2000–2020.

The per capita CO_2 emissions and CO_2 intensity indicators exclude the influences of city size and total economic volume; thus, they can more objectively compare the environmental pressures brought by socioeconomic activities (Figure 2). The overall CO_2 intensities of PRD cities show fluctuating declines from 2000 to 2020, indicating that the environmental pressure brought by economic development has decreased. and the development mode gradually changed from high emissions and high pollution to low-carbon and efficient development. The CO_2 intensities of Guangzhou, Shenzhen, and Foshan cities declined more slowly than those of the other six PRD cities. However, the effect of energy consumption due to technological progress may have caused the CO_2 intensity of cities such as Jiangmen, Zhaoqing, Dongguan, and Foshan to rebound after 2018. As of 2020, Guangzhou and Shenzhen have reached low levels of carbon emission intensity at 0.32 t per million yuan and 0.16 t per million yuan, respectively. This indicates that these two cities have reduced carbon emissions through efficient use of energy and application of carbon-reducing technologies while experiencing steady economic growth. Zhaoqing is at the highest level in the PRD region with 1.12 t per million yuan.

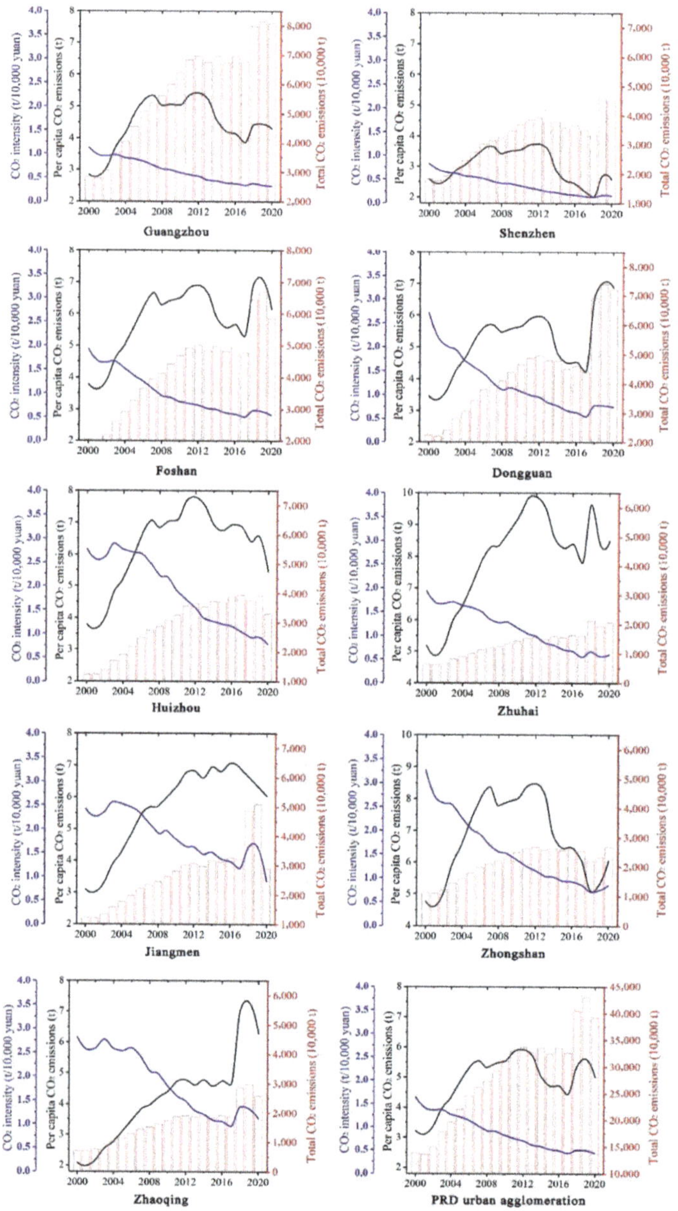

Figure 2. Trends of total CO_2 emissions, CO_2 intensity (blue lines), and per capita CO_2 emissions (black lines) in the Pearl River Delta (PRD) urban agglomeration, 2000–2020.

Figures 3 and 4 depict the spatial and temporal evolutions of the total CO_2 emissions, per capita CO_2 emissions, and CO_2 intensities of the PRD from 2000 to 2020. The gaps among the total CO_2 emissions of the cities in the PRD have widened significantly, and the polarization is more serious by 2020 (Figure 3). The overall spatial distribution revolves around the three cities of Guangzhou, Foshan, and Dongguan, presenting a high center and low periphery pattern indicating a close correlation between carbon emissions and the level of urban economic development with central aggregation.

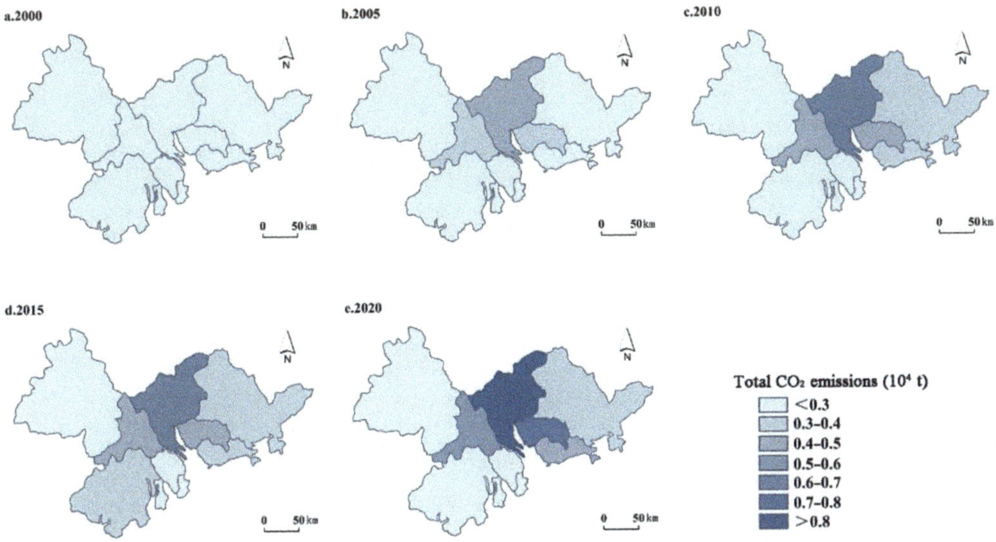

Figure 3. Spatiotemporal evolution of total CO_2 emissions in the Pearl River Delta (PRD) urban agglomeration, 2000–2020.

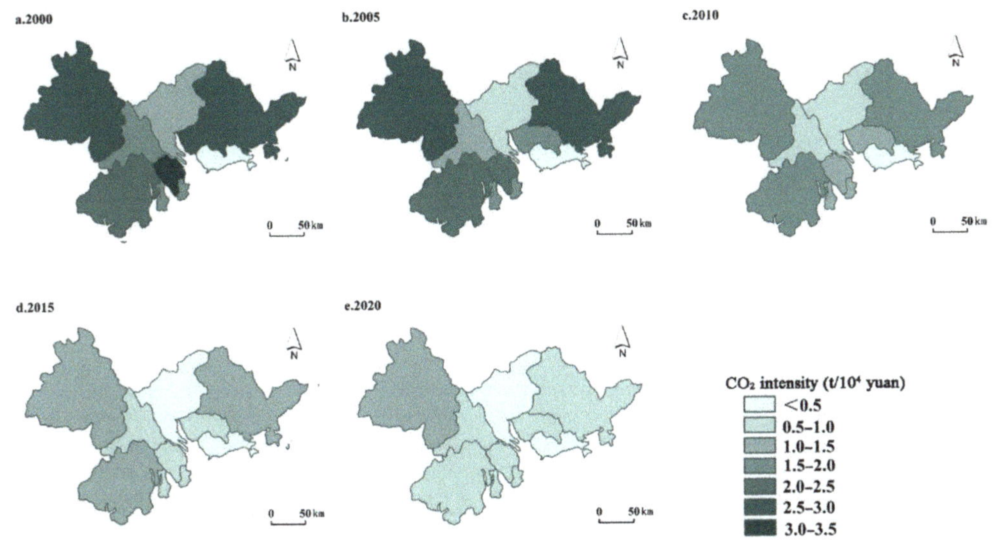

Figure 4. Spatiotemporal evolution of CO_2 intensity in the Pearl River Delta (PRD) urban agglomeration, 2000–2020.

The spatial distribution of carbon emission intensity (Figure 4) is opposite to that of the total carbon emissions, with low values in the center and high values on the periphery. Comparing the spatial distribution patterns of 2000 and 2020, this depression-like feature tends to weaken, indicating a narrowing of regional differences in carbon emission intensity and a more balanced impact of economic development on environmental pressure within the region.

3.2. Decoupling Statuses of Economic Growth from CO_2 Emissions

In recent years, the topic of regional economic development and its correlation with carbon emissions has become a focal point of research. This study undertakes a profound investigation into the decoupling statuses and their transformations over the period of 2000–2020, shedding light on the dynamics and underlying factors steering the decoupling trajectory in the PRD region.

Our analysis, encapsulated in Table 1, reveals marked fluctuations in the decoupling statuses between the PRD's economic development and CO_2 emissions over the study period. This trend of fluctuating decoupling statuses resonates with other research, including a study by Zhao et al. (2017) [37], which also identified variations in decoupling states across different time spans in China. Within the study period of 2000–2020, both weak decoupling and strong decoupling can be observed between economic growth and CO_2 emissions. The majority of years experienced weak decoupling, indicating insufficient decoupling between CO_2 emissions and economic development. These fluctuations can be traced back to various factors such as high urbanization, policy shifts favoring rapid industrialization, which sometimes heightened CO_2 emission [38], and strides in technological advancements and renewable energy adoption contributing to phases of strong decoupling [26].

Table 1. The decoupling of CO_2 emissions from economic development and the trend of decoupling elasticity in the whole Pearl River Delta (PRD) region.

Time Period	ΔC/C	ΔGDP/GDP	ΔEC/EC	$E_{C, GDP}$	Decoupling State	$E_{C, EC}$	$E_{EC, GDP}$
The whole PRD region							
2000–2005	0.105	0.171	2.097	0.614	Weak decoupling	0.050	12.286
2006–2010	0.025	0.094	0.379	0.271	Weak decoupling	0.067	4.055
2011–2015	−0.031	0.041	0.078	−0.769	Strong decoupling	−0.402	1.912
2016–2020	0.012	0.028	0.195	0.442	Weak decoupling	0.064	6.938

The years 2011–2015 notably demonstrated strong decoupling, a positive indication of a shift towards a more sustainable economic model, less reliant on energy-intensive industries. This transformation hints at a burgeoning commitment to sustainable practices within the region, spurred by technological innovations and policy directives aimed at fostering energy efficiency and curtailing carbon footprints. Comparative analysis with preceding studies illustrates a consistent pattern of evolving towards a more sustainable model, a critical step in mitigating the adverse effects of economic growth on the environment.

However, the progress appears somewhat stalled in the period 2016–2020, where a reversion to weak decoupling was observed. This transition signals persistent challenges in sustaining energy conservation and emissions reduction efforts, warranting a deeper scrutiny of the underlying causal factors. The period saw a resurgence in the reliance on energy-consuming industries, dampening the gains achieved in the previous years.

In general, the decoupling of carbon emissions from economic development in the PRD improved during 2000–2015, followed by a slight deterioration during 2016–2020. Specifically, strong decoupling marked the period of 2011–2015, as carbon emissions decreased under economic growth. This indicates that during this period, the economic development of the PRD region has reduced its dependence on energy-intensive and carbon-intensive industries and has transformed towards a low-carbon and sustainable economic development model. However, during the 2016–2020, the relationship between CO_2 emissions and economic development transformed from strong decoupling in the previous period to weak decoupling. This suggests that energy conservation and emissions reduction in the PRD are still challenging.

The trends of energy-saving elasticity and emissions reduction elasticity show a significant gap. Energy-saving elasticity has high volatility, with a particularly sharp decrease in 2010–2015. Compared with the period of 2000–2015, the energy-saving elasticity has picked up in recent years, indicating that the industry in the PRD is still heavily reliant on energy consumption, and energy efficiency should increase. Expansive negative decoupling dominated the emissions reduction elasticity during all periods, indicating that the economic growth of the PRD city cluster still relies on extensive energy consumption at the cost of environmental degradation. However, during the study period, the emissions reduction elasticity has decreased, indicating that the PRD cities are placing greater emphasis on social and economic development while seeking a more low-carbon and environmentally friendly development path. They are improving energy efficiency through technological innovation.

There is significant heterogeneity in the decoupling statuses among the nine cities within the PRD (Table 2). From 2011 to 2015, seven out of nine cities in the PRD achieved strong decoupling, indicating that economic development had the least environmental pressure, and an increase in economic output was accompanied by a decrease in CO_2 emissions. This phenomenon was particularly marked in cities experiencing rapid urban expansion, where strategies to mitigate environmental impacts were more pronounced. However, with further technological advances, a rebound effect has begun to appear, in which the growth rate of the economy was larger than the increase in CO_2 emissions, leading to a worsening decoupling status from strong decoupling in 2016–2020 to weak decoupling (or expansive negative decoupling) in 2016–2020 in Guangdong, Zhuhai, Foshan, and Dongguan. This reiterates the persistent challenge of maintaining an environmental equilibrium amidst booming economic growth, a focal point of discussion in numerous previous analyses [39].

Further, the analysis of the energy-saving elasticity revealed an encouraging trend of decreased dependency on energy consumption for economic development, a sentiment echoed in other contemporary studies focusing on China's sustainable urban development trajectory [40]. Particularly in cities like Guangzhou and Shenzhen, the significant role of energy-saving elasticity in fostering sustainable economic development has been highlighted, showcasing similar patterns to other metropolises globally, where energy efficiency has emerged as a linchpin for sustainable growth [29].

The patterns of decoupling statuses observed within the PRD region serve as a significant indicator of the larger narrative of China's rapid urbanization and strategic regional development. This era of swift urban expansion is marked by an ambitious drive towards economic growth, intricately coupled with sustainable environmental practices. Within the context of the PRD region, the nuanced interplay between escalating urbanization rates, expanding city sizes, intricacies of the social system, and the pace of technological progress serves as central determinants steering both economic and environmental developments.

Within this landscape, cities like Guangzhou and Shenzhen have emerged as frontrunners, exemplifying how advancements in energy efficiency can spearhead sustainable growth, a trend echoed in other global metropolises. The variances in decoupling statuses across different cities within the PRD region underscore the necessity for nuanced approaches in policy formulation and implementation, adapting to the unique characteristics and developmental phases of each city.

Looking forward, the trajectory of the PRD region will be significantly influenced by the intertwining forces of technological advancements and green policy initiatives. The challenge lies in harnessing these dynamic forces effectively to pave a path towards a more sustainable urban development, where economic growth harmonizes with environmental conservation. Thus, leveraging the insights gained from the fluctuations in decoupling statuses, it is incumbent upon policymakers and urban planners to craft strategies that synergize economic growth objectives with sustainable environmental practices, fostering a future where prosperity and ecology exist in harmony.

Table 2. The decoupling of CO_2 emissions from economic development and the trend of decoupling elasticity in the nine cities of the Pearl River Delta (PRD) urban agglomeration.

	ΔC/C	ΔGDP/GDP	ΔEC/EC	$E_{C, GDP}$	Decoupling State	$E_{C, EC}$	$E_{EC, GDP}$
Guangzhou							
2000–2005	0.144	0.160	1.935	0.897	Expansive coupling	0.074	12.088
2006–2010	0.010	0.052	0.200	0.188	Weak decoupling	0.048	3.872
2011–2015	−0.039	0.027	−0.016	−1.436	Strong decoupling	2.412	−0.595
2016–2020	0.005	0.018	0.166	0.301	Weak decoupling	0.032	9.387
Shenzhen							
2000–2005	0.084	0.083	1.804	1.013	Expansive coupling	0.047	21.679
2006–2010	0.020	0.062	0.307	0.314	Weak decoupling	0.064	4.944
2011–2015	−0.111	0.013	−0.180	−8.466	Strong decoupling	0.619	−13.681
2016–2020	0.006	0.009	−0.040	0.640	Strong decoupling	−0.145	−4.416
Zhuhai							
2000–2005	0.053	0.105	1.713	0.500	Weak decoupling	0.031	16.295
2006–2010	0.032	0.095	0.772	0.337	Weak decoupling	0.041	8.130
2011–2015	−0.012	0.030	0.222	−0.422	Strong decoupling	−0.056	7.515
2016–2020	0.003	0.003	0.037	0.992	Expansive coupling	0.087	11.464
Foshan							
2000–2005	0.086	0.162	1.329	0.530	Weak decoupling	0.065	8.203
2006–2010	0.024	0.095	0.279	0.252	Weak decoupling	0.086	2.948
2011–2015	−0.026	0.045	0.084	−0.567	Strong decoupling	−0.306	1.855
2016–2020	0.017	0.034	0.130	0.490	Weak decoupling	0.129	3.797
Huizhou							
2000–2005	0.094	0.203	3.870	0.460	Weak decoupling	0.024	19.039
2006–2010	0.029	0.145	1.132	0.200	Weak decoupling	0.026	7.822
2011–2015	−0.007	0.071	0.497	−0.097	Strong decoupling	−0.014	7.039
2016–2020	−0.035	0.052	0.541	−0.676	Strong decoupling	−0.065	10.326
Dongguan							
2000–2005	0.098	0.284	2.024	0.345	Weak decoupling	0.049	7.117
2006–2010	0.014	0.134	0.099	0.108	Weak decoupling	0.146	0.743
2011–2015	−0.045	0.063	−0.035	−0.720	Strong decoupling	1.274	−0.565
2016–2020	0.077	0.063	0.299	1.216	Expansive negative decoupling	0.256	4.756
Zhongshan							
2000–2005	0.074	0.242	2.366	0.305	Weak decoupling	0.031	9.761
2006–2010	0.011	0.110	0.296	0.098	Weak decoupling	0.036	2.689
2011–2015	−0.030	0.044	0.056	−0.688	Strong decoupling	−0.538	1.279
2016–2020	−0.010	0.045	0.217	−0.224	Strong decoupling	−0.046	4.866
Jiangmen							
2000–2005	0.117	0.273	2.503	0.427	Weak decoupling	0.047	9.162
2006–2010	0.049	0.186	0.917	0.263	Weak decoupling	0.053	4.925
2011–2015	0.012	0.096	0.763	0.121	Weak decoupling	0.015	7.912
2016–2020	−0.019	0.047	0.377	−0.399	Strong decoupling	−0.050	8.049
Zhaoqing							
2000–2005	0.157	0.429	6.504	0.367	Weak decoupling	0.024	15.169
2006–2010	0.096	0.347	3.834	0.276	Weak decoupling	0.025	11.063
2011–2015	0.008	0.175	1.104	0.048	Weak decoupling	0.008	6.313
2016–2020	0.059	0.089	0.370	0.668	Weak decoupling	0.161	4.157

4. Conclusions and Policy Implications

4.1. Conclusions

This study estimated the imbalances of CO_2 emissions and emission intensity in the PRD urban agglomeration from 2000 to 2020, and discussed the decoupling states between CO_2 emissions and economic growth. Urban agglomerations play important roles as China's regional economic development engines and are an important geographical unit driving coordinated regional development, as well as a key node in China's low-carbon sustainable transition development. This work provides a reference for regional emission-reduction policies and low-carbon development. The main conclusions are as follows.

Firstly, the unbalanced development of the PRD urban agglomeration is a major concern, as it leads to significant disparities in economic development and CO_2 emissions. The spatial pattern of total CO_2 emissions is high in the center and low on the periphery, indicating that urban areas in the center of the region emit more CO_2 than those on the outskirts. In contrast, CO_2 intensity is lower in the center and higher on the periphery.

Secondly, the decoupling analysis show that there is significant volatility in the decoupling of economic development and CO_2 emissions in the PRD. Weak decoupling occurred in most years, indicating economic growth leading to an increase in CO_2 emissions. Overall, the decoupling between CO_2 emissions and economic development in the PRD shows an improving trend from 2000 to 2015, meaning that economic growth and growth in carbon emissions are becoming less correlated, but the decoupling then worsens slightly from 2016 to 2020, especially in Guangdong, Zhuhai, Foshan, and Dongguan. This indicates that the PRD region still faces challenges in energy conservation and emission reduction.

Thirdly, a detailed examination of energy-saving and emission-reduction elasticities underscore a clear divergence in trends. The energy-saving elasticity has high volatility, with an especially significant decline from 2010 to 2015. Compared with 2000–2015, the energy-saving elasticity has rebounded in recent years, indicating that the dependence of industries in the PRD region on energy consumption is still high, and energy efficiency needs to improve. Emission-reduction elasticities have declined over the study period, indicating that the PRD cities are emphasizing social and economic development while seeking a low-carbon and environmentally friendly development path. In Guangzhou and Shenzhen, the trends of energy-efficiency elasticity and decoupling elasticity are the same, indicating that energy-efficiency elasticity plays a more critical role in decoupling economic development from carbon emissions. In other cities, the decoupling elasticity is more important than the energy-efficiency elasticity.

Given the dynamic and multi-faceted nature of economic and environmental developments, future research can take a deeper dive into analyzing the intricate relationship between CO_2 emission decoupling and various influential variables such as urbanization rate, city size, social system, and technological progress. In addition, it would be immensely beneficial to verify the findings of this study through the application of regression models. Such analytical approaches can provide a more granular understanding of the factors influencing decoupling statuses in the PRD region, aiding in the creation of more nuanced and effective policies. Furthermore, studies could explore the potential synergies between technological innovations and urban policies in fostering sustainable growth, drawing from a wider pool of data that encompasses recent developments and trends in the region.

4.2. Policy Implications

To forge a resilient and sustainable path for urban agglomerations, it is critical to focus on actionable strategies that can significantly influence the trajectory of environmental sustainability while fostering economic growth.

Firstly, the establishment of a data-driven policy-making framework would serve as a cornerstone in crafting policies that are sensitive to regional discrepancies [41–44]. This involves not only monitoring but also leveraging insights from a centralized database that keeps track of environmental parameters across various cities. This dynamic repository can foster intelligent decision-making, enabling cities to adapt strategies that focus on economic agglomeration effects and industrial restructuring based on real-time data and trends. Furthermore, it can serve as a blueprint for other cities in identifying sectors ripe for low-carbon transformations.

Secondly, in fostering collaboration, the creation of regional knowledge platforms can be a vital asset [45–47]. Regular forums and dialogue platforms can be institutionalized to foster a culture of knowledge exchange and collaboration. These platforms should be equipped to facilitate technical know-how exchange, and sharing of successful case studies, thereby fostering a culture of collaborative growth and learning.

Thirdly, to streamline regional cooperation, a regional coordination entity should be established, tasked with the orchestration of collaborative projects, overseeing efficient resource allocation, and maximizing synergies [48–50]. This entity could act as a nerve center, standardizing tools and methodologies for monitoring and management of carbon emission initiatives across various regions, promoting a cohesive approach to sustainable development.

Lastly, policy formulation should be fine-tuned to the unique attributes of individual cities within urban conglomerates, taking into cognizance their respective economic stages, energy consumption patterns, and carbon emission levels. Governments could develop comprehensive guidelines and training programs to aid local policy-makers in crafting policies that resonate with both local conditions and overarching regional emission-reduction objectives, fostering a harmonized approach to sustainable development.

Author Contributions: Conceptualization, Z.W. and F.L.; methodology, F.L. and Z.X.; software, F.L. and Z.X.; validation, Z.W. and F.L.; formal analysis, Q.L. and Y.Z.; investigation, Q.L and M.D.; resources, Z.W. and F.L.; data curation, Z.X.; writing-original draft preparation, Z.W. and F.L.; supervision, Z.X.; project administration, Z.X.; funding acquisition, Q.L and Y.Z. All authors have read and agreed to the published version of the manuscript.

Funding: This work was supported by National Natural Science Foundation of China (No. 42001147), the Guangdong Province Natural Science Fund (No. 2022A1515011728), Special Funds for the Cultivation of Guangdong College Students' Scientific and Technological Innovation. ('Climbing Program' Special Funds) (pdjh2022a0153), the Guangdong Provincial Key Laboratory of Intelligent Urban Security Monitoring and Smart City Planning (Research on soil health assessment and intelligent monitoring and early warning technology in the context of "dual carbon").

Institutional Review Board Statement: Not applicable.

Informed Consent Statement: Not applicable.

Data Availability Statement: Data is contained within the article.

Conflicts of Interest: The authors declare no conflict of interest.

References

1. Wang, Q.; Su, M. The effects of urbanization and industrialization on decoupling economic growth from carbon emission—A case study of China. *Sustain. Cities Soc.* **2019**, *51*, 101758.
2. Hu, M.M.; Wang, Y.F.; Xia, B.C.; Jiao, M.Y.; Huang, G.H. How to balance ecosystem services and economic benefits?—A case study in the Pearl River Delta, China. *J. Environ. Manag.* **2020**, *271*, 110917.
3. Wang, S.; Zeng, J.; Liu, X. Examining the multiple impacts of technological progress on CO_2 emissions in China: A panel quantile regression approach. *Renew. Sustain. Energy Rev.* **2019**, *103*, 140–150.
4. Li, L.X.; Li, Y.N.; Ye, F.; Zhang, L. Carbon dioxide emissions quotas allocation in the Pearl River Delta region: Evidence from the maximum deviation method. *J. Clean. Prod.* **2018**, *177*, 207–217.
5. Wu, R.; Wang, J.Y.; Wang, S.J.; Feng, K.S. The drivers of declining CO_2 emissions trends in developed nations using an extended STIRPAT model: A historical and prospective analysis. *Renew. Sustain. Energy Rev.* **2021**, *149*, 111328.
6. Dong, F.; Long, R.Y.; Bian, Z.F.; Xu, X.H.; Yu, B.L.; Wang, Y. Applying a Ruggiero three-stage super-efficiency DEA model to gauge regional carbon emission efficiency: Evidence from China. *Nat. Hazards* **2017**, *87*, 1453–1468.
7. Fang, C.L.; Yu, D.L. Urban agglomeration: An evolving concept of an emerging phenomenon. *Landsc. Urban Plan.* **2017**, *162*, 126–136. [CrossRef]
8. Shan, Y.L.; Fang, S.; Cai, B.F.; Zhou, Y.; Li, D.; Feng, K.S.; Hubacek, K. Chinese cities exhibit varying degrees of decoupling of economic growth and CO_2 emissions between 2005 and 2015. *One Earth* **2021**, *4*, 124–134.
9. Zheng, H.R.; Meng, J.; Mi, Z.F.; Song, M.L.; Shan, Y.L.; Ou, J.M.; Guan, D.B. Linking city-level input-output table to urban energy footprint: Construction framework and application. *J. Ind. Ecol.* **2019**, *23*, 781–795.
10. Yu, X.; Wu, Z.Y.; Zheng, H.R.; Li, M.Q.; Tan, T.L. How urban agglomeration improve the emission efficiency? A spatial econometric analysis of the Yangtze River Delta urban agglomeration in China. *J. Environ. Manag.* **2020**, *260*, 110061.
11. Zheng, H.R.; Zhang, Z.Y.; Zhang, Z.K.; Li, X.; Shan, Y.L.; Song, M.L.; Mi, Z.F.; Meng, J.; Ou, J.M.; Guan, D.B. Mapping Carbon and Water Networks in the North China Urban Agglomeration. *One Earth* **2019**, *1*, 126–137. [CrossRef]

12. Chen, X.C.; Li, F.; Li, X.Q.; Hu, Y.H.; Wang, Y. Mapping ecological space quality changes for ecological management: A case study in the Pearl River Delta urban agglomeration, China. *J. Environ. Manag.* **2020**, *267*, 110658. [CrossRef] [PubMed]
13. Li, X.; Li, Y.P.; Li, G.H.; Lv, J.; Ma, Y.; Li, Y.F. A multi-scenario input-output economy-energy-environment nexus management model for Pearl River Delta urban agglomeration. *J. Clean. Prod.* **2021**, *317*, 128402. [CrossRef]
14. Zhou, X.; Bai, L.; Bai, J.; Tian, Y.; Li, W. Scenario prediction and critical factors of CO_2 emissions in the Pearl River Delta: A regional imbalanced development perspective. *Urban Clim.* **2022**, *44*, 101226. [CrossRef]
15. Li, L.; Shan, Y.L.; Lei, Y.L.; Wu, S.M.; Yu, X.; Lin, X.Y.; Chen, Y.P. Decoupling of economic growth and emissions in China's cities: A case study of the Central Plains urban agglomeration. *Appl. Energy* **2019**, *244*, 36–45. [CrossRef]
16. Xiao, H.J.; Duan, Z.Y.; Zhou, Y.; Zhang, N.; Shan, Y.L.; Lin, X.Y.; Liu, G.S. CO_2 emission patterns in shrinking and growing cities: A case study of Northeast China and the Yangtze River Delta. *Appl. Energy* **2019**, *251*, 113384. [CrossRef]
17. Zhou, X.Y.; Zhou, D.Q.; Wang, Q.W.; Su, B. Who shapes China's carbon intensity and how? A demand-side decomposition analysis. *Energy Econ.* **2020**, *85*, 104600. [CrossRef]
18. Zhou, D.Q.; Zhou, X.Y.; Xu, Q.; Wu, F.; Wang, Q.W.; Zha, D.L. Regional embodied carbon emissions and their transfer characteristics in China. *Struct. Chang. Econ. Dyn.* **2018**, *46*, 180–193. [CrossRef]
19. Adams, S.; Adedoyin, F.; Olaniran, E.; Bekun, F.V. Energy consumption, economic policy uncertainty and carbon emissions; causality evidence from resource rich economies. *Econ. Anal. Policy* **2020**, *68*, 179–190. [CrossRef]
20. Chontanawat, J.; Hunt, L.C.; Pierse, R. Does energy consumption cause economic growth?: Evidence from a systematic study of over 100 countries. *J. Policy Model.* **2008**, *30*, 209–220. [CrossRef]
21. Bella, G.; Massidda, C.; Mattana, P. The relationship among CO_2 emissions, electricity power consumption and GDP in OECD countries. *J. Policy Model.* **2014**, *36*, 970–985. [CrossRef]
22. Ntanos, S.A.G.; Milioris, K.; Chalikias, M.; Lalou, P. Energy Consumption and CO_2 Emissions on a Global Level. In Proceedings of the 4th International Conference: Quantitative and Qualitative Methodologies in the Economic & Administrative Sciences (ICQQMEAS 2015), Athens, Greece, 21–22 May 2015; Technological Education Institute of Athens: Athens, Greece, 2015; pp. 251–260.
23. de Freitas, L.C.; Kaneko, S. Decomposing the decoupling of CO_2 emissions and economic growth in Brazil. *Ecol. Econ.* **2011**, *70*, 1459–1469. [CrossRef]
24. Tapio, P. Towards a theory of decoupling: Degrees of decoupling in the EU and the case of road traffic in Finland between 1970 and 2001. *Transp. Policy* **2005**, *12*, 137–151. [CrossRef]
25. Luo, Y.S.; Long, X.L.; Wu, C.; Zhang, J.J. Decoupling CO_2 emissions from economic growth in agricultural sector across 30 Chinese provinces from 1997 to 2014. *J. Clean. Prod.* **2017**, *159*, 220–228. [CrossRef]
26. Wang, Q.; Jiang, R. Is China's economic growth decoupled from carbon emissions? *J. Clean. Prod.* **2019**, *225*, 1194–1208. [CrossRef]
27. Raza, M.Y.; Lin, B.Q. Decoupling and mitigation potential analysis of CO_2 emissions from Pakistan's transport sector. *Sci. Total. Environ.* **2020**, *730*, 139000. [CrossRef]
28. Lin, S.J.; Beidari, M.; Lewis, C. Energy Consumption Trends and Decoupling Effects between Carbon Dioxide and Gross Domestic Product in South Africa. *Aerosol Air Qual. Res.* **2015**, *15*, 2676–2687. [CrossRef]
29. Wang, Q.; Su, M. Drivers of decoupling economic growth from carbon emission—An empirical analysis of 192 countries using decoupling model and decomposition method. *Environ. Impact Asses* **2020**, *81*, 106356. [CrossRef]
30. Huo, T.F.; Ma, Y.L.; Yu, T.; Cai, W.G.; Liu, B.S.; Ren, H. Decoupling and decomposition analysis of residential building carbon emissions from residential income: Evidence from the provincial level in China. *Environ. Impact Assess. Rev.* **2021**, *86*, 106487. [CrossRef]
31. Zhao, X.C.; Jiang, M.; Zhang, W. Decoupling between Economic Development and Carbon Emissions and Its Driving Factors: Evidence from China. *Int. J. Environ. Res. Public Health* **2022**, *19*, 2893. [CrossRef] [PubMed]
32. Zhou, X.; Zhang, M.; Zhou, M.H.; Zhou, M. A comparative study on decoupling relationship and influence factors between China's regional economic development and industrial energy-related carbon emissions. *J. Clean. Prod.* **2017**, *142*, 783–800. [CrossRef]
33. Shuai, C.Y.; Chen, X.; Wu, Y.; Zhang, Y.; Tan, Y.T. A three-step strategy for decoupling economic growth from carbon emission: Empirical evidences from 133 countries. *Sci. Total. Environ.* **2019**, *646*, 524–543. [CrossRef]
34. Elvidge, C.D.; Cinzano, P.; Pettit, D.R.; Arvesen, J.; Sutton, P.; Small, C.; Nemani, R.; Longcore, T.; Rich, C.; Safran, J.; et al. The Nightsat mission concept. *Int. J. Remote. Sens.* **2007**, *28*, 2645–2670. [CrossRef]
35. Chen, J.D.; Gao, M.; Cheng, S.L.; Hou, W.X.; Song, M.L.; Liu, X.; Liu, Y.; Shan, Y.L. County-level CO_2 emissions and sequestration in China during 1997–2017. *Sci. Data* **2020**, *7*, 391. [CrossRef] [PubMed]
36. Zhang, B.R.; Li, J.; Wang, M.G.; Duan, P.; Li, C. Using DMSP/OLS and NPP/VIIRS Images to Analyze the Expansion of 21 Urban Agglomerations in Mainland China. *J. Urban. Plan. Dev.* **2021**, *147*, 04021024. [CrossRef]
37. Zhao, X.; Zhang, X.; Li, N.; Shao, S.; Geng, Y. Decoupling economic growth from carbon dioxide emissions in China: A sectoral factor decomposition analysis. *J. Clean. Prod.* **2017**, *142*, 3500–3516. [CrossRef]
38. Shang, M.; Luo, J. The Tapio Decoupling Principle and Key Strategies for Changing Factors of Chinese Urban Carbon Footprint Based on Cloud Computing. *Int. J. Environ. Res. Public Health* **2021**, *18*, 2101. [CrossRef] [PubMed]
39. Wang, Q.; Su, M.; Li, R.R. Toward to economic growth without emission growth: The role of urbanization and industrialization in China and India. *J. Clean. Prod.* **2018**, *205*, 499–511. [CrossRef]

40. Xie, P.J.; Gao, S.S.; Sun, F.H. An analysis of the decoupling relationship between CO_2 emission in power industry and GDP in China based on LMDI method. *J. Clean. Prod.* **2019**, *211*, 598–606. [CrossRef]
41. Wang, S.; Liu, X.; Zhou, C.; Hu, J.; Ou, J. Examining the impacts of socioeconomic factors, urban form, and transportation networks on CO_2 emissions in China's megacities. *Appl. Energy* **2017**, *185*, 189–200. [CrossRef]
42. Fang, C.; Wang, S.; Li, G. Changing urban forms and carbon dioxide emissions in China: A case study of 30 provincial capital cities. *Appl. Energy* **2015**, *158*, 519–531. [CrossRef]
43. Wang, S.; Fang, C.; Wang, Y.; Huang, Y.; Ma, H. Quantifying the relationship between urban development intensity and carbon dioxide emissions using a panel data analysis. *Ecol. Indic.* **2014**, *49*, 121–131. [CrossRef]
44. Wang, S.; Fang, C.; Wang, Y. Spatiotemporal variations of energy-related CO_2 emissions in China and its influencing factors: An empirical analysis based on provincial panel data. *Renew. Sustain. Energy Rev.* **2016**, *55*, 505–515. [CrossRef]
45. Wang, S.; Shi, C.; Fang, C.; Feng, K. Examining the spatial variations of determinants of energy-related CO_2 emissions in China at the city level using Geographically Weighted Regression Model. *Appl. Energy* **2019**, *235*, 95–105. [CrossRef]
46. Wang, S.; Zeng, J.; Huang, Y.; Shi, C.; Zhan, P. The effects of urbanization on CO_2 emissions in the Pearl River Delta: A comprehensive assessment and panel data analysis. *Appl. Energy* **2018**, *228*, 1693–1706. [CrossRef]
47. Wang, S.; Wang, J.; Fang, C.; Li, S. Estimating the impacts of urban form on CO_2 emission efficiency in the Pearl River Delta, China. *Cities* **2019**, *85*, 117–129. [CrossRef]
48. Wang, S.; Fang, C.; Sun, L.; Su, Y.; Chen, X.; Zhou, C.; Feng, K.; Hubacek, K. Decarbonizing China's Urban Agglomerations. *Ann. Am. Assoc. Geogr.* **2019**, *109*, 266–285. [CrossRef]
49. Li, S.; Zhou, C.; Wang, S. Does modernization affect carbon dioxide emissions? A panel data analysis. *Sci. Total. Environ.* **2019**, *663*, 426–435. [CrossRef] [PubMed]
50. Zhou, C.; Wang, S.; Feng, K. Examining the socioeconomic determinants of CO_2 emissions in China: A historical and prospective analysis. *Resour. Conserv. Recycl.* **2018**, *130*, 1–11. [CrossRef]

land

Article

Impact of Land-Use Changes on Climate Change Mitigation Goals: The Case of Lithuania

Renata Dagiliūtė and Vaiva Kazanavičiūtė *

Department of Environmental Sciences, Vytautas Magnus University, Universiteto Str. 10, Akademija, LT-53361 Kaunas, Lithuania; renata.dagiliute@vdu.lt
* Correspondence: vkazanaviciute@gmail.com

Abstract: The land-use, land-use change and forestry (LULUCF) sector is receiving increasing attention in climate change mitigation and greenhouse gas (GHG) emission offsetting. The sector itself and measures applied to mobilize this sector in order to tackle climate change are dominant in nationally determined contributions under the Paris Agreement as well as in national strategies, as in the case of Lithuania. Lithuania has set the goal of becoming a carbon-neutral country in 2050, reducing GHGs by 80% compared to 1990 and offsetting the remaining 20% through the LULUCF sector. Therefore, this paper aims at analyzing historical land-use changes in 1990–2021, as reported for the United Nations Framework Convention on Climate Change (UNFCCC) secretariat, and LULUCF's potential to achieve climate change mitigation goals, taking into account different land-use change scenarios (business as usual, forest development, forest development + additional measures and forest land 40% + additional measures) for 2030 and 2050 in Lithuania. The scenarios are based on historical and potential future policy-based land-use changes. Projections of GHG emissions/removals for different scenarios are prepared according to the Good Practice Guidance and Uncertainty Management in National Greenhouse Gas Inventories (2006) by the Intergovernmental Panel on Climate Change (IPCC). The results indicate that land-use changes over the period 1990–2021 remained rather stable, with some increases in forest area and grassland at the expense of cropland. The whole LULUCF sector acted as a carbon sink in most cases, forests being a key category for removal. However, reaching climate neutrality in 2050 might be challenging, as the goal to offset 20% of remaining GHG emission compared to 1990 through LULUCF would not be met in any of the scenarios analyzed, even the scenario of maximal forest-area development and additional measures. Considering the high historical GHG-removal fluctuations and the uncertainties of the sector itself, caution should be taken when relying on LULUCF's potential to reach the set goals.

Keywords: LULUCF; climate change; mitigation; policy; forestry; GHG

1. Introduction

In 2021, pursuing the global sustainability agenda [1], particularly goal 13 on climate action, and in line with the Paris Agreement [2] as well as the European Union aims on climate neutrality [3], Lithuania adopted the National Climate Change Management Agenda [4], setting ambitious goals for GHG reduction in the short term and the long term. Compared to 1990, the national agenda foresees reduction of national net GHG emissions by 70% by 2030 and reduction of net GHG emissions by 100% by 2050 [4]. Both goals strongly rely on the LULUCF sector and its carbon removals: the 2030 reductions include absorption in the LULUCF sector (no specific target in the national agenda, but according to EU Regulation 2023/839 [5], removal should amount at least to 4.633 million t CO_2 eq for Lithuania); the 2050 goals anticipate an 80% reduction due to various climate change mitigation measures applied in different economic sectors, while the remaining 20% will be offset by the LULUCF sector [4]. It means that 9.558 million t CO_2 eq must be absorbed by the LULUCF sector in 2050 in Lithuania.

Lithuania is not the only country to rely significantly on the LULUCF sector while aiming at carbon neutrality. Under the Paris Agreement, countries were required to undertake national commitments for greenhouse gas (GHG) emissions reduction, known as nationally determined contributions (NDC), and prepare long-term low greenhouse gas emission development strategies to achieve climate change policy goals. A recent UN report [6] on NDC indicates that most of the countries included the LULUCF sector and corresponding mitigation measures in their NDC. Most often, this covers afforestation, reforestation and revegetation (48%). Such reliance on the LULUCF sector is supported by a number of studies showcasing high historical carbon removals [7–9] and/or potential net carbon removal [10–12] particularly by forests. For example, Finnish forests could outweigh carbon emissions from the other sectors no later than 2040, as indicated by Kallio et al. [13]. It is also estimated that forest expansion alone could contribute from 6% to 10% of the EU GHG emission reduction target [14]. In addition, the forestry sector plays a role in GHG emission reduction via sequestration not only in biomass but also in forest products due to both storage and substitution effects [15–20]. For example, it is suggested [13] that use of wood for bioenergy would significantly reduce emissions and play an important role in reaching Finland's 80% emission-reduction target by 2050. Nevertheless, the results of a study performed for European forests [21] indicate that to achieve climate neutrality, the EU forests' net carbon removals should increase about 25%, while the combined EU + UK forest sink is projected to decline; therefore, additional efforts are needed.

Hence, next to the forest related measures, other measures might also be significant for carbon removal in the LULUCF sector. Usually, wetland restoration, soil carbon sequestration, bioenergy with carbon capture and storage (BECCS) and agroforestry are analyzed as important land management measures for GHG removals [22,23]. Still, forest-related measures cover the largest share of mitigation potential, followed by peatland restoration and soil organic carbon enhancement on agricultural lands (cropland and grassland) [23]. In general, it is estimated that land-use-based measures have the potential to contribute approximately 20–30% to the 1.5° temperature target before 2050 [24–27]. Though policies and measures in land-use-related sectors in 2009–2019 have contributed only to 0.5% of total emission reduction [23], reliance on the LULUCF sector might even increase, especially if GHG reductions in other sectors fail to be achieved [23,26]. It may also be the case that a focus on LULUCF will decrease efforts to reduce emissions in other sectors [28].

Lithuania has not provided a separate NDC but is represented in the EU's joint NDC [29]. As mentioned, national climate neutrality goals are set in the National Climate Change Management Agenda [24], setting GHG reduction targets for 2030 and 2050, which include LULUCF GHG removals. To ensure the implementation of these commitments, measures for different sectors, including LULUCF, are listed in Policies and Measures and Projections of Greenhouse Gas emissions in Lithuania [30]. Measures for the LULUCF sector include biomass sink enhancement, such as afforestation and reforestation, restoration of damaged forests and redevelopment of shrubs as well as various soil carbon stock enhancement measures: wetland restoration, grassland management in locations with organic soils, promotion of perennial crops, promotion of cultivation of cover crops, and promotion of no-tillage agricultural practices. To date, the main factors and drivers of land-use-related GHG emissions and removals in Lithuania are considered to be land-use changes due to political and economic factors. Land-use changes were induced by the restructuring of the agricultural sector after the restoration of independence, followed by support allocated for rural development after joining the EU, intensive afforestation of abandoned land or land not suitable for agriculture [31,32], strict governmental control of deforestation and preservation of domestic forest resources [33,34].

Therefore, the main research questions of this paper are (i) whether Lithuania can rely on LULUCF and (ii) whether foreseen measures are sufficient to increase GHG removals in the LULUCF sector to reach desirable levels. For that purpose, historical land-use changes and net removals/emissions by the LULUCF sector in Lithuania in 1990–2021 are analyzed, and based on land-use scenarios and planned measures, the potential of the sector to contribute to the climate change mitigation goals for 2030 and 2050 is estimated.

The paper is structured as follows. The Section 2 introduces the time frames and data sources as well as descriptions of selected land-use scenarios and measures included in the estimations of LULUCF's GHG removal potential. Section 3 presents the main findings on land-use changes, the sector's net GHG emissions and its potential to contribute to the national climate change mitigation goals in the long run. The paper closes with a discussion and conclusions.

2. Methods

2.1. Land-Use Changes and National GHG Emissions

Historical analysis of land-use changes covers 1990–2021 period and is based on the data collected by the State Forest Service in executing the National Forest inventory (NFI), which serves as the main database for national greenhouse gas inventory and provides data on annual area and its changes covering all land uses—forest land, cropland, grassland, wetlands, settlements and other land. A matrix of land-use changes is developed from the monitoring of more than 16,000 sampling plots on the 4 × 4 km grid of the NFI, covering the whole country area and all land uses, including afforestation, which considers national criteria for forest land—minimum area, height of trees (at maturity), crown cover, etc. [35]. Each sampling plot represents nearly 400 ha of country area. National sectoral emissions, including those of the LULUCF sector, also cover the 1990–2021 period and are obtained from national greenhouse gas inventory (as of 2023) [36] prepared according to the IPCC Good Practice Guidelines [36].

2.2. Projections of Greenhouse Gas Emissions and Removals under Different Scenarios

Projections of GHG emissions and removals are calculated using the same methodology as for the national GHG inventory under the United Nations Framework Convention on Climate Change requirements, applying IPCC Good Practice Guidelines [37]. GHG projections are estimated according to 4 different land-use scenarios for the period of 2021–2025 and the years 2030 and 2050, taking into account the LULUCF accounting rules provided in the LULUCF Regulation No EU 2018/841 [38] and its amendment No. EU 2023/839 [5] for 2021–2025. Land-use area changes are projected according to either historical changes or policy documents and established goals related to land-use change (Table 1). Analyzed scenarios include the following:

- The business-as-usual (BAU) scenario (scenario I) contains the assumption that the recently observed forest area will increase to reach 34.4% forest-area coverage in 2030 and 34.5% forest-area coverage in 2050 (3200 ha annually) according to the national forestry sector development plan for 2012–2020 [39];
- The forestry development scenario (scenario II) includes the assumption of a significant forest-area increase (from 34.1% in 2021 to 35.1% in 2030 and 35.3% in 2050)—8000 ha annually, including both human-induced afforestation and natural forest expansion, as indicated in Policies and Measures and Projections of Greenhouse Gas Emissions in Lithuania [30];
- The forestry development + additional measures scenario (scenario III) makes additions to scenario II, including preliminary measures for increasing GHG removals and decreasing GHG emissions from the LULUCF sector as indicated in the Integrated National Energy and Climate Plan [40] and Policies and Measures and Projections of Greenhouse Gas Emissions in Lithuania [30]. All additional measures under this scenario are dedicated to the cropland, wetland and grassland categories (Table 2);
- The forest area 40% + additional measures scenario (scenario IV) takes into consideration a more pronounced afforestation rate according to the project of the National Forest Agreement [41], aiming at 40% forested land to be achieved by 2050 and the same measures as in scenario III (Table 2). This nonbinding forest-land expansion

could be achieved with a 13,200 ha annual forest-land increase and represents more ambitious employment of the LULUCF sector for climate neutrality goals.

To project GHG emissions and removals for forest land, projections of growing stock volume, increment and mortality on forest land remaining forest land as prepared by State Forest Service and applied in the Policies and Measures and Projections of Greenhouse Gas Emissions in Lithuania [30] are used. The aforementioned projections by the State Forest Service consider the age-class distribution in Lithuanian forests in the future and, due to the relatively large share of old forest stands, foresees a nearly 3% decrease in the growing stand volume increment in 2050, as well as an 8% increase in forest harvests in 2030 and 10% in 2050 [30]. These ratios of change are applied for projecting GHG emissions/removals for forest land and for estimation of carbon-stock changes in harvested wood products. To project growing stock volume changes in afforested land, areas are multiplied by the annual growing stock volume change, according to the function applied in the National GHG Inventory (as of 2021) [42].

The same criteria for the areas that could be converted to forest land in all scenarios are used. They include the fertility rate of agricultural areas (only nonfertile or abandoned agricultural areas can be afforested), limitations regarding existing drainage systems in agriculture, etc. [43].

Table 1. Description of scenarios.

<table>
<tr><th>Land-Use Category</th><th>BAU</th><th>Forestry Development</th><th>Forestry Development + Additional Measures</th><th>Forest Land 40% + Additional Measures</th></tr>
<tr><td rowspan="2">Forest land (remaining)</td><td>2.25×10^6 ha in 2030;
2.30×10^6 ha in 2050</td><td colspan="2">2.29×10^6 ha in 2030;
2.45×10^6 ha in 2050</td><td>2.34×10^6 ha in 2030;
2.59×10^6 ha in 2050</td></tr>
<tr><td colspan="4">Growing stock increment: 19.76×10^6 m^3 in 2030, 19.48×10^6 m^3 in 2050;
Growing stock change: 4.95×10^6 m^3 in 2030, 5.05×10^6 m^3 in 2050;
Felling: 11.38×10^6 m^3 in 2030, 11.54×10^6 m^3 in 2050
Policies and Measures and Projections of Greenhouse Gas emissions in Lithuania [30]</td></tr>
<tr><td>Land converted to forest land</td><td>3.2 kha annually from grassland to forest land</td><td colspan="2">4 kha annually from grassland to forest land;
4 kha annually from cropland to forest land</td><td>6 kha annually from cropland to forest land;
7.2 kha annually from grassland to forest land</td></tr>
<tr><td>Cropland (remaining)</td><td colspan="2">16.25 kha of perennial cropland (as of 2019);
2.29 kha annual increase in certified organic cropland (2010–2019 average);
4.29 kha annual increase in no-tillage cropland (2010–2019 average)</td><td colspan="2">Additional measures, covering perennial, certified organic and no-tillage cropland, as described in Table 2</td></tr>
<tr><td>Land converted to cropland</td><td colspan="4">33.74 kha annually from grassland to cropland (2010–2019 average)</td></tr>
<tr><td>Grassland (remaining)</td><td colspan="4">Organic drained soils comprise 6.2% of total grassland area [42]</td></tr>
<tr><td>Land converted to grassland</td><td colspan="2">36.66 kha annually from cropland to grassland (2010–2019 average)</td><td colspan="2">Additional measures, covering cropland conversions to grassland, as described in Table 2</td></tr>
<tr><td>Wetlands (remaining)</td><td colspan="4">13.83 kha of peat extraction (as of 2019)</td></tr>
<tr><td>Land converted to wetlands</td><td colspan="2">No new conversions projected</td><td colspan="2">Additional measures, covering cropland conversion to wetlands, as described in Table 2.</td></tr>
<tr><td>Settlements (land converted to settlements)</td><td colspan="4">0.4 kha annually from grassland to settlements</td></tr>
</table>

Table 1. *Cont.*

Land-Use Category	BAU	Forestry Development	Forestry Development + Additional Measures	Forest Land 40% + Additional Measures
Other land (land converted to other land)	No new conversions projected			
Harvested wood products	8% increase by 2030, 10% increase by 2050; same ratio as in 2019 among the categories of sawn wood, wood-based panels and paper products			

Harvested wood products (HWPs) are projected by applying a first-order decay function, as specified in the IPCC Guidelines [44], meaning that all wood products (sawn wood, wood-based panels, paper and paperboard), once produced, enter the HWP pool as an input (whole amount of CO_2 sequestered) and then gradually decay each subsequent year. The half-period of decay is 35 years for sawn wood, 25 years for wood-based panels and 2 years for paper products, meaning that each subsequent year after production, 1/35 of (remaining) sawn wood's CO_2, 1/25 of (remaining) wood-based panels' CO_2 and ½ of paper products' CO_2 is released back to the atmosphere. HWP carbon stock change is a balance between CO_2 input (with new products) and output (from the decay of previous products). HWP input includes both domestically consumed and exported products (sawn wood, wood-based panels, paper and paperboard) produced from domestically harvested wood; exported roundwood is not included in the calculations. The carbon stock balance in HWP for all 4 scenarios is projected by applying the same ratio among harvested wood products as in 2019 and considering projections of harvested wood volume. According to FAO [45], harvested wood products in Lithuania in 2019 consisted of 1.27×10^6 m^3 of sawn wood (56%), 0.85×10^6 m^3 of wood-based panels (37%) and 0.16×10^6 m^3 of paper products (7%) produced from a total of 6.67×10^6 m^3 of roundwood.

According to the IPCC Guidelines [37], conversion from one land use to another is considered to be effective for 20 years; therefore, at a certain time, the result of measures shifts from, for example, afforested land to managed forest land. A 20-year transition period is applied for all changes in the land-use categories in the projections; therefore, the effect of measures applied to increase carbon stocks might decline if new conversions are not projected.

Table 2. Additional measures included in projections of GHG emissions under scenarios III and IV, according to the Policies and Measures and Projections of Greenhouse Gas Emissions in Lithuania [30].

Description of Measure	Affected Land-Use Category	Annual Area	Period Affected
Promotion of no-tillage crop management	Cropland	Gradually increasing to 800,000 ha in 2040	2021–2050, with the same ratio of area increase applied to 2041–2050
Restoration of wetlands on arable peatlands and protection of perennial grass cover	Cropland, wetlands	Gradually increasing to 20,000 ha in 2040 Cropland converted to wetlands	2021–2050, with the same ratio of area increase applied to 2041–2050
Promotion of perennial crops (shrubs and trees)	Cropland	Gradually increasing to 26,300 ha in 2040	2021–2050, the same ratio of area increase applied to 2041–2050
Promotion of perennial grassland management on organic soils	Cropland, grassland	Gradually increasing to 40,000 ha in 2040 Cropland converted to grassland	2021–2050, the same area of grasslands on organic soils (converted from cropland) as in 2040 applied to 2041–2050
Promotion of green bedding in agricultural land, planting of landscape elements on agricultural land	Cropland, grassland	Gradually increasing to 178,000 ha in 2040 (10% of arable land) Cropland converted to grassland	2021–2050, the same area of grasslands (converted from cropland) as in 2040 applied to 2041–2050

Projected GHG emissions and removals are compared to the 2030 and 2050 GHG reduction targets as set in EU regulation (EU 2023/839) and the Lithuanian Climate Change Management Agenda [4]. It should be acknowledged that economic growth and other factors, such as the influence of climate change, are not addressed in the analysis. Potentially increasing biomass consumption due to bioeconomy development is partly covered in the projections in the form of increased harvest volume in 2030 and 2050, projected by the State Forest Service [30].

3. Results

3.1. GHGs Emissions, Removals and Land-Use Changes in 1990–2021

Over the period under analysis, overall GHG emissions in Lithuania decreased significantly (Figure 1). This more than double decrease has been mainly the result of significantly dropped energy consumption due to transitional decline, reforms and market restructuring after the country regained independence in 1990 [46]. In 1995, emissions reached 43% of the 1990 level. However, afterwards, the trend of national emissions shows no significant reductions, and a rather stable level of GHG emissions should be acknowledged.

As in the beginning of the analyzed period (67%), energy-related GHGs continued to constitute the largest share, though a decreasing share, of total countrywide GHG emissions over the period (in 2021—61%). There are some reasons behind this. First, until the pandemic situation and the energy crisis, final energy consumption had trended slightly upwards since the last economic crisis. Second, while the share of renewable energy sources increased from 17.2% in 2004 to 26.8% in 2020 in Lithuania [47], GHG emissions related to the energy sector decreased only 4.4% in the same period, mostly due to significantly increasing emissions from transport. From 2004 to 2020, transport GHG emissions increased by 55.5%, transport being the largest source of emissions in the energy sector—54.1% [36]. According to Eurostat data, renewables account for 21.28% of electricity and 46.63% in heating and cooling, but in the transport sector, renewables amount to only 6.46% as of 2021.

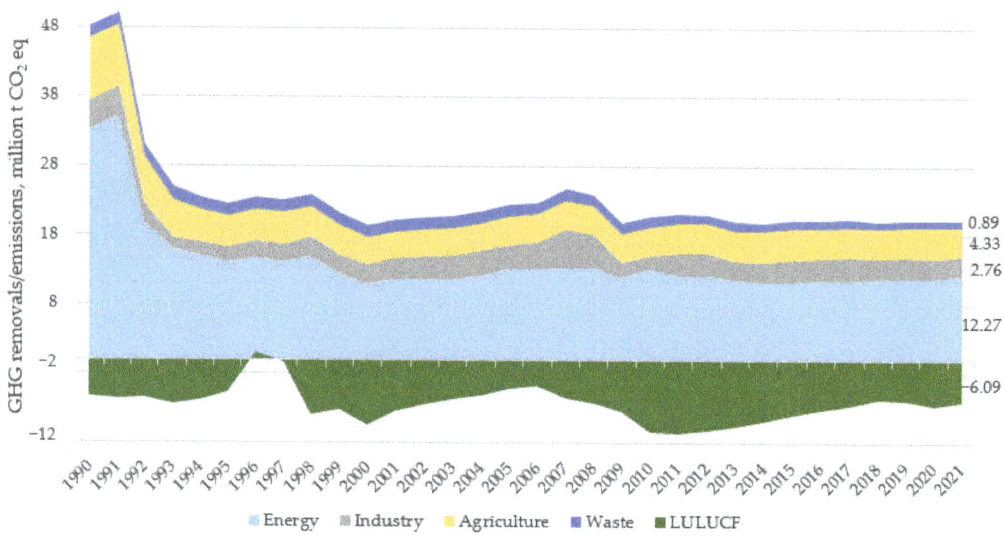

Figure 1. GHG emissions/removals in the LULUCF sector and total emissions in Lithuania during 1990–2021, million tons CO_2 eq (based on data from National GHG Inventory Report 2023).

The LULUCF sector has been a net sink of greenhouse gas emissions in Lithuania for almost the whole reporting period (1990–2021), except for 1996 and 1997, when, due to adverse natural conditions, LULUCF was recorded to be a net source of emissions

(Figure 1). Emissions in 1996 and 1997 are the result of repetitive droughts and consequent invasion by pests (e.g., *Ips Typographus*), which caused massive damage and death of spruce stands in Lithuania and therefore biomass losses from its forest land [42]. In addition to this, high emissions due to the drainage of organic soils (especially in cropland) also had an impact on the overall sector's net emissions. Emissions from drained organic soils varied from 1.9 in 1990 to 1.5 million tons of CO_2 eq in 2021 [36] due to conversions between land uses and different emissions factors (EFs) applied for different land uses. Though a sector's removal potential generally varies according to the natural conditions, economic factors are also of importance as they drive afforestation rates and use of agricultural land, as well as the volume of harvested wood products. Hence, an increasing area of grasslands converted from croplands increased GHG removals (starting after 2005); changing harvesting levels had an impact on both increasing and decreasing GHG removals, while an increasing growing stand volume increment, to some extent, compensates for the impact of increasing harvest levels. Though, in general, growing stand volume and harvest showed increasing trends over the analyzed period, the pattern of changes was different. Growing stock volume increased by 21.5% from 2007 to 2012, while afterwards, only 3.4% growth (2012–2020) was observed. The harvest level decreased by 16.8% from 2007 to 2012 and afterwards increased by 28.3% until 2020. Currently, 36% of wood is used for energy and 64% for materials (estimated from data in the National Greenhouse Gas Inventory [42] and FAO [45]).

The biggest change in the land-use categories was recorded for croplands (Figure 2), whose share shrank from 37% in 1990 to 31% in 2021. Correspondingly, forest area increased from 31% to 34% and grassland from 20% to 23%. Hence, land-use changes indicate some higher potential for GHG removal, as the area of cropland, which usually acts as a source, has decreased, initially due to the abandonment of cropland areas (which were gradually converted to grassland) after the restoration of independence and the subsequent economic recession.

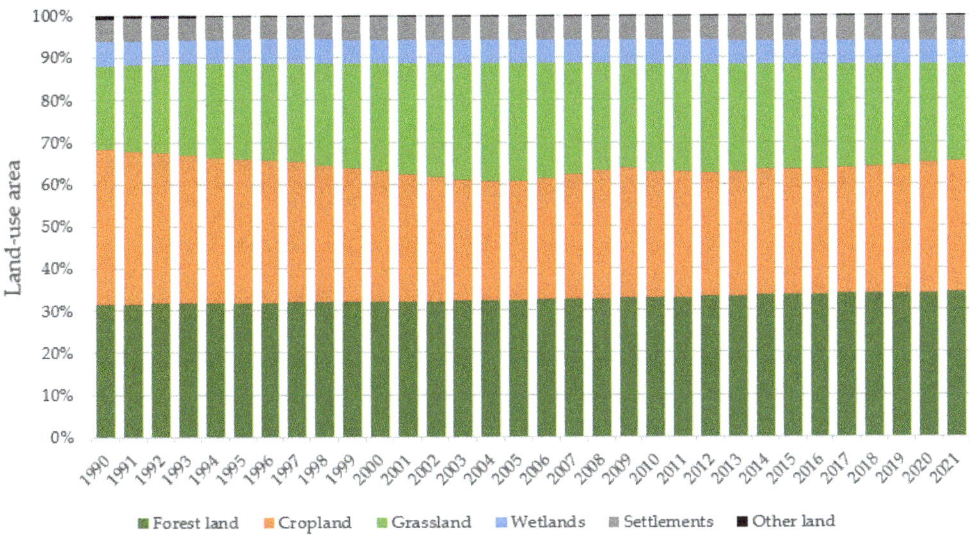

Figure 2. Land-use changes in Lithuania in 1990–2021 (based on data from National GHG Inventory Report 2023).

During the whole period of analysis, cropland acted as a source of GHGs, constituting 81% of the sector's GHG emissions at the beginning and 30% at the end of the period (Figure 3). The decrease in GHG emissions from cropland is related to the shift from traditional intensive agricultural practice to larger areas of no tillage crop practice, organic agriculture, which is reported to increase soil organic carbon due to reduced soil disturbance and increased input of organic matter [36]. Decreasing overall emissions from cropland are partially outweighed by increasing emissions from wetlands, settlements and other land. The results also show the significance of the forest land category, as it provides the highest share of removals in the overall LULUCF balance (Figure 3). The forest category counterbalances emissions from cropland, wetlands and settlements and provides the potential to counterbalance other sectors' emissions altogether with harvested wood products (HWPs) and grassland. In 1990, forests accounted for 89% of the sector's removals; in 2021, 77%. The total amounts reached a maximum in 2011 with removal of some 10.173 million tons of CO_2 eq Though forests dominate removals, over the period of analysis, the share of grassland and harvested wood products in GHG removals also slightly increased.

If the years 1996 and 1997 are excluded, the land-use-related sector in Lithuania absorbed some 11–46% of the country's yearly emissions in 1990–2021 (Figure 1). This indicates that the LULUCF sector's foreseen offsetting potential (9.558 million tons of CO_2 eq), as needed for 2050, has been reached already in 2010, 2011 and 2012. Even bearing in mind the sector's uncertainties, this suggests some possibilities to reach the target set.

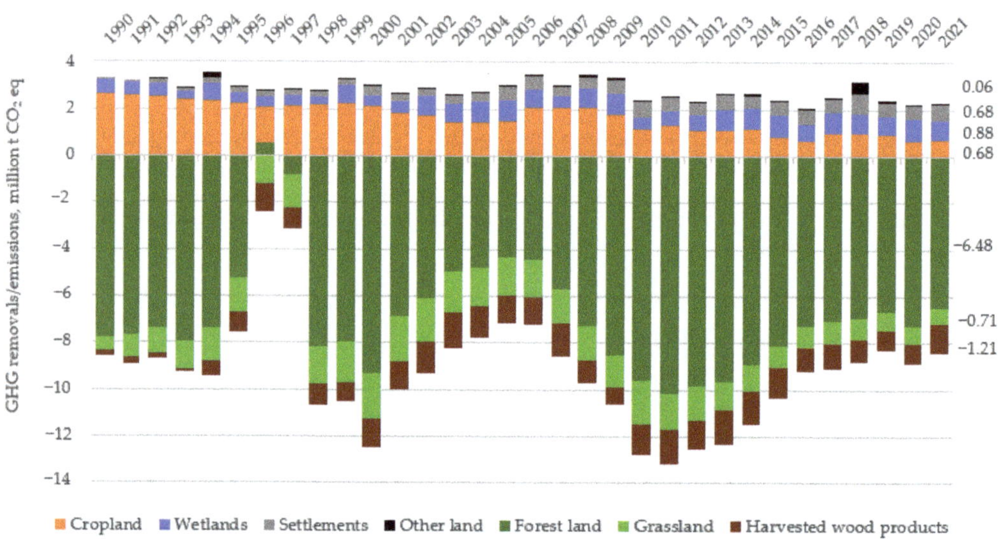

Figure 3. LULUCF GHG balance in 1990–2021 (million tons of CO_2 eq) (based on data from National GHG Inventory Report 2023).

3.2. GHG Projections for the 2021–2025 Period According to the Different Land-Use Scenarios

Though historical removals look favorable, after the EU LULUCF accounting rules (EU Regulation 2023/841 and its amendment 2023/839) [5,38] are applied, this aforementioned offsetting looks less promising. The maximum LULUCF credit allowance for Lithuania equals 6.5 million tons of CO_2 eq for 2021–2030 (EU Regulation 2023/857)) [48] (Table 3); however, only removals accounted for in 2021–2025 can be used for this purpose [5].

In the case of the BAU scenario (scenario I), 2.33 million tons of CO_2 eq in accounted-for removals by the LULUCF sector could be generated in 2021–2025, which is only 35.8% of the removals allotted to offset other sectors' GHG emissions. Accounted-for removals from LULUCF sector could amount to 2.41 and 4.53 million tons of CO_2 eq in the forest development (scenario II) and forest development + additional measures (Scenario III) scenarios during 2021–2025, correspondingly. It is evident that additional measures included in scenario III could have a significant impact on the accounted-for LULUCF GHG balance in 2021–2025. However, even in scenario III, the achieved accounted-for removals are not sufficient to offset the desired amount of GHG emission from the other sectors. Furthermore, in this case, LULUCF could provide only some 70% of the amount allowed for offsetting, even considering that the amount allocated for offsetting (6.5 million tons of CO_2 eq) seems quite low—it is less than average annual sector's removals during the last 10 years of inventory. In addition, according to the unpublished data of the Ministry of the Environment of Lithuania, a shortage of approximately 6 million tons of CO_2 eq of annual emission allocations (AEAs) is expected in 2021–2030 if no additional measures for GHG emission reduction in non-ETS sectors are applied [49]. Hence, the results indicate that LULUCF might play a crucial role, despite being insufficient, for the implementation of climate change mitigation goals in the case of Lithuania in the short term to conform to the EU level commitments.

In addition, it is apparent that the LULUCF sector's contribution to climate change mitigation target achievement is significantly determined by the managed forest land category (Table 3), since the estimated managed forest reference level (−5164 kt CO_2 eq annually and −25,820 kt CO_2 eq for 5 years) for Lithuania [50] is higher than the projected forest land remaining forest land removals in 2021–2025. Therefore, application of the reference level leads to the accounted-for GHG emissions from this category in 2021–2025. Despite that, the afforested land category balances emissions with net removals in forest land (Table 3). The forest land 40% + measures scenario (scenario IV) shows that if the National Forest Agreement [41] were implemented, the LULUCF sector could provide additional GHG removals even in the short term (73.5% of the allowed offsetting). However, the greatest potential lies in the future. Afforestation plays a crucial role not only due to its high GHG removal potential but also because all GHG removals by afforestation can be accounted for as removals for offsetting (EU Regulation 2018/841 and its amendment 2023/839) [5,37] until the end of the conversion period of 20 years, when afforested areas are shifted to the managed forest land category. However, afforestation rates may be limited by the Nature Restoration Regulation [51] in areas where different land-use restoration may be required; thus, reported and accounted-for GHG removals from afforested land could be smaller.

Table 3. Reported and accounted-for 2021–2025 GHG emissions/removals (per EU Regulation 2023/839) [5] and LULUCF flexibility limit for 2021–2030 (per EU Regulation 2018/842 and its amendment 2023/857) [48,52], kt CO_2 eq ("IE": included elsewhere—in managed forest land reference level).

Accounting Category	Reference Values (Annual)	BAU Scenario	Forestry Development Scenario	Forestry Development + Measures Scenario	Forest Land 40% + Measures Scenario
		2021–2025	2021–2025	2021–2025	2021–2025
Managed forest land	−5164.64	−24,716.13	−24,530.24	−24,530.24	−24,530.24
Afforested land	-	−5417.84	−5668.39	−5668.39	−5906.76
Deforested land	-	590.67	590.67	590.67	590.67
Managed cropland	841.9653	4249.97	4237.89	2119.67	2119.67
Managed grassland	−1210.13	−837.05	−833.12	−843.28	−865.84
Managed wetlands	791.9271	4083.10	4083.10	4087.50	4087.50
Harvested wood products	IE	−3988.86	−3988.86	−3988.86	−3988.86
Balance (accounted-for GHG)		−2331.73	−2404.54	−4528.52	−4778.93
Limit for offsetting		−6500	−6500	−6500	−6500

3.3. GHG Projections until 2030 and 2050 According to the Different Land-Use Scenarios

To assess the potential of LULUCF for climate mitigation in Lithuania until 2030 and 2050, projections based on the LULUCF reporting guidelines set forth in the IPCC Good Practice Guidelines [37] and the LULUCF accounting rules set forth by the EU in LULUCF Regulation [5,38] were carried out.

The results (Table 4) show that projections of GHG balance vary significantly and are sensitive to land-use changes, except for harvested wood products. Since forest stands in newly afforested areas will not reach maturity for harvest until 2050, projected forest land expansion does not have an impact on carbon sequestration in harvested wood products. It is evident that newly afforested areas (due to afforestation and natural forest expansion) not only constitute a significant sink but also play a significant role in offsetting emissions from agricultural land uses. It could be stated that if Lithuania is able to maintain a stable land-use change pattern as observed in recent years (small areas of deforestation, large afforested/reforested areas (32,000 ha annually) and increased conversion from cropland to grassland), a total of 5.74 and 6.31 million tons of CO_2 eq could be sequestered in the LULUCF sector correspondingly in 2030 and 2050. This could ensure compliance with and overachievement of the EU target for 2030 but could only partly (66%) reach the national target for 2050. In 2050, those numbers could amount to 7.70 and 8.1 million tons of CO_2 eq correspondingly in the forest development and forest development + additional measures scenarios. The latter implies that the amount expected to be offset by the LULUCF sector in 2050 would be closer to the target only if the additional measures included in scenarios II and III were applied (81% and 85% respectively). Meanwhile, the 2030 LULUCF targets would be exceeded in all four scenarios (Table 4).

Table 4. Projected GHG emissions/removals (kt CO_2 eq) and targets for 2030 (per Regulation 2023/839 (EU)) [5] and 2050 (per Lithuania's National Climate Change Management Agenda [4]).

Land-Use Category	**BAU Scenario**		**Forestry Development Scenario**		**Forestry Development + Measures Scenario**		**Forest Land 40% + Measures Scenario**	
	2030	**2050**	**2030**	**2050**	**2030**	**2050**	**2030**	**2050**
Forest land	−6403.0	−7531.8	−6466.2	−8907.9	−6466.2	−8907.9	−6681.7	−9865.5
Cropland	912.4	1351.7	904.3	1327.5	83.3	1045.8	80.5	1034.9
Grassland	−623.9	−583.4	−621.7	−575.8	−628.9	−749.2	−636.3	−754.4
Wetlands	816.6	816.6	816.6	816.6	836.7	878.7	836.7	878.7
Settlements	324.4	131.2	329.9	133.3	329.9	133.3	329.9	133.3
Other land	12.3	0.0	12.3	0.0	12.3	0.0	12.3	0.0
Harvested wood products	−779.4	−497.3	−779.4	−497.3	−779.4	−497.3	−779.4	−497.3
Balance	−5740.7	−6312.9	−5804.2	−7703.6	−6612.3	−8096.7	−6838.0	−9070.3
Target	−4633	−9558	−4633	−9558	−4633	−9558	−4633	−9558
% of the target	124	66	125	81	143	85	148	95

Although the targets for 2050 will be not reached, in all scenarios, higher total removals will be achieved in the long run (2050) except in the cropland category, where higher emission reduction will be achieved in the short term, and in the harvested wood product category, with declining removals in the long term (Table 4). The most significant effect of the measures applied can be observed in the cropland and grassland categories in 2030 (Table 4). Conversion of conventional agricultural land to no-tillage crops will be the most intense until 2030, therefore causing the most significant effect on carbon sequestration (in soils) until 2030. In addition, restoration of wetlands will result in a slight decrease in emissions for the cropland category at the expense of restored drained areas but will contribute to the wetland GHG source. Restoration of wetlands via CH_4 would additionally result in 25 kt CO_2 eq in 2030 and 62 kt CO_2 eq in 2050. This suggests that measures should be thoroughly accounted for and considered before implementation, also taking into account different time perspectives regarding LULUCF's climate change mitigation potential. It should also be considered that the growing stock increment in mature stands is shrinking, and the areas of mature stands will increase in upcoming decades. The declining sink in old forest stands and the remaining high emissions from drained forest organic soils will be counterbalanced by significant removals in young forests—areas recently shifted from land converted to forest land to forest land remaining forest land. Therefore, expected afforestation might be not enough, and additional measures (or an increase in their volumes) for increasing carbon sinks or reducing emissions in other land-use categories (where available) are needed to rely fully on the sector's potential for 2050. This is clearly seen in the case of scenario IV (forest land 40% + measures), which indicates that, at least, much more pronounced afforestation levels are needed to reach higher removals in the LULUCF sector by 2050 to approach the climate neutrality target. If the National Forest Agreement were implemented [41], it could be expected that some 95% of the target for LULUCF GHG removals would be achieved (Table 4). Hence, even in the case of the most significant forest expansion, a shortage of approximately 0.5 million tons of CO_2 eq in GHG removals (Figure 4) in the LULUCF sector is expected compared to what would be necessary to reach the climate neutrality goal. In all cases, this shortage must be covered either by the sector itself taking additional measures or with more pronounced reductions of GHG emissions in the other sectors.

Figure 4. Shortage of GHG removals (million tons of CO_2 eq) in LULUCF in 2050.

4. Discussion

Since 1990, Lithuania has reduced its national GHG emissions more than 60%, but the challenge remains to meet the intended goals for 2050. Most of the decrease in GHGs resulted from a transformational decline in 1990–1994 after the fall of the Eastern Bloc, as energy consumption dropped more than twofold [46]. It should be admitted that not much more action has been taken than general EU regulations and market pressure have demanded. International and other commitments regarding GHGs have been achieved without significant efforts due to the very high emissions in the reference year. Hence, reducing GHGs in so-called non-ETS sectors remains the most problematic. The transport sector prevails as the most complicated one with the largest share in energy consumption and GHG emissions in Lithuania [36], with only 6.5% renewables (Eurostat data) and a relatively old car fleet in the EU [53].

In addition to the other economic sectors to be addressed to reduce GHGs, according to the National Climate Change Management Agenda [4], it is expected that LULUCF will remove at least 6.5 million tons of CO_2 eq over the period 2021–2030 and, in 2050, will offset 20% of the other sectors' remaining emissions compared to 1990. Historical trends provide some optimistic perspectives, as, over the total period of the analysis, LULUCF acted as a sink for 26% of national CO_2 eq emissions on average annually. However, though removals by the LULUCF sector are accounted for in all scenarios for 2021–2025 to offset emissions from the other sectors, application of the EU accounting rules results in much lower GHG removal potential for offsetting. According to the different scenarios, accounted-for removals can reach only up to 36%, 37%, 70% and 74% of total amount allowed for offsetting by EU Regulation 2023/857 for 2021–2030. The role of accounting rules and the cap in achieving climate change goals also is acknowledged by other scholars [13,54], and the need for substantial changes in accounting rules is discussed [55]. Schlamadinger et al. [56] suggest that a fixed cap for forest-management accounting does not encourage countries to improve forest management unless a country is below the cap or faces such a risk. Considering that a significant share of Lithuanian forest stands are relatively old or will become old in the very near future, higher harvest rates are obvious; hence, the set forest reference level is not favorable in the case of Lithuania, as it encourages a reduction in harvest intensity (at least for 2021–2025, as set in EU Regulation 2018/841) [38] in order to preserve larger GHG removals in forest biomass. On one hand, a lower harvest rate would be preferable for forest-land carbon sink enhancement. On the other hand, it could be a solution for the short term only, since old forest stands have lower GHG removal potential due to their lower yield [57] or might even become a GHG source in the future. Therefore, the newest updates in the rules (EU Regulation 2023/839) [5], maintaining no specific accounting categories or reference values since 2026 and setting an overall GHG removal goal for the LULUCF sector for the first time, seems to be more beneficial for Lithuania. The results show that the 2030 mitigation targets could even be exceeded, and Lithuania could rely on LULUCF at least until 2030.

However, reaching the 2050 target remains more challenging. Lithuania would achieve only some 81% and 85% of the desirable 20% offsetting of 1990's GHG levels via LULUCF in 2050 in the forest development and forest development + measures scenarios. Hence, additional measures or changes in their volumes are still needed either to increase LULUCF potential or reduce emissions more significantly from the other economic sectors. The effect and the continuity of the measures proposed should be considered to yield a substantial number of credits in the future because the given measures might have different results in the short and long term, as our results indicate. Additionally, different environmental goals might intervene, such as wetlands restoration and climate change mitigation.

Forest land is the main reporting category for carbon removal in Lithuania. It could play a more pronounced role in climate mitigation if the National Forest Agreement [41] is implemented and forest land area reaches 40% of country area by 2050. Projections show a significant input of afforestation for climate change mitigation. The scenario including more extensive forest development (40% of total country area) indicates that, in this case, climate mitigation goals for 2050 could be achieved by 95% for LULUCF in Lithuania. Hence, potentially more ambitious goals for afforestation should be set instead of the current 3200 ha or 8000 ha planned annually. Other studies also report the significant influence of afforestation on carbon sequestration (for e.g., [11]). Nevertheless, taking into account the recently decreasing ratio of forest expansion (both natural forest expansion and afforestation) in Lithuania, the business-as-usual scenario may also be challenging to maintain. Natural forest expansion has been the most important source of the forest coverage increase in Lithuania, and according to the results of the State Audit [58], there are still large areas of natural forest expansion that are not included under forest land and are at risk of being clear-felled and used again for agricultural purposes. In addition, as historical trends indicate, natural disturbances or economic factors might also influence forest development and the potential to remove GHGs.

Climate change and bioeconomic inconsistencies might also play an important role in reliance on the LULUCF sector's climate mitigation potential. Trade-offs between bioeconomy (forest biomass harvesting) and carbon sequestration [59] should be considered in the light of climate mitigation goals, as an increase in biomass removals might lead to forests becoming a carbon source rather than a sink in the future [60]. Nevertheless, Kauppi et al. [61] suggest that timber harvesting and carbon sequestration can be aligned if proper management of forests is applied. Though the influence of bioeconomic development is not analyzed in detail in this paper, results show that a projected 10% harvest removal increase will have no negative impact on GHG removals as growing stand volume is still increasing in the case of Lithuania.

Some other uncertainties within the sector also should be considered [62], and saturation of sequestration capacity and the vulnerability of the sector should be taken into account when relying solely on forests and LULUCF in general [13]. For example, the forest management cycle [63] and other forest and agroecosystem management decisions [64] may have an impact on carbon sequestration and, thus, climate change mitigation potential in the LULUCF sector as well. Forests' climate change mitigation potential might also be affected negatively by climate change (see, for example, [65]), and forests might become carbon sources instead of sinks [66,67].

5. Conclusions

Despite this sector's vulnerability to natural disturbances and uncertainties, the LULUCF sector in Lithuania shows significant potential for carbon sequestration, with most of the removals occurring in forest land, followed by harvested wood products and sequestration in grassland. GHG removals by forest land not only ensure this large sector's removal potential but also ensure its ability to counterbalance other sectors' emissions. Though the "no debit" rule set in LULUCF regulations is to be met and it will even be possible to generate accounted-for removals in the LULUCF sector during 2021–2025 as well as the 2030 target for LULUCF, reaching carbon neutrality in 2050 will be challenging for Lithuania. Along with afforestation, the analyzed measures for increasing LULUCF potential are insufficient and must be considered while taking into account the benefits in the short and long term—additional measures in line with planned forestry development could achieve only some 85% of the needed GHG removals for 2050. In addition, even rather ambitious afforestation goals (an increase in forest land to 40% of total country area) are not enough and, with other measures applied, could ensure only 95% of the national climate neutrality target for LULUCF. Moreover, afforestation rates could be limited by the lack of areas suitable for afforestation due to national criteria. Additionally, the requirements set by Nature Restoration Regulation (2022) might influence land-use change patterns and the GHG balance of the LULUCF sector or separate land-use categories. Hence, reconsideration of afforestation targets and of other measures is needed if LULUCF is to be relied on for climate neutrality implementation. In addition, the sector's peculiarities and uncertainties are of importance and must be considered by policy makers while pursuing national climate neutrality and climate change mitigation in general.

Author Contributions: Conceptualization, R.D. and V.K.; methodology, V.K., R.D.; software, V.K.; validation, R.D., V.K.; formal analysis, R.D.; investigation, V.K.; resources, V.K.; data curation, V.K., R.D; writ-ing—original draft preparation, V.K.; writing—review and editing, R.D.; visualization, R.D.; su-pervision, R.D. All authors have read and agreed to the published version of the manuscript.

Funding: This research received no external funding.

Data Availability Statement: Data used in the research will be made available upon request.

Conflicts of Interest: The authors declare that they have no known competing financial interests or personal relationships that could have appeared to influence the work reported in this paper.

References

1. UN. *Adoption of the Paris Agreement; Decision 1/CP.21*; FCCC/CP/2015/10/Add; UN: New York, NY, USA, 2015.
2. UN. *Transforming Our World: The 2030 Agenda for Sustainable Development*; Resolution Adopted by the General Assembly on 25 September 2015; UN: New York, NY, USA, 2015; Volume 42809, pp. 1–13.
3. EU Green Deal. Communication from the Commission to the European Parliament, the European Council, the Council, the European Economic and Social Committee and the Committee of the Regions the European Green Deal. COM/2019/640 Final. 2019. Available online: https://eur-lex.europa.eu/legal-content/en/ALL/?uri=CELEX:52019DC0640 (accessed on 1 June 2023).
4. Parliament of the Republic of Lithuania. National Climate Change Management Agenda (in Lithuanian). *Legal Act.* **2021**. Available online: https://e-seimas.lrs.lt/portal/legalAct/lt/TAD/7eb37fc0db3311eb866fe2e083228059?positionInSearchResul (accessed on 1 May 2022).

5. EU. Regulation (EU) 2023/839 of the European Parliament and of the Council of 19 April 2023 amending Regulation (EU) 2018/841 as regards the scope, simplifying the reporting and compliance rules, and setting out the targets of the Member States for 2030, and Regulation (EU) 2018/1999 as regards improvement in monitoring, reporting, tracking of progress and review. *Off. J. Eur. Union* **2023**.
6. UN. *Nationally Determined Contributions under the Paris Agreement*; Synthesis Report by the Secretariat FCCC/PA/CMA/2021/8; UN: New York, NY, USA, 2021.
7. Lobianco, A.; Caurla, S.; Delacote, P.; Barkaoui, A. Carbon mitigation potential of the French forest sector under threat of combined physical and market impacts due to climate change. *J. For. Econ.* **2016**, *23*, 4–26. [CrossRef]
8. Tzamtzis, I.; Ganatsas, P. Land use, land-use change and their effect on greenhouse gas emissions and removals from Greek forests. *Int. J. Glob. Warm.* **2020**, *22*, 111–131. [CrossRef]
9. Njana, M.A.; Mbilinyi, B.; Eliakimu, Z. The role of forests in the mitigation of global climate change: Emprical evidence from Tanzania. *Environ. Chall.* **2021**, *4*, 100170. [CrossRef]
10. Bhan, M.; Sharma, D.; Ashwin, A.S.; Mehra, S. Policy forum: Nationally-determined climate commitments of the BRICS: At the forefront of forestry-based climate change mitigation. *For. Policy Econ.* **2017**, *85*, 172–175. [CrossRef]
11. Qiu, Z.; Feng Zh Song, Y.; Li, M.; Zhang, P. Carbon sequestration potential of forest vegetation in China from 2003 to 2050: Predicting forest vegetation growth based on climate and the environment. *J. Clean. Prod.* **2020**, *252*, 119715. [CrossRef]
12. Zhao, M.; Yang, J.; Zhao, N.; Xiao, X.; Yue, T.; Wilson, J.P. Estimation of the relative contributions of forest areal expansion and growth to China's forest stand biomass carbon sequestration from 1977 to 2018. *J. Environ. Manag.* **2021**, *300*, 113757. [CrossRef] [PubMed]
13. Kallio, A.M.I.; Salminen, O.; Sievänen, R. Forests in the Finnish low carbon scenarios. *J. For. Econ.* **2016**, *23*, 45–62. [CrossRef]
14. Ovando, P.; Caparrós, A. Land use and carbon mitigation in Europe: A survey of the potentials of different alternative. *Energy Policy* **2009**, *37*, 992–1003. [CrossRef]
15. Sathre, R.; O'Connor, J. Meta-analysis of greenhouse gas displacement factors of wood product substitution. *Environ. Sci. Policy* **2010**, *13*, 104–114. [CrossRef]
16. Lundmark, T.; Bergh, J.; Hofer, P.; Lundström, A.; Nordin, A.; Poudel, C.B.; Sathre, R.; Taverna, R.; Werner, F. Potential roles of Swedish forestry in the context of climate change mitigation. *Forests* **2014**, *5*, 557–578. [CrossRef]
17. Gustavsson, L.; Haus, S.; Lundblad, M.; Lundström, A.; Ortiz, C.A.; Sathre, R.; Truong, N.L.; Wikberg, P.-E. Climate change effects of forestry and substitution of carbon-intensive materials and fossil fuels. *Renew. Sustain. Energy Rev.* **2017**, *67*, 612–624. [CrossRef]
18. Gustavsson, L.; Nguyen, T.; Sathre, R.; Tettey, U.Y.A. Climate effects of forestry and substitution of concrete buildings and fossil energy. *Renew. Sustain. Energy Rev.* **2021**, *136*, 110435. [CrossRef]
19. Leskinen, P.; Cardellini, G.; González-García, S.; Hurmekoski, E.; Sathre, R.; Seppälä, J.; Smyth, C.; Stern, T.; Verkerk, P.J.; European Forest Institute. *Substitution Effects of Wood-Based Products in Climate Change Mitigation*; European Forest Institute (EFI): Joensuu, Finland, 2018.
20. Ellison, D.; Breidenbach, J.; Petersson, H.; Korhonen, K.T.; Henttonen, H.M.; Wallerman, J.; Appiah Mensah, A.; Gobakken, T.; Næsset, E.; Astrup, R. Europe's Forest Sink Obsession (v19.05.2022). *Zenodo* **2022**. [CrossRef]
21. Pilli, R.; Alkama, R.; Cescatti, A.; Kurz, W.A.; Grassi, G. The European forest carbon budget under future climate conditions and current management practices. *Biogeosciences* **2022**, *19*, 3263–3284. [CrossRef]
22. Smith, P.; Adams, J.; Beerling, D.J.; Beringer, T.; Calvin, K.V.; Fuss, S.; Griscom, B.; Hagemann, N.; Kammann, C.; Kraxner, F.; et al. Land-Management Options for Greenhouse Gas Removal and Their Impacts on Ecosystem Services and the Sustainable Development Goals. *Annu. Rev. Environ. Resour.* **2019**, *44*, 255–286. [CrossRef]
23. Roe, S.; Streck, C.; Obersteiner, M.; Frank, S.; Griscom, B.; Drouet, L.; Fricko, O.; Gusti, M.; Harris, N.; Hasegawa, T.; et al. Contribution of the land sector to a 1.5 °C world. *Nat. Clim. Chang.* **2019**, *9*, 817–828. [CrossRef]
24. Griscom, B.W.; Adams, J.; Ellis, P.W.; Houghton, R.A.; Lomax, G.; Miteva, D.A.; Schlesinger, W.H.; Shoch, D.; Siikamäki, J.V.; Smith, P.; et al. Natural climate solutions. *Proc. Natl. Acad. Sci. USA* **2017**, *114*, 11645. [CrossRef]
25. UNEP. *The Emissions Gap Report 2017*; UNEP: Nairobi, Kenya, 2017. [CrossRef]
26. Jia, G.; Shevliakova, E.; Artaxo, P.; Noblet-Ducoudré, N.D.; Houghton, R.; House, J.; Kitajima, K.; Lennard, C.; Popp, A.; Sirin, A.; et al. Land–climate interactions. In *Climate Change and Land: An IPCC Special Report on Climate Change, Desertification, Land Degradation, Sustainable Land Management, Food Security, and Greenhouse Gas Fluxes in Terrestrial Ecosystems*; IPCC: Geneva, Switzerland, 2019.
27. Roe, S.; Streck, C.; Beach, R.; Busch, J.; Chapman, M.; Daioglou, V.; Deppermann, A.; Doelman, J.; Emmet-Booth, J.; Engelmann, J.; et al. Land-based measures to mitigate climate change: Potential and feasibility by country. *Glob. Chang. Biol.* **2021**, *27*, 6025–6058. [CrossRef]
28. Böttcher, H.; Graichen, J. *Impacts on the EU 2030 Climate Target of Including LULUCF in the Climate and Energy Policy Framework*; Report; Öko-Institut e.V.: Berlin, Germany, 2015.
29. EU. The update of the nationally determined contribution of the European Union and its Member States. Submission By Germany and The European Commission on Behalf of The European Union and Its Member States. Berlin. *Off. J. Eur. Union* **2020**.

30. Ministry of Environment. Policies and Measures and Projections of Greenhouse Gas emissions in Lithuania. 2020. Available online: https://am.lrv.lt/lt/veiklos-sritys-1/klimato-kaita/sesd-apskaitos-ir-prognoziu-ataskaitos-nacionaliniai-pranesimai (accessed on 10 March 2022).
31. Jukneliené, D.; Kazanavičiūtė, V.; Valčiukiené, J.; Atkočevičiené, V.; Mozgeris, G. Spatiotemporal Patterns of Land-Use Changes in Lithuania. *Land* **2021**, *10*, 619. [CrossRef]
32. Gerulytė, P.; Stravinskienė, V. Žemės ūkio naudmenų kaita Lietuvos ir Europos agrarinėse teritorijose [The change of agricultural lands in the agrarian areas of Lithuania and Europe]. *Miškininkystė Ir Kraštotvarka. Kauno Miškų Ir Aplink. Inžinerijos Kolegija* **2017**, *2*, 95–99.
33. Makrickienė, E.; Mozgeris, G.; Brukas, V.; Brodrechtova, Y.; Sedmak, R.; Salka, J. From command-and-control to good forest governance: A critical comparison between Lithuania and Slovakia. *For. Policy Econ.* **2019**, *109*, 102024. [CrossRef]
34. Mozgeris, G.; Jukneliené, D. Modelling Future Land Use Development: A Lithuanian Case. *Land* **2021**, *10*, 360. [CrossRef]
35. Forest Law. Forest Law of the Republic of Lithuania (in Lithuanian). 1994. Available online: https://e-seimas.lrs.lt/portal/legalAct/lt/TAD/TAIS.6036/KZzWEnfumI (accessed on 9 January 2024).
36. Ministry of Environment, 2023. Lithuania's National Inventory Report 2023. Available online: https://am.lrv.lt/lt/veiklos-sritys-1/klimato-politika/klimato-kaita/sesd-apskaitos-ir-prognoziu-ataskaitos-nacionaliniai-pranesimai/lietuvos-nacionalines-sesd-apskaitos-ataskaitos-anglu-k/ (accessed on 1 September 2023).
37. IPCC. *2006 IPCC Guidelines for National Greenhouse Gas Inventories, Prepared by the National Greenhouse Gas Inventories Programme*; Eggleston, H.S., Buendia, L., Miwa, K., Ngara, T., Tanabe, K., Eds.; IGES: Hayama, Japan, 2006.
38. EU. Regulation (EU) 2018/841 of the European Parliament and the Council of the European Union on the inclusion of greenhouse gas emissions and removals from land use, land use change and forestry in the 2030 climate and energy framework, and amending Regulation (EU) No 525/2013 and Decision No 529/2013/EU. *Off. J. Eur. Union* **2018**.
39. The Government of the Republic of Lithuania. Resolution on the Approval of the National Forestry Sector Development Program 2012–2020 (in lithuanian). Vilnius. 2012. Available online: https://e-seimas.lrs.lt/portal/legalAct/lt/TAD/TAIS.425608 (accessed on 10 March 2022).
40. NECP. National Energy and Climate Action Plan of the Republic of Lithuania for 2021–2030. 2018. Available online: https://commission.europa.eu/energy-climate-change-environment/implementation-eu-countries/energy-and-climate-governance-and-reporting/national-energy-and-climate-plans_en (accessed on 2 April 2022).
41. National Forest Agreement. Project for the National Forest Agreement. 2022. Available online: https://nacionalinismiskususitarimas.lt/kvieciame-susipazinti-su-galutiniu-nacionalinio-susitarimo-del-misku-susitarimo-dokumento-projektu/ (accessed on 1 May 2022).
42. Ministry of Environment. Lithuania's National Inventory Report 2021. 2021. Available online: https://am.lrv.lt/lt/veiklos-sritys-1/klimato-politika/klimato-kaita/sesd-apskaitos-ir-prognoziu-ataskaitos-nacionaliniai-pranesimai/lietuvos-nacionalines-sesd-apskaitos-ataskaitos-anglu-k/ (accessed on 10 March 2022).
43. Order of the Minister of Agriculture and of the minister of Environment. Approval of Procedural Description of Afforestation of Private Non-Forest Land (in Lithuanian). 2004. Available online: https://e-seimas.lrs.lt/portal/legalAct/lt/TAD/TAIS.230808/IfpSqkkyrt (accessed on 9 January 2024).
44. IPCC. *2013 Revised Supplementary Methods and Good Practice Guidance Arising from the Kyoto Protocol*; Hiraishi, T., Krug, T., Tanabe, K., Srivastava, N., Baasansuren, J., Fukuda, M., Troxler, T.G., Eds.; IPCC: Geneva, Switzerland, 2014.
45. FAO. Forestry Production and Trade. Database. 2021. Available online: https://www.fao.org/faostat/en/#data/FO (accessed on 30 March 2022).
46. Juknys, R.; Dagiliūtė, R.; Miškinis, V. From Transition to Sustainability: A Comparative Study. *Environ. Res. Eng. Manag.* **2008**, *1*, 61–68.
47. Eurostat. Share of Energy from Renewable Sources [NRG_IND_REN__custom_4020440]. 2022. Available online: https://ec.europa.eu/eurostat/data/database (accessed on 1 December 2022).
48. EU. Regulation (EU) 2023/857 of the European Parliament and of the Council of 19 April 2023 amending Regulation (EU) 2018/842 on binding annual greenhouse gas emission reductions by Member States from 2021 to 2030 contributing to climate action to meet commitments under the Paris Agreement, and Regulation (EU) 2018/1999 (Text with EEA relevance). *Off. J. Eur. Union* **2023**.
49. Ministry of Environment, 2018. Statement on Lithuania's Commitments to the EU on Reduction of Greenhouse Gas (GHG) and Tasks for Sectoral Ministries.
50. EC. Commission Delegated Regulation (EU) 2021/268 of 28.10.2020 amending Annex IV to Regulation (EU) 2018/841 of the European Parliament and of the Council as regards the forest reference levels to be applied by the Member States for the period 2021–2025. *Off. J. Eur. Union* **2020**, *L60*, 21–23.
51. EC 2022. European Commission Proposal for a Regulation of the European Parliament and of the Council on Nature Restoration, 2022/0195 (COD), Brussels, Belgium, p. 1–79. Available online: https://eur-lex.europa.eu/legal-content/EN/TXT/?uri=CELEX:52022PC0304 (accessed on 9 January 2024).
52. EU. Regulation (EU) 2018/842 of the European Parliament and the Council of the European Union on binding annual greenhouse gas emission reductions by Member States from 2021 to 2030 contributing to climate action to meet commitments under the Paris Agreement and amending Regulation (EU) No 525/2013. *Off. J. Eur. Union* **2018**.

53. ACEA. *Vehicles in Use, Europe 2023. Report*; European Automobile Manufacturers' Association: Brussels, Belgium, 2023; p. 21.
54. Kazanavičiūtė, V.; Dagiliūtė, R. Impact of LULUCF Accounting Rules for Climate Change Mitigation Goals: Winning or Losing? *J. Environ. Eng. Landsc. Manag.* **2023**, *31*, 19466. [CrossRef]
55. Liu, S.; Wilkes, A.; Li, Y.; Gao, Q.; Wan, Y.; Ma, X.; Qin, X. Contribution of different sectors to developed countries' fulfillment of GHG emission reduction targets under the first commitment period of the Kyoto Protocol. *Environ. Sci. Policy* **2016**, *61*, 143–153. [CrossRef]
56. Schlamadinger, B.; Johns, T.; Ciccarese, L.; Braun, M.; Sato, A.; Senyaz, A.; Stephens, P.; Takahashi, M.; Zhang, X. Options for including land use in a climate agreement post-2012: improving the Kyoto Protocol approach. *Environ. Sci. Policy* **2007**, *10*, 295–305. [CrossRef]
57. Mozgeris, G.; Kazanavičiūtė, V.; Jukneliene, D. Does Aiming for Long-Term Non-Decreasing Flow of Timber Secure Carbon Accumulation: A Lithuanian Forestry Case. *Sustainability* **2021**, *13*, 2778. [CrossRef]
58. National Audit Office. Conservation of Lithuanian Forest Resources (in Lithuanian). *Audit Report.* 2022. Available online: https://www.valstybeskontrole.lt/EN/Product/24091/protection-of-lithuanian-forestry-resources (accessed on 15 July 2022).
59. Lin, B.; Ge, J. To harvest or not to harvest? Forest management as a trade-off between bioenergy production and carbon sink. *J. Clean. Prod.* **2020**, *268*, 122219. [CrossRef]
60. MacLean, M.G.; Duveneck, M.J.; Plisinski, J.; Morreale, L.L.; Laflower, D.; Thompson, J.R. Forest carbon trajectories: Consequences of alternative land-use scenarios in New England. *Glob. Environ. Chang.* **2021**, *69*, 102310. [CrossRef]
61. Kauppi, P.E.; Stål, G.; Arnesson-Ceder, L.; Hallberg Sramek, I.; Hoen, H.F.; Svensson, A.; Wernick, I.K.; Högberg, P.; Lundmark, T.; Nordin, A. Managing existing forests can mitigate climate change. *For. Ecol. Manag.* **2022**, *513*, 120186. [CrossRef]
62. Peltoniemi, M.; Palosuo, T.; Monni, S.; Mäkipää, R. Factors affecting the uncertainty of sinks and stocks of carbon in Finnish forests soils and vegetation. *For. Ecol. Manag.* **2006**, *232*, 75–85. [CrossRef]
63. Jonard, M.; Nicolas, M.; Coomes, D.A.; Caignet, I.; Saenger, A.; Ponette, Q. Forest soils in France are sequestering substantial amounts of carbon. *Sci. Total Environ.* **2017**, *574*, 616–628. [CrossRef] [PubMed]
64. Lugato, E.; Panagos, P.; Bampa, F.; Jones, A.; Montanarella, M. A new baseline of organic carbon stock in European agricultural soils using a modelling approach. *Glob. Chang. Biol.* **2014**, *20*, 313–326. [CrossRef]
65. Li, X.; Wang, Y.-P.; Lu, X.; Yan, J. Diagnosing the impacts of climate extremes on the interannual variations of carbon fluxes of a subtropical evergreen mixed forest. *Agric. For. Meteorol.* **2021**, *307*, 108507. [CrossRef]
66. Bradshaw, C.J.; Warkentin, I.G. Global estimates of boreal forest carbon stocks and flux. *Glob. Planet. Chang.* **2015**, *128*, 24–30. [CrossRef]
67. Walden, L.L.; Fontaine, J.B.; Ruthrof, K.X.; Matusick, G.; Harper, R.J.; Hardy, G.E.S.J. Carbon consequences of drought differ in forests that resprout. *Glob. Chang. Biol.* **2019**, *25*, 1653–1664. [CrossRef] [PubMed]

land

MDPI

Article

Unveiling the Impact of Urbanization on Net Primary Productivity: Insights from the Yangtze River Delta Urban Agglomeration

Jing Gao [1], Min Liu [2,3] and Xiaoping Wang [4,*]

1 School of Teacher Education, Northwest Normal University, Lanzhou 730070, China; gaojing@nwnu.edu.cn
2 Shanghai Key Laboratory for Urban Ecological Processes and Eco-Restoration, School of Ecological and Environmental Sciences, East China Normal University, Shanghai 200241, China; mliu@re.ecnu.edu.cn
3 Technology Innovation Center for Land Spatial Eco-Restoration in Metropolitan Area, Ministry of Natural Resources, Shanghai 200241, China
4 State Key Laboratory of Soil Erosion and Dryland Farming on the Loess Plateau, College of Natural Resources and Environment, Northwest A&F University, Xianyang 712100, China
* Correspondence: wxp4911@nwafu.edu.cn; Tel.: +86-130-5752-5010

Abstract: Urbanization has significantly altered the carbon cycle of the terrestrial environment, particularly in relation to net primary productivity (NPP). Gaining a more comprehensive comprehension of how NPP is affected by urbanization is crucial for obtaining fresh perspectives on sustainable urban landscape design and decision making. While there is a significant body of research examining the geographical and temporal patterns of NPP supply capacity, there are only a few studies that have investigated the spatial relationships between NPP and urbanization, particularly at the grid scale. This research investigated the temporal and geographical features and patterns of NPP and their impact mechanisms. In order to estimate NPP and the level of urbanization in the Yangtze River Delta Urban Agglomeration (YRDUA), we used a combination of different models and datasets. To evaluate the geographical correlations and dependence between NPP and urbanization, we utilized local bivariate autocorrelation methods and spatial regression models to describe and visualize these relationships. The findings revealed that there was a consistent negative relationship between NPP and urbanization on a global scale from 1990 to 2020. However, when examining the local scale, the geographical correlations could be classified into four distinct categories: areas with both low NPP and low urbanization, areas with high NPP and high urbanization, areas with low NPP and high urbanization, and areas with high NPP and low urbanization. Our analysis showed that spatial regression models are more suitable for quantifying the spatial relationship between NPP and urbanization due to their ability to include the impacts of spatial Moran's I techniques. Due to the growing urbanization, the highest NPP value was recorded in 2005, followed by 2000, 2020, and 2010. Conversely, the smallest association was observed in 2015. Examining the geographical connection between NPP and urbanization offers theoretical and practical insights for urban planning that prioritizes human needs and promotes sustainable development. It also aids in the development of reasonable methods for organizing ecological functional systems.

Keywords: NPP; urbanization; spatial dependence; spillover effects; urban agglomeration

1. Introduction

Urban agglomerations have caused the conversion of natural ecosystems into ecosystems that are either dominated by humans or closely connected to human activities [1]. The process of urbanization is primarily driven by population concentration, economic growth, and urban expansion. These factors are recognized as the key drivers of changes in NPP within urban agglomerations [2]. Urbanization often leads to transformations in land use and land cover (LULC), affecting not just metropolitan areas but also their surrounding regions. For instance, the substitution of vegetation areas and the implementation of urban greening may directly change the composition of local terrestrial ecosystems [3,4]. Furthermore, urbanization has significantly impacted the environment for plant growth, including factors such as temperature, soil texture, and atmospheric conditions. Net primary production (NPP) is a common consequence of urbanization and has always been a subject of significant study interest [5]. Vegetation NPP refers to the total amount of organic matter produced by photosynthesis, minus the organic matter consumed by respiration; it represents the total amount of organic matter accumulated by vegetation per unit area and per unit time [6]. The dynamic changes in NPP, as a key parameter of terrestrial ecological processes and an important indicator reflecting the regional ecological conditions, can reflect the impact of climate change and human activities on ecosystems [7]. Studying the spatial–temporal patterns and driving factors of vegetation NPP is of great significance for the protection of regional ecological environments and sustainable development [8]. Human activities have a significant role in ecological management, since actions such as irrigation, pruning, and tree cutting may have substantial impacts [9]. In addition, urbanization may significantly alter terrestrial ecosystems, particularly the carbon cycle systems within them, due to the aforementioned effects [10]. Over the last several decades, there has been a significant increase in urbanization worldwide, making it a crucial aspect of global transformation [11]. Gaining a deeper understanding of how urbanization affects NPP of terrestrial ecosystems is crucial in this specific context [12].

Ecosystem services refer to all the benefits that humans derive from ecosystems, which are categorized into four distinct groups: supply services (such as providing food and pure water), regulatory services (such as controlling floods and diseases), cultural services (such as entertainment and cultural benefits), and support services (such as maintaining nutrient cycling) [13,14]. The regulatory services act as a conduit that links the NPP of the environment with the welfare of humans. It mostly pertains to the functions of climate control, such as carbon fixation, oxygen release, and cooling impacts. The global NPP relies on regulatory services as a crucial component and essential connection, which contribute significantly to the overall global carbon equilibrium [15]. Hence, within the framework of global climate change, investigating alterations in vegetation NPP has immense importance in comprehending the interplay between variations in plant productivity and climate [16,17]. Historically, the study of NPP has mostly relied on quantitative methods, such as biometric assessments including sample surveys and field measurements [18]. Nevertheless, these conventional measures conducted in the field often require a significant amount of time and effort, making them challenging to implement on a large scale to estimate NPP. Models have been extensively used in recent decades to obtain more precise NPP estimates on broader temporal and spatial scales; these models include statistical [19], process-based [20], and light energy utilization [21] models. Researchers have used NPP simulation models to study the effects of urbanization and LULC changes on NPP. Imhoff et al. used the Carnegie Ames Stanford Approach (CASA) model to examine the consequences of urban land conversion in the United States. Their findings indicate that urbanization has significantly and detrimentally affected NPP [22]. Paz-Kagan et al. used NPP as a measure to evaluate the impact of land-use changes on the ecosystems in semi-arid regions of Israel [23]. In China, many scholars have used the CASA model to assess the temporal and geographical NPP patterns and the influence of urban growth on NPP [24–26].

The changes in NPP in terrestrial ecosystems are a clear indicator of the impact of both human activities and global climate change on vegetation. These changes have a significant effect on the global carbon cycle and climate change. The capacity of the earth to support life and the sustainable evolution of terrestrial ecosystems can be evaluated by using this indicator [27]. Zhao et al. utilized the Moderate Resolution Imaging Spectrometer MOD13A2 Enhanced Vegetation Index (EVI) product to quantify the changes in NPP and found that plant growth in most Chinese cities saw substantial improvements as a result of indirect factors [28]. This improvement offset approximately 40% of the losses resulting from direct effects. Peng et al. used spatial regression to quantify the linear correlation between NPP changes and the three indicators of urbanization. They also identified the threshold at which NPP changes respond to these indicators [29]. Su et al. used the spatial lag model (SLM) to enhance the visualization of the non-stationary correlation between environmental services and urbanization [30]. While these studies attempted to examine the correlation between NPP and urbanization, several elements remain unexplored. There is a lack of consideration for the spatial relationship between NPP and urbanization, particularly at the regional level. Hence, other statistical methods must be used to address spatial autocorrelations. Furthermore, the previous research mostly concentrated on a single urban area, often using administrative districts to represent spatial entities. This level of study is insufficient to capture the spatial phenomena occurring at the meso or macro level, such as those occurring in towns, counties, and cities. This might restrict the practical feasibility of incorporating the NPP impact into comprehensive regional landscape design and the industrial arrangements of urban agglomerations.

The YRDUA is one of China's three main urban agglomerations and has the greatest economic growth rate and population density in the country. Over the last several decades, urbanization has caused significant changes in the land-cover conditions in the YRDUA, altering the structure and function of its terrestrial ecosystems. This process has significantly impacted the carbon budget of the area [31]. Hence, it is crucial to conduct more research on the impact of urbanization in the YRDUA on its NPP. The MOD17A3HGF V061 data products obtained from the data distribution system of the National Aeronautics and Space Administration (NASA) website offer NPP datasets with a resolution of 500 m. These datasets cover the period from 2000 to 2020 and fulfill the requisite criteria for both temporal duration and spatial precision.

The purpose of this study was to (1) use various models and multi-source data to quantify and map the degree of comprehensive urbanization, and analyze its spatial–temporal evolution pattern; (2) examine the relationship between urbanization and NPP using bivariate global and local Moran's I approaches; and (3) investigate the geographical relationship between urbanization and NPP, as well as other relevant parameters, using spatial regression models such as ordinary least squares (OLS) regression models and geographic weighted regression (GWR) models.

2. Materials and Methods

2.1. Study Area

Our study chose 16 prefecture-level cities as the research object, which are the core area of the YRDUA (Figure 1). The Lower Yangtze River, which borders both the East China Sea and Yellow Sea, is home to the central region of the YRDUA, which is located at 118° E–123° E, 28° N–33° N, and has an area of 167 thousand km^2, accounting for 1.74% of the total national land area. It is a segment of the alluvial plain near the point where the Yangtze River flows into the ocean, with an altitude of more than 10 m and low hills scattered between 200 and 300 m. The gross domestic product (GDP) of the 16 cities in the YRDUA's central area reached CNY 9.47 trillion in 2020 or 11.43% of the country's GDP. At this time, there were 119 million people living there, making up 9.68% of the entire population of the country. According to statistics data, the energy consumption of the 16 prefecture-level cities in the YRDUA core region surpassed 6869 Mt in 2020, constituting 15.92% of the overall energy usage in China.

2.2. Data Sources

We combined a variety of data sources that were diverse in nature, including both geographical and statistical feature data, into our analysis. More precisely, the datasets used were (1) yearly net primary production (NPP) data from the MOD17A3HGF V061 products with a spatial resolution of 500 m, obtained from the National Aeronautics and Space Administration (NASA) (https://lpdaac.usgs.gov/, accessed on 10 February 2021) [32];

(2) Land-use/land-cover (LULC) data from 1990 to 2020, compiled using a conventional interpretation method that analyzed Landsat Thematic Mapper (TM) and Landsat 8 OLI remote-sensing imagery with a 30 m resolution. The LULC data were derived from Landsat scenes covering path/row designations 118–120 and 37–40, and they achieved a classification accuracy of over 95% based on confusion matrix and Kappa coefficient testing [33,34]; (3) Gridded datasets of GDP and population at a 1 km spatial resolution, obtained from the Resource and Environment Science and Data Center (RESDC) of the Chinese Academy of Sciences (https://www.resdc.cn/, accessed on 12 June 2021) [35]; (4) Meteorological datasets including annual air temperature and annual rainfall from 1990 to 2020, interpolated to a 1 km grid from observations at 83 weather stations situated across and around the Yangtze River Delta Urban Agglomeration (YRDUA) region. These meteorological data were acquired from the China Meteorological Data Service Center (http://data.cma.cn, accessed on 18 August 2021) [36]; (5) Digital elevation model (DEM) data at 90 m resolution, resampled to 1 km resolution, obtained from the Geospatial Data Cloud platform of the Computer Network Information Center, Chinese Academy of Sciences (https://www.gscloud.cn/, accessed on 18 April 2021) [37].

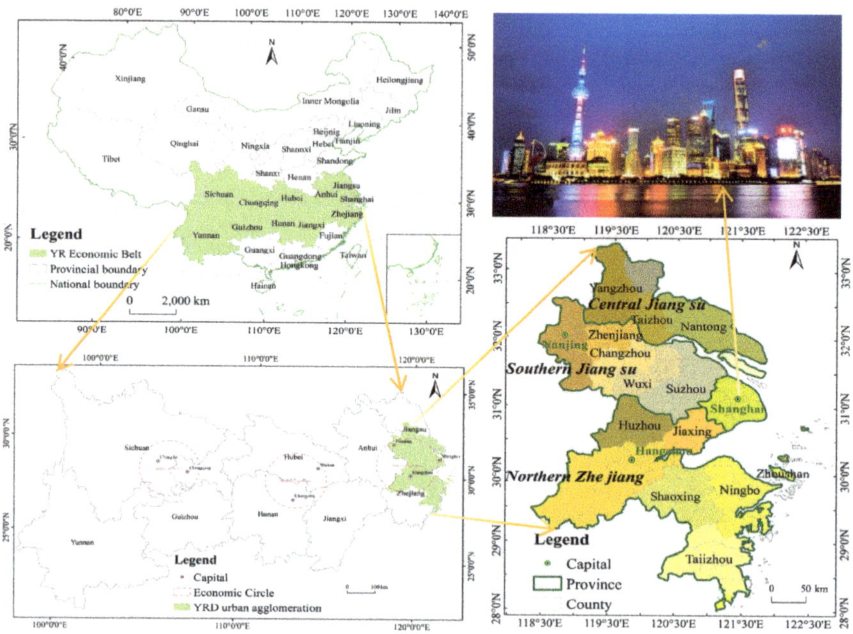

Figure 1. Map of the study area. **Top left** panel: the national boundary of China (green area is the location of the YR Economic Belt in China); **bottom left** panel: the administrative boundary of the YR Economic Belt (green area is the YRDUA in the YR Economic Belt); **bottom right** panel: the administrative boundary of the YRDUA (administrative districts); and **top right** panel: a representative city (Shanghai) in the YRDUA.

2.3. Data Analyses and Methods

The process of determining the NPP response to urbanization mainly included the following three steps: (1) quantitative characterization of the degree of comprehensive urbanization; (2) spatial correlation analysis between the changes in NPP and urbanization; and (3) spatial regression analysis of NPP and urbanization. A flow chart of the procedure is shown in Figure 2.

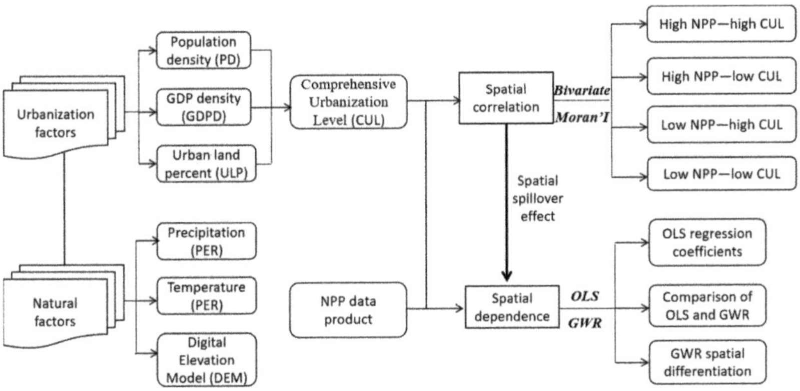

Figure 2. The process for determining the spatial relationship between NPP and urbanization.

2.3.1. Urbanization Assessment

The urbanization process can be generally characterized by the growth of the population, increase in the total economy, continuous improvement in the quality of life, and rapid growth of urban construction areas. In view of the fact that social urbanization data are not easy to collect and the indicators are relatively complex, we did not consider these data here; thus, the degree of urbanization was measured through the three other aspects. More precisely, population density (PD) was used as a metric to quantify the degree of urbanization in terms of population, gross domestic product density (GDPD) was selected to reflect the economic development level, and urban land percentage (ULP) was utilized to gauge the extent of urbanization in terms of land usage. Due of the high similarity in the geographical patterns of PD, GDPD, and ULP, these three variables were combined into a single indicator known as comprehensive urbanization level (CUL). The various indices were subjected to range standardization in order to convert their values into a uniform range of 0 to 1. These standardized values were then averaged to obtain the CUL value. The range standardization method (Equation (1)) and CUL calculation (Equation (2)) are as follows:

$$U'_{i,j} = \frac{U_{i,j} - U_{i,min}}{U_{i,max} - U_{i,min}} \tag{1}$$

$$CUL_j = (PD_j + GDPD_j + ULP_j)/3 \tag{2}$$

where $U'_{i,j}$ represents the normalized value of $U_{i,j}$; $U_{i,j}$ is the i-th urbanization indicator (PD, GDPD, or ULP) in the j-th raster, relative to the original value; $U_{i,max}$ and $U_{i,min}$ represent the highest and lowest values, respectively, of the i-th urbanization indicator over all grids; CUL_j represents the urbanization level of the i-th grid; and PD_j, $GDPD_j$, and ULP_j represent the population density, GDP density, and urban land proportion, respectively, of the j-th grid after standardization. The rationale for using Equation (2) to calculate the CUL is that it provides a balanced and holistic measure of urbanization by equally weighting the three key aspects [38].

2.3.2. Spatial Correlation Measure

The bivariate Moran's I statistic was utilized to identify any geographical clustering or discontinuous link between the comprehensive urbanization level (CUL) and vegetation net primary production (NPP). Global and local bivariate Moran's I are two strategies that can be used for this purpose. The formulae that were used are as follows (Equations (3a) and (3b)):

$$I_{cu} = \frac{N\sum_i^N \sum_{j\neq i}^N W_{ij} z_i^c z_j^u}{(N-1)\sum_i^N \sum_{j\neq i}^N W_{ij}}, \tag{3a}$$

$$I'_{cu} = z^c \sum_{j=i}^N W_{ij} z_j^u, \tag{3b}$$

Here, I_{cu} and I'_{cu} are the global and local bivariate Moran's I of NPP and CUL, respectively, and N represents the aggregate number of spatial grid cells. In the model, the parameters obtained are W_{ij}, which represents an $N \times N$ weighted matrix that was used to detect the correlation between the i-th and j-th grids. The spatial unit is a 4×4 matrix generated based on the first-order neighborhood in the weight adjacent to the queen [39]. The input data were z_i^c and z_j^u; z_i^c represents the i-th standardized NPP grid value obtained by using Equation (1), and z_j^u represents the j-th standardized CUL unit value calculated using Equation (1) [40,41]. The output result is I_{cu}/I'_{cu}, where the range of values for I_{cu}/I'_{cu} is -1 to 1. The computed p-value for the regional connection between the NPP impact and CUL was below 0.05, indicating statistical significance [42]. NPP and CUL were readjusted to a 1×1 km raster map using the mean value approach in ArcGIS 10.5. Next, the NPP and CUL data of all grids were entered into GeoDa 1.12 (https://geodacenter.github.io/, accessed on 16 August 2021) for execution, and spatial correlation analyses were conducted [43].

Bivariate spatial autocorrelation can determine whether two variables are spatially correlated and evaluate the strength and direction of the correlation. It can help us in exploring the laws of geographical phenomena and spatial distributions, providing a scientific basis for decision making [44]. However, bivariate spatial autocorrelation analysis also has some limitations, since it is sensitive to data distribution biases, spatial scale effects, spatial connections, and causal relationships. Therefore, we should take these limitations into consideration when interpreting the results [45,46].

2.3.3. Spatial Regression Test

1. Analysis of global spatial regression

Ordinary least squares (OLS) can generate predictions when performing global linear regression, or model a dependent variable and a set of explanatory variables to detect the influence relationship. Anselin provides the general form of the spatial regression equation for raster data, taking into account the spatial correlation between independent variables and dependent variables [47] (Equations (4) and (5)):

$$Y = \rho W_1 Y + X\beta + \varepsilon, \tag{4}$$

$$\varepsilon = \lambda W_2 + \mu, \mu \sim N(0, \Omega), \Omega_{ii} = h_i(za), \tag{5}$$

where ρ represents the coefficient of the geographical lag variable W_1Y; β represents the $k \times 1$ parameter vector associated with the independent variable X; ε is the vector representing the random error term; the weight matrix W_1 represents the geographical pattern of the variable; the order weight matrix W_2 represents an $n \times n$ matrix; the normal distribution is denoted by N; the exogenous variable is represented by z, while Ω denotes the variance matrix, its diagonal elements are Ω_{ii}, h_i is the functional relationship, and the constant term is represented by a; and the spatial autonomy is denoted by λ. The coefficients of the regression structure W_2 should generally be $0 \leq \rho < 1$, $0 \leq \lambda < 1$, and μ is a random error vector of a normal distribution. The regression equation of the whole grid data space is subject to 3 parameters: ρ, λ, and a.

2. Analysis of local spatial regression

Spatial regression technology was used to study the spatial dependence of the effect of urbanization on NPP (that is, how NPP changes in response to the process of urbanization). Geographically weighted regression (GWR) is a type of regression that adds regional ordinary least squares (OLS) to improve the model [48]; the expression of the model is as follows (Equation (6)):

$$ZFJG_i = \beta_o(u_i, v_i) + \sum_{i=1}^{T} x_u \beta_i(u_i, v_i) + \varepsilon_i, \quad (6)$$

where $\beta_o(u_i, v_i)$ is a constant term; $\beta_i(u_i, v_i)$ is the characteristic elastic coefficient of the i-th sample point. The elastic coefficient of every point (u_i, v_i) in the sample region is determined using a weighted least square multiplication method; the calculation formula is as follows (Equation (7)):

$$\hat{\beta}(u_i, v_i) = \left(\hat{\beta}_\rho(u_i, v_i), \hat{\beta}_\tau(u_i, v_i)\cdots, \hat{\beta}_\gamma(u_i, v_i)\right)^T = \left(X^\gamma W(u_i, v_i)X\right)^{-1} X^\gamma W(u_i, v_i) ZFJG_i \quad (7)$$

where X represents the matrix of independent variables and $W(u_i, v_i)$ represents the spatial weight matrix. The spatial weight matrix is constructed using a monotonically decreasing function that calculates the geographical distance between the location to be estimated and the surrounding observation sites. Different function forms can be used. Our study used the Gauss kernel function; its expression is as follows (Equations (8) and (9)):

$$W(u_i, v_i) = \text{diag}\left(K(d_{io}/h), K(d_{i\tau}/h)\cdots, K(d_{i\gamma}/h)\right), \quad (8)$$

$$K(t)\frac{1}{\sqrt{2\pi}}\exp\left(-\frac{1}{2}t^2\right), \quad (9)$$

where d_i is the Euclidean distance between each sample point and h is the optimal bandwidth, which can be determined using the cross-determination method to minimize h (Equation (10)).

$$CV(h) = \frac{1}{n}\sum_{i=1}^{n}\left(ZFJG_i - Z\hat{F}JG_{(-i)}(h)\right)^2, \quad (10)$$

where $Z\hat{F}JG_{(-i)}(h)$ is the simulated predicted value of the NPP at point i obtained by simulation after the i-th observation value is discarded under h, and $ZFJG_i$ is the actual observed value of the NPP at point i.

3. Results

3.1. Spatial CUL Patterns in the YRDUA

According to the analysis of Figures 3 and 4 and Supplementary Materials, from 1990 to 2000, the comprehensive urbanization level of the YRDUA showed a predominant pattern of circular expansion around the major cities. During the two five-year periods from 1990 to 1995 and 1995 to 2000, Shanghai, as the main city in the urban agglomeration, saw significant urbanization development. The comprehensive urbanization level increased from 67.37 to 105.75, with growth rates of 6.05% and 12.22%, respectively. The comprehensive urbanization level developed slowly in the YRDUA.

After 2000, the YRDUA achieved axial expansion, i.e., expansion of the circle around core cities. The urbanization level of the growth poles and key cities on the Nanjing–Hangzhou Expressway increased significantly. The growth pole city Hangzhou had an added value of urbanization of 130.86, with a growth rate of 67.11%. The comprehensive urbanization growth values of key cities such as Ningbo, Huzhou and Shaoxing were 66.21, 21.55, and 51.73, and the growth rates were 120.32%, 41.66%, and 87.43%, respectively. The process of urbanization developed rapidly.

After 2010, as the development of urban agglomerations continued to increase, cities within urban agglomerations gradually obtained their own independent development space. The linear connection mode was gradually replaced by the expand-around mode, and the evolution of urban agglomerations was transformed into a network mode. Due to the changes in macro policy and concept aspects, such as adjusting industry structures and promoting upgrades, Shanghai and Hangzhou have maintained a steady growth rate due to the development of their emerging high-tech industries [49]. The added values of urbanization were 104.78 and 73.93, respectively, with growth rates of 9.35% and 17.39%; thus, the urbanization process has grown steadily.

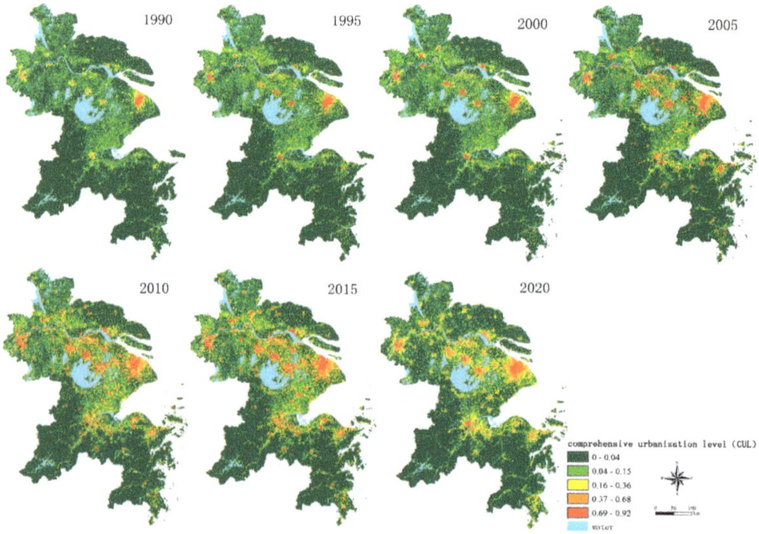

Figure 3. Spatial pattern of comprehensive urbanization level (CUL) in the YRDUA from 1990 to 2020.

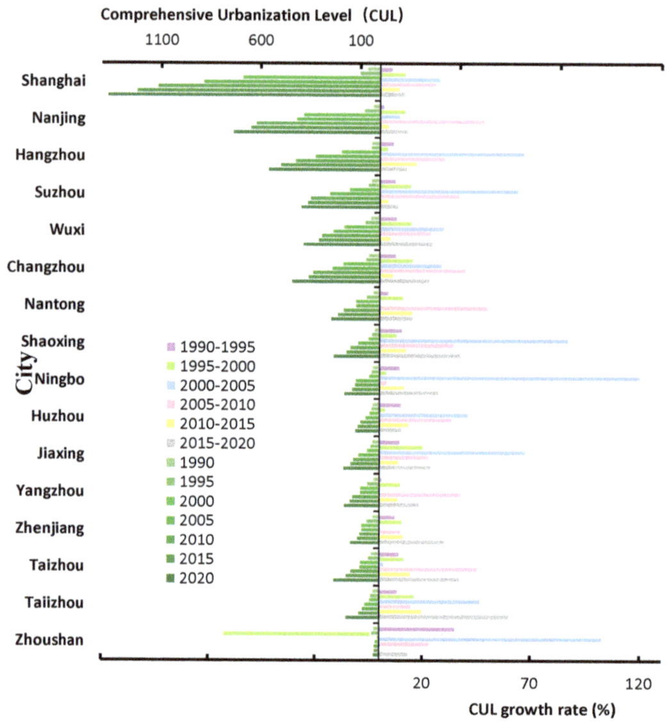

Figure 4. Comparison of comprehensive urbanization level (CUL) across the 16 cities in the YRDUA during 1990–2020.

3.2. Geographical Links between Urbanization and NPP

Moran's I analysis revealed a notable negative geographical association between the NPP and urbanization, irrespective of the year (Figure 5). It is known that the growth of built-up land led to a decline in NPP at the global level. Nevertheless, the extent of the negative association varies depending on the various phases of urban agglomeration growth. Here, the association between NPP and urbanization was the strongest in 2010 (Moran's I: −0.2492), followed by 2020 (Moran's I: −0.1937), 2015 (Moran's I: −0.1841), 1990 (Moran's I: −0.1685), 2000 (Moran's I: −0.1685), and 2005 (Moran's I: −0.1470). The weakest correlation was in 1995 (Moran's I: −0.1234). The results of the global spatial autocorrelation analysis, to some extent, showed a spatial correlation between NPP and CUL, and overall, the negative correlation increased over time.

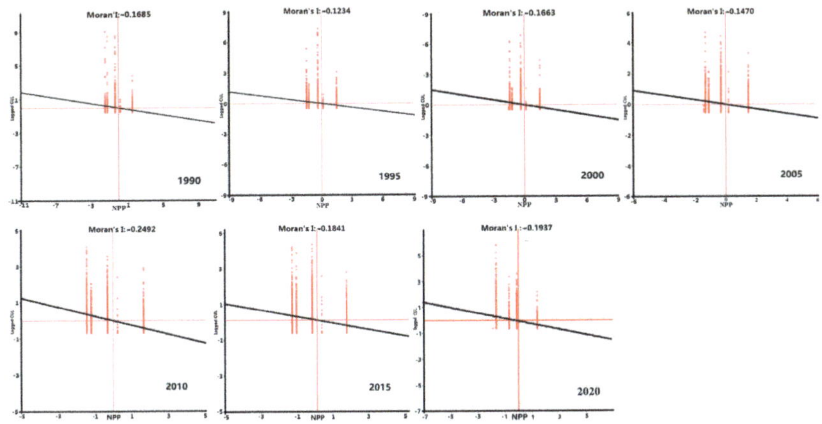

Figure 5. Spatial–temporal correlations between NPP and CUL (global bivariate Moran's I autocorrelation).

Local bivariate spatial autocorrelation is a statistical method used to analyze the spatial correlation between two variables at the local level; it introduces the spatial concept into the autocorrelation analysis, allowing us to determine whether two variables are spatially correlated and the strength and direction of the correlation, which are usually represented by the local indicators of spatial association (LISA) [50]. The LISA diagram shows the four possible geographical correlations that exist between urbanization and NPP (Figure 6): the high–high (HH) type represents the clustering of high NPP and high CUL values; the low–low (LL) type represents the clustering of low NPP and low CUL values; the low–high (LH) clustering represents the clustering of low NPP and high CUL values; and the high–low (HL) clustering represents the clustering of high NPP and low CUL values [51]. Using a seven-year sample, we saw distinct similarities in the way NPP and urbanization were clustered in different regions. The places with the highest elevation are mostly located in the central regions of the urban land of the YRDUA. With the expansion of urban land, HH areas also increased. The LH regions were mostly dispersed over the whole HH region and concentrated around the HH areas. The low–high regions were mostly concentrated in the northern region of the urban agglomeration, whereas low–high regions were absent from the southern region. The HL area occupied a large area in the south, concentrated in the mountains in southwest Hangzhou and the lush vegetation areas in the southern mountainous areas of Taizhou, Ningbo, and Shaoxing, which had relatively low levels of urbanization due to being restricted by natural conditions such as topography and landforms. The LL area did not show any changes in its spatial pattern over time. In 2005, 2015, and 2020, the LL areas appeared in the coastal areas of Hangzhou, Shaoxing, and Ningbo, while in the other four years, the LL areas did not show any obvious spatial characteristics or LL areas did not appear.

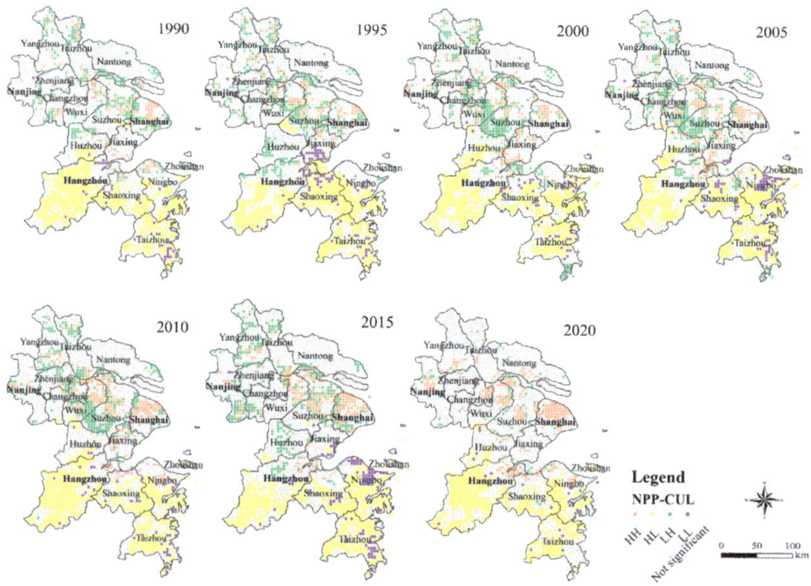

Figure 6. Spatial–temporal correlations between NPP and CUL (LISA diagram).

3.3. *Spatial NPP Pattern Dependence on Urbanization*

Changes in the ecosystem are fundamental impacts of climate change; therefore, climatic factors (including temperature, precipitation, and digital elevation models (DEM)) were analyzed in order to reveal the driving factors affecting the regional NPP. The results simulated by the OLS model showed that the regression coefficients of precipitation in all years were positive, indicating that precipitation and the NPP were positively correlated (Table 1). Except for the regression coefficients of PD in 2000 and ULP in 2005 that showed positive correlations, all factors from 1990 to 2020 were negatively correlated with NPP. From 1990 to 2010, the absolute value of the regression coefficient of the CUL was always greater than that of the other factors (PD: −0.48; GDPD: −0.61; ULP: −0.53; TEM: −0.86; PRE and DEM: −0.56). GDPD ranked second in 1990 and 2000, with coefficients of −0.55 and −0.48, and PD in 1995 and 2005 ranked second with coefficients of 0.27 and −0.61, respectively. As the pace of urbanization stabilized, the population and economic growth in the YRDUA reached a state of relative saturation, and the influence of urbanization factors on the NPP diminished. In 2010, the coefficients of temperature and precipitation were relatively large at −0.56 and 0.36. In 2020, the regression coefficients of PD, GDPD, ULP, and CUL continued to decrease compared with 2015 and 2010, and they were still smaller than the meteorological factors, showing a relatively weak degree of influence.

Table 2 shows the R^2, adjusted R^2, AIC, and Moran's I values from the OLS and GWR models. The R^2 (adjusted R^2) values for the GWR model ranged from 0.42 to 0.53, surpassing those of the OLS model. Meanwhile, Moran's I and AIC from the GWR model surpassed those from the OLS model, suggesting that the GWR model is superior to the OLS model in examining the variables influencing the NPP. GWR is more appropriate for spatial regression analyses than OLS. Since the CUL regression coefficient is the largest, we analyzed the regional differences and evolution trend of the CUL regression coefficients and residual and explored whether CUL increasingly affects the spatial pattern of the NPP.

Table 1. OLS model analysis results of NPP-influencing factors from 1990 to 2020.

Year	Variable	Coefficient	Standard Deviation	t/z Value	p-Value (>|t|)
1990	(Intercept)	0.23	0.03	19.79	0.00 **
	PD	0.02	0.17	0.09	0.93 *
	GDPD	−0.55	0.14	−3.93	0.00
	ULP	0.35	0.12	2.83	0.01 **
	CUL	−0.48	0.34	−3.16	0.01 ***
	TEM	−0.37	0.04	−10.23	0.00 **
	PRE	0.36	0.02	22.05	0.05 **
	DEM	0.35	0.06	14.20	0.00 **
1995	(Intercept)	0.27	0.03	19.08	0.00 ***
	PD	0.27	0.12	2.26	0.02 *
	GDPD	−0.05	0.17	−6.03	0.00
	ULP	0.10	0.10	1.10	0.27
	CUL	−0.61	0.21	−2.82	0.01 ***
	TEM	−0.37	0.04	−9.78	0.00 ***
	PRE	0.32	0.02	18.37	0.05 ***
	DEM	0.25	0.03	9.42	0.00 ***
2000	(Intercept)	0.24	0.03	21.00	0.00 ***
	PD	0.22	0.13	1.73	0.08 *
	GDPD	−0.48	0.15	−3.22	0.00
	ULP	0.12	0.09	0.02	0.98
	CUL	−0.53	0.22	−2.47	0.01 ***
	TEM	−0.43	0.04	−12.08	0.00 ***
	PRE	0.33	0.02	20.20	0.00 ***
	DEM	0.41	0.05	11.02	0.00 ***
2005	(Intercept)	0.38	0.03	24.73	0.00 ***
	PD	−0.61	0.24	−2.55	0.01 ***
	GDPD	−0.36	0.07	−5.28	0.00 ***
	ULP	0.17	0.08	2.09	0.04 **
	CUL	−0.86	0.17	−5.19	0.00 ***
	TEM	−0.42	0.04	−11.77	0.00 ***
	PRE	0.22	0.02	14.07	0.00 ***
	DEM	0.51	0.06	10.28	0.00 ***
2010	(Intercept)	0.29	0.02	25.28	0.00 ***
	PD	−0.15	0.10	−1.48	0.14 *
	GDPD	−0.32	0.17	−2.42	0.02 ***
	ULP	−0.20	0.15	−1.33	0.08 *
	CUL	−0.46	0.25	−0.66	0.11 *
	TEM	−0.56	0.03	−13.91	0.00 ***
	PRE	0.36	0.02	23.02	0.00 ***
	DEM	0.42	0.01	16.35	0.00 ***
2015	(Intercept)	0.25	0.02	25.25	0.00 ***
	PD	−0.13	0.13	−1.00	0.32
	GDPD	−0.13	0.05	−2.71	0.01 ***
	ULP	−0.09	0.05	−1.72	0.08 *
	CUL	−0.29	0.13	−1.40	0.16 *
	TEM	−0.41	0.03	−12.83	0.00 ***
	PRE	0.23	0.02	10.51	0.00 ***
	DEM	0.49	0.05	9.24	0.00 ***
2020	(Intercept)	0.31	0.01	22.13	0.00 ***
	PD	−0.11	0.25	−1.03	0.28
	GDPD	−0.15	0.08	−2.34	0.01 ***
	ULP	−0.16	0.06	−1.65	0.059 *
	CUL	−0.32	0.11	−1.58	0.14 *
	TEM	−0.52	0.04	−11.54	0.01 **
	PRE	0.31	0.01	10.36	0.00 ***
	DEM	0.52	0.06	8.87	0.00 ***

Note: *, **, and *** indicate significance at 90%, 95%, and 99% confidence levels, respectively. Abbreviations: population density (PD); GDP density (GDPD); urban land percent (ULP); comprehensive urbanization level (CUL); temperature (TEM); precipitation (PRE); digital elevation model (DEM).

Figure 7 shows the spatial impact (regression coefficients and residual) of CUL on NPP from 1990 to 2015. In 2000, the regression coefficient of CUL was roughly centered on the western part of Hangzhou, the regression coefficient becomes smaller as it goes outward, and the influence of comprehensive urbanization level (CUL) on NPP gradually decreases. In 1995, the regression coefficient for both the high-value area and low-value area shifted slightly northward towards the traffic line connecting Shanghai and Nanjing, compared to 1990. This indicates that Hangzhou's e-commerce, Internet development, and other tertiary industries gradually influenced the surrounding areas such as Shanghai and Nanjing [52]. The spatial pattern of the regression coefficients in 2000 was basically the same as that in 2005. The areas of Taizhou and Wenling in the southeast corner changed from median regression coefficients to high values. This is because the degree of urbanization in Taizhou and Wenling was strengthened and it had begun to have a greater impact on the spatial NPP pattern.

Table 2. OLS and GWR model results of NPP-influencing factors from 1990 to 2020.

Parameter	Model	1990	1995	2000	2005	2010	2015	2020
AIC	OLS	−1794.51	−1499.15	−1892.90	−1484.85	−1220.41	−1592.63	−1975.63
	GWR	−1880.94	−1619.63	−2063.51	−1648.74	−1311.01	−1687.76	−1653.41
R^2	OLS	0.46	0.43	0.48	0.47	0.50	0.42	0.46
	GWR	0.30	0.48	0.55	0.54	0.55	0.46	0.43
Adjusted R^2	OLS	0.45	0.42	0.47	0.47	0.50	0.42	0.44
	GWR	0.49	0.47	0.53	0.52	0.53	0.45	0.50
Moran's I	OLS	−0.04	−0.06	−0.03	−0.05	−0.01	−0.03	−0.02
	GWR	−0.16	−0.17	−0.15	−0.17	−0.18	−0.17	−0.14

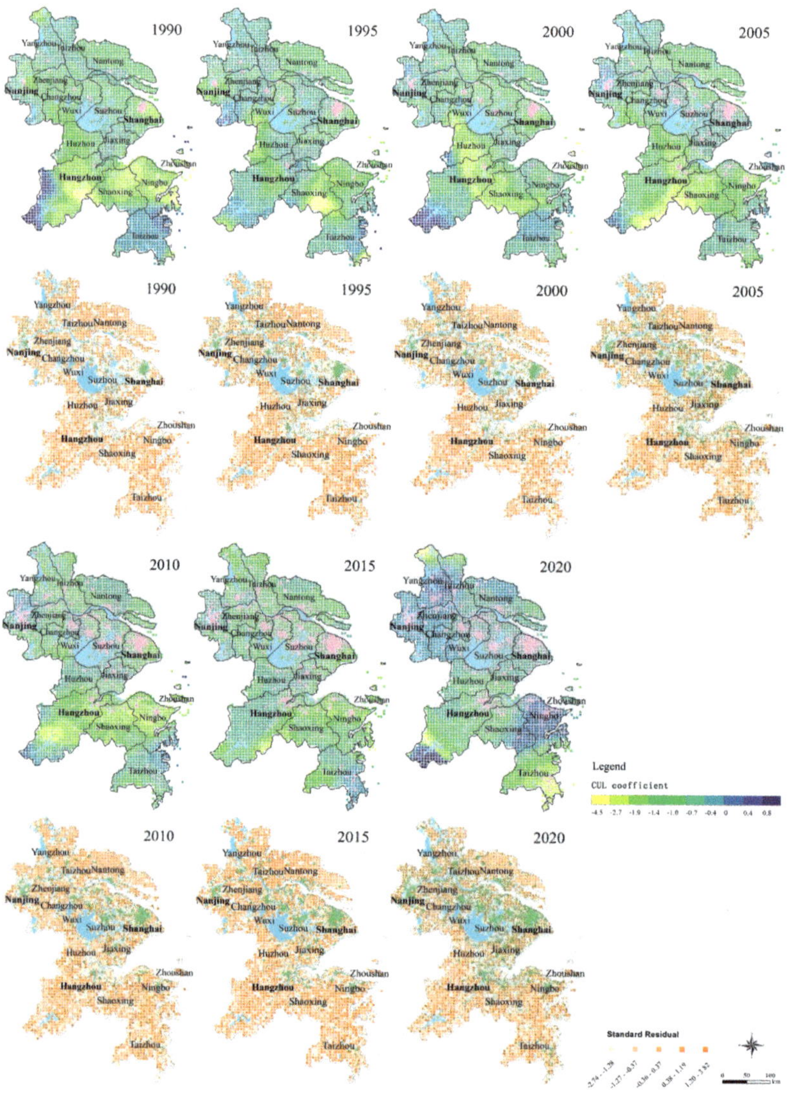

Figure 7. Regression coefficient and standard residual distribution of CUL from 1990 to 2020.

4. Discussion

4.1. Geographical Spillover Consequences in the Correlation between Urbanization and NPP

Geographical spillovers occur when the proximity of one unit to its nearby units influences its benefits or costs [53]. Statistically significant bivariate global Moran's I values were observed in all cases (Figure 5). This suggests that there was a significant spillover effect on the geographical correlation between urbanization and NPP in the YRDUA. Additionally, all bivariate global Moran's I values were negative, indicating that urbanization resulted in negative externalities for NPP. The findings of the bivariate LISA analysis (Figure 6) indicate that the spatial spillover process was not spatially independent. Furthermore, the transmission of the spatial spillover impact across grid cells was severely limited by the regional context [54]. In other words, when a grid is next to a highly urbanized grid, there is a higher likelihood of its NPP dropping. Conversely, when it is next to a grid with low urbanization, the reverse scenario occurs. The bivariate LISA analysis revealed distinct clustering patterns of high–high and low–low associations, indicating the need for more investigation into the relationship between NPP and urbanization. It is important to include other characteristics such as plant cover, water cover, terrain, and soil, since these may also have an impact. These variables, in conjunction with urbanization, influenced the alterations in the regional NPP [55].

The bivariate LISA diagram displays many conspicuous characteristics (Figure 6).

1. Examining the correlation between NPP and urbanization, it was evident that from 1990 to 2020, the geographical arrangement of regions deemed as not significant remained consistent. These areas are mostly located on the outskirts of metropolitan agglomerations. This is due to the fact that, compared to the Shanghai, Suzhou–Wuxi–Changzhou, and Nanjing metropolitan areas, these places exhibited a lower level of urbanization activities, such as population concentration, economic investment, and land development. Hence, urbanization is not the primary influence on NPP on the outskirts of metropolitan agglomerations;
2. In 2010, the geographical correlations of NPP and urbanization were highest in the high–high and low–high areas. The regions at the highest elevations are mostly located in the inner region, while the regions at lower elevations are found around the urban built-up areas. Prior to 2010, the rate of urban expansion exhibited a consistent and steady increase. Since 2010, there has been a growing awareness at both the national and regional levels of the rapid expansion of urban agglomerations and the environmental pollution issues associated with economic development and high-energy-consuming industries. These factors have significantly contributed to global climate change and the degradation of the ecological environment. Consequently, regulations have been implemented to regulate the unrestricted expansion of urban areas and reconfigure energy-intensive businesses in order to transition and enhance the use of clean energy sources. Hence, starting in 2015, the association between NPP and urbanization seemed to diminish;
3. The geographical distribution patterns of NPP and urbanization in 2000 and 2010 exhibited a significant degree of similarity. This outcome aligns with the findings of Qiu's study. He discovered that the decade spanning from 2000 to 2010 had the highest rate of urbanization and the greatest stability for the urban agglomeration of the Yangtze River Delta. During this time, the urban agglomeration underwent a phase of creation and development, with a consistent and continuous increase in the level of urbanization [56]. Simultaneously, the decline in NPP failed to attract attention and recognition for its effect and repercussions. The decline in NPP and the rise in anthropogenic carbon emissions are not being effectively managed and regulated. NPP and urbanization exhibited a strong geographical correlation, indicating a consistent pattern of spatial clustering.

4.2. NPP and CUL Spatial Link Implications for Urban Agglomeration Development Programs

Using an OLS model, our research quantitatively analyzed the influence mechanism of the spatial–temporal evolution of NPP in the YRDUA. The findings indicate that the impact of different variables on the spatial–temporal development of NPP varied across different time periods, and the influence of CUL on NPP showed an inverted "U" pattern. Coordinating regional development, adapting measures to local conditions, maximizing the benefits of regional development, and achieving a balanced, coordinated, and sustainable urban agglomeration regional development model are important goals for the progress of the YRDUA region. Prioritizing the low-NPP areas that are most vulnerable to urbanization is crucial when embarking on urban development projects. Any region exhibiting a substantial NPP should be designated as an ecological reserve to prevent or limit urban expansion [57,58].

The regression analysis of the CUL in 2015 revealed that the regions with high regression coefficients were concentrated in the western border region of the urban agglomeration, as seen from the distribution patterns of Shanghai and Hangzhou. This is mostly due to the urbanization growth of the urban agglomeration, which has started expanding towards the west. As a result, it has begun to link with and influence the urbanization process of the western strip region of the urban agglomeration [59]. As far as the actual situation is concerned, in the "Yangtze River Delta Urban Agglomeration Development Plan (2015–2030)", 10 prefecture-level cities under the jurisdiction of Anhui Province, which is close to Jiangsu, have also been assigned to the YRDUA, so that the YRDUA has a more solid development foundation and geographical space, which can better realize the improvement of quality and efficiency and the integrated development of large regions [60].

The study findings can more accurately align with the current state of development in the YRDUA. Additionally, the research conclusions may serve as a foundation for the creation of regional development policies for the YRDUA. The analytical framework not only emphasizes the application of new methods but also pays attention to the dynamic space of regional cooperation and its interconnections [61]. It aims to provide support for further coordinating regional balanced development and strengthening regional exchanges and cooperation by identifying and quantifying the spillover effect between regions. This research approach can also provide new research ideas and methods for other domestic regional economic development research [62].

4.3. Limitations of the Applied Method

However, there are still some limitations in this study. One problem is that the R^2 values are relatively low for both the OLS and GWR models, suggesting that other factors affecting NPP were not fully explored. Further research should develop improved models based on the characteristics of the study area to analyze the spatial–temporal evolution of NPP, such as spatiotemporal weighted regression (STWR) models. The STWR model is characterized by using a novel "time distance" for weighting to capture fine spatiotemporal heterogeneity, as opposed to the traditional geographically and temporally weighted regression (GTWR) approach [63]. These advanced spatiotemporal modeling techniques could help to better elucidate the relationship between NPP and its driving factors, including the role of past carbon storage data to determine the relationship between social behavior and natural resources. Another issue is that the geospatial data layer should be further refined in future research. It is worth noting that the assessment of ecosystem services depends on the choice of proxy indicators. In our study, the NPP indicator, which characterizes carbon fixation and oxygen release, was selected to reflect the regulatory function of ecosystem services. If the NPP data based on MODIS data cannot be accurately spatially predicted, or the collected data do not have sufficiently high spatial and temporal resolutions, there may be errors in the results of the correlation analyses and regression fittings. This issue also arises in proxies used for evaluating urbanization, and all three levels have inherent issues concerning the quality and accuracy that may impact the study's findings.

The findings indicate that GWR outperforms the standard regression method (OLS) in explaining the relationship between urbanization and the responsiveness of ecosystem services, especially for NPP. The results derived from the methodologies used in this investigation are very reliable and satisfactory, indicating that spatially explicit modeling approaches could be valuable for decision making and policy formulation. If the data used in this investigation can be gathered for other geographical areas, the technique could be applied in those areas to evaluate indicators of urbanization and NPP and perform a geographical examination of the interconnections between these two aspects. The purpose of this research is to assist governments in making informed choices that will contribute to the long-term sustainability of urban agglomeration areas, including the economic, environmental, and sociocultural aspects.

5. Conclusions

Our research investigated the relationship between ecosystem services (with NPP representing regulatory functions) and urbanization from a geographical standpoint, taking into account the spatial correlations and dependencies. The findings of our investigation led to the following conclusions: (1) The bivariate global Moran's I of urbanization and NPP from 1990 to 2020 exhibited negative values, suggesting a global negative connection between the two variables. From a local standpoint, there was a geographic disparity in the correlations between CUL and NPP. The bivariate LISA approach identified and presented four different types of local correlations (namely, high–high, high–low, low–high, and low–low) between NPP and urbanization. (2) The spatial regression analysis revealed that urbanization and other influencing variables have varying effects on NPP. Due to the growing urbanization, NPP reached its peak in 2005, with 1995 and 2010 following closely after. The correlation was lowest in 2015. (3) Aside from urbanization, environmental services are also influenced by other variables such as climate and geography. When accounting for spillover effects in the regression analysis, the influence of urbanization on ecosystem services showed a steady increase from 1990, followed by a gradual decline after 2010, in contrast to the findings obtained using the OLS method. However, our study has certain limitations. First, we used NPP as a proxy for ecosystem services, which may not fully capture all aspects of regulatory functions. Second, the spatial regression models employed may not account for all potential confounding factors and complex interactions. Third, our analysis focused on the national scale, and finer-scale local variations may exist. Despite these limitations, our findings offer practical implications for urban planning and industrial site selection, highlighting the importance of considering spatial dependencies and trade-offs between urbanization and ecosystem services. Future research could explore alternative measures of ecosystem services, incorporate additional explanatory variables, and conduct multi-scale analyses to further refine our understanding of this complex relationship. We recommend that urban planners and policymakers consider the spatial patterns and correlations identified in our study when developing urbanization strategies and industrial zoning plans. Balancing economic growth with ecosystem conservation requires a holistic approach that accounts for geographic variations and spatial spillover effects. Integrating these considerations into decision-making processes can promote sustainable urban development and environmental stewardship.

Supplementary Materials: The following supporting information can be downloaded at: https://www.mdpi.com/article/10.3390/land13040562/s1, Figure S1: (a) Spatial pattern of population (POP) in the YRDUA from 1990 to 2020. (b) Spatial pattern of gross domestic product (GDP) in the YRDUA from 1990 to 2020. (c) Spatial pattern of urban land percentage (ULP) in the YRDUA from 1990 to 2020.

Author Contributions: Conceptualization, methodology, writing—review and editing, J.G. and M.L.; formal analysis, writing—original draft preparation, visualization, J.G.; supervision, M.L. and X.W. All authors have read and agreed to the published version of the manuscript.

Funding: This work was supported by the National Natural Science Foundation of China (42371219), Gansu Province Philosophy and Social Science Planning Project (2023QN010), and the Young Teachers' Research Ability Enhancement Program Project (NUNW-SKQN2023-24).

Data Availability Statement: Data are contained within the article.

Acknowledgments: The authors are grateful to the editor and reviewers for their valuable comments and suggestions.

Conflicts of Interest: The authors declare no conflicts of interest.

References

1. Qiu, B.K.; Li, H.L.; Zhou, M.; Zhang, L. Vulnerability of ecosystem services provisioning to urbanization: A case of China. *Ecol. Indic.* **2015**, *57*, 505–513. [CrossRef]
2. Seto, K.C.; Guneralp, B.; Hutyra, L.R. Global forecasts of urban expansion to 2030 and direct impacts on biodiversity and carbon pools. *Proc. Natl. Acad. Sci. USA* **2012**, *109*, 16083–16088. [CrossRef]
3. Yan, Y.C.; Liu, X.P.; Wang, F.Y.; Li, X.; Ou, J.P.; Wen, Y.Y.; Liang, X. Assessing the impacts of urban sprawl on net primary productivity using fusion of Landsat and MODIS data. *Sci. Total Environ.* **2018**, *613*, 1417–1429. [CrossRef]
4. Li, J.; Bi, M.; Wei, G. Investigating the Impacts of Urbanization on Vegetation Net Primary Productivity: A Case Study of Chengdu–Chongqing Urban Agglomeration from the Perspective of Townships. *Land* **2022**, *11*, 2077. [CrossRef]
5. Guan, X.; Shen, H.; Li, X.; Gan, W.; Zhang, L. A long-term and comprehensive assessment of the urbanization-induced impacts on vegetation net primary productivity. *Sci. Total Environ.* **2019**, *669*, 342–352. [CrossRef]
6. Li, H.; Zhang, H.; Li, Q.; Zhao, J.; Guo, X.; Ying, H.; Deng, G.; Rihan, W.; Wang, S. Vegetation Productivity Dynamics in Response to Climate Change and Human Activities under Different Topography and Land Cover in Northeast China. *Remote Sens.* **2021**, *13*, 975. [CrossRef]
7. Qin, K.; Li, J.; Liu, J.; Yan, L.; Huang, H. Setting conservation priorities based on ecosystem services—A case study of the Guanzhong-Tianshui Economic Region. *Sci. Total Environ.* **2018**, *650*, 3062–3074. [CrossRef]
8. Wang, H.; Liu, G.; Li, Z.; Zhang, L.; Wang, Z. Processes and driving forces for changing vegetation ecosystem services: Insights from the Shaanxi Province of China. *Ecol. Indic.* **2020**, *112*, 106105. [CrossRef]
9. Wu, S.H.; Zhou, S.L.; Chen, D.X.; Wei, Z.Q.; Dai, L.; Li, X.G. Determining the contributions of urbanisation and climate change to NPP variations over the last decade in the Yangtze River Delta, China. *Sci. Total Environ.* **2014**, *472*, 397–406. [CrossRef]
10. Pei, F.S.; Li, X.; Liu, X.P.; Lao, C.H.; Xia, G.R. Exploring the response of net primary productivity variations to urban expansion and climate change: A scenario analysis for Guangdong Province in China. *J. Environ. Manag.* **2015**, *150*, 92–102. [CrossRef] [PubMed]
11. Yao, R.; Zhang, S.; Sun, P.; Bian, Y.; Yang, Q.; Guan, Z.; Zhang, Y. Diurnal Variations in Different Precipitation Duration Events over the Yangtze River Delta Urban Agglomeration. *Remote Sens.* **2022**, *14*, 5244. [CrossRef]
12. Paolini, L.; Araoz, E.; Gioia, A.; Powell, P.A. Vegetation productivity trends in response to urban dynamics. *Urban For. Urban Green.* **2016**, *17*, 211–216. [CrossRef]
13. Costanza, R.; d'Arge, R.; de Groot, R.; Farber, S.; Grasso, M.; Hannon, B.; Limburg, K.; Naeem, S.; O'Neill, R.V.; Paruelo, J.; et al. The value of the world's ecosystem services and natural capital. *Nature* **1997**, *387*, 253–260. [CrossRef]
14. Keeler, B.L.; Polasky, S.; Brauman, K.A.; Johnson, K.A.; Finlay, J.C.; O'Neill, A.; Kovacs, K.; Dalzell, B. Linking water quality and well-being for improved assessment and valuation of ecosystem services. *Proc. Natl. Acad. Sci. USA* **2012**, *109*, 18619–18624. [CrossRef]
15. Cramer, W.; Kicklighter, D.W.; Bondeau, A.; Moore, B.; Churkina, G.; Nemry, B.; Ruimy, A.; Schloss, A.L. Comparing global models of terrestrial net primary productivity (NPP): Overview and key results. *Glob. Change Biol.* **1999**, *5*, 1–15. [CrossRef]
16. Piao, S.L.; Fang, J.Y.; Chen, A.P. Seasonal dynamics of terrestrial net primary production in response to climate changes in China. *Acta Bot. Sin.* **2003**, *45*, 269–275.
17. Duan, H.; Xue, X.; Wang, T.; Kang, W.; Liao, J.; Liu, S. Spatial and Temporal Differences in Alpine Meadow, Alpine Steppe and All Vegetation of the Qinghai-Tibetan Plateau and Their Responses to Climate Change. *Remote Sens.* **2021**, *13*, 669. [CrossRef]
18. Betts, R.A.; Boucher, O.; Collins, M.; Cox, P.M.; Falloon, P.D.; Gedney, N.; Hemming, D.L.; Huntingford, C.; Jones, C.D.; Sexton, D.M.H.; et al. Projected increase in continental runoff due to plant responses to increasing carbon dioxide. *Nature* **2007**, *448*, 1037–1041. [CrossRef]
19. Grimm, N.B.; Grove, J.M.; Pickett, S.T.A.; Redman, C.L. Integrated approaches to long-term studies of urban ecological systems. *Bioscience* **2000**, *50*, 571–584. [CrossRef]
20. Imhoff, M.L.; Bounoua, L.; DeFries, R.; Lawrence, W.T.; Stutzer, D.; Tucker, C.J.; Ricketts, T. The consequences of urban land transformation on net primary productivity in the United States. *Remote Sens. Environ.* **2004**, *89*, 434–443. [CrossRef]
21. McRoberts, R.E. Probability- and model-based approaches to inference for proportion forest using satellite imagery as ancillary data. *Remote Sens. Environ.* **2010**, *114*, 1017–1025. [CrossRef]
22. Tian, G.J.; Qiao, Z. Assessing the impact of the urbanization process on net primary productivity in China in 1989–2000. *Environ. Pollut.* **2014**, *184*, 320–326. [CrossRef]

23. Paz-Kagan, T.; Shachak, M.; Zaady, E.; Karnieli, A. Evaluation of ecosystem responses to land-use change using soil quality and primary productivity in a semi-arid area, Israel. *Agric. Ecosyst. Environ.* **2014**, *193*, 9–24. [CrossRef]
24. Qi, S.; Chen, S.; Long, X.; An, X.; Zhang, M. Quantitative contribution of climate change and anthropological activities to vegetation carbon storage in the Dongting Lake basin in the last two decades. *Adv. Space Res.* **2023**, *71*, 845–868. [CrossRef]
25. Shi, S.; Yu, J.; Wang, F.; Wang, P.; Zhang, Y.; Jin, K. Quantitative contributions of climate change and human activities to vegetation changes over multiple time scales on the Loess Plateau. *Sci. Total Environ.* **2021**, *755*, 142419. [CrossRef]
26. Zhang, M.; Yuan, N.; Lin, H.; Liu, Y.; Zhang, H. Quantitative estimation of the factors impacting spatiotemporal variation in NPP in the Dongting Lake wetlands using Landsat time series data for the last two decades. *Ecol. Indic.* **2022**, *135*, 108544. [CrossRef]
27. Field, C.B.; Behrenfeld, M.J.; Randerson, J.T.; Falkowski, P. Primary production of the biosphere: Integrating terrestrial and oceanic components. *Science* **1998**, *281*, 237–240. [CrossRef]
28. Zhao, S.Q.; Liu, S.G.; Zhou, D.C. Prevalent vegetation growth enhancement in urban environment. *Proc. Natl. Acad. Sci. USA* **2016**, *113*, 6313–6318. [CrossRef]
29. Peng, J.; Shen, H.; Wu, W.H.; Liu, Y.X.; Wang, Y.L. Net primary productivity (NPP) dynamics and associated urbanization driving forces in metropolitan areas: A case study in Beijing City, China. *Landsc. Ecol.* **2016**, *31*, 1077–1092. [CrossRef]
30. Su, S.L.; Li, D.L.; Hu, Y.N.; Xiao, R.; Zhang, Y. Spatially non-stationary response of ecosystem service value changes to urbanization in Shanghai, China. *Ecol. Indic.* **2014**, *45*, 332–339. [CrossRef]
31. Fang, J.Y.; Chen, A.P.; Peng, C.H.; Zhao, S.Q.; Ci, L. Changes in forest biomass carbon storage in China between 1949 and 1998. *Science* **2001**, *292*, 2320–2322. [CrossRef]
32. Guerschman, J.P.; Mcvicar, T.R.; Vleeshower, J.; Niel, T.G.V.; Pea-Arancibia, J.L.; Chen, Y. Estimating actual evapotranspiration at field-to-continent scales by calibrating the CMRSET algorithm with MODIS, VIIRS, Landsat and Sentinel-2 data. *J. Hydrol.* **2022**, *605*, 127318. [CrossRef]
33. Liu, J.; Liu, M.; Tian, H.; Zhuang, D.; Zhang, Z.; Zhang, W.; Tang, X.; Deng, X. Spatial and temporal patterns of China's cropland during 1990–2000: An analysis based on Landsat TM data. *Remote Sens. Environ.* **2005**, *98*, 442–456. [CrossRef]
34. Liu, J.; Zhang, Z.; Xu, X.; Kuang, W.; Zhou, W.; Zhang, S.; Li, R.; Yan, C.; Yu, D.; Wu, S. Spatial patterns and driving forces of land use change in China during the early 21st century. *J. Geogr. Sci.* **2010**, *20*, 483–494. [CrossRef]
35. Liu, H.; Jiang, D.; Yang, X.; Luo, C. Spatialization Approach to 1km Grid GDP Supported by Remote Sensing. *Geo-Inf. Sci.* **2005**, *7*, 120–123.
36. Wei, S.; Dong, Y.; Qiu, Y.; Li, B.; Li, S.; Dong, C. Temporal and spatial analysis of vegetation cover change in the Yellow River Delta based on Landsat and MODIS time series data. *Environ. Monit. Assess.* **2023**, *195*, 1057. [CrossRef]
37. Wolock, D.M.; Mccabe, G.J. Differences in topographic characteristics computed from 100- and 1000-m resolution digital elevation model data. *Hydrol. Process.* **2015**, *14*, 987–1002. [CrossRef]
38. Jianxin, Z.; Lufeng, D. Measure on the Development Relationship of China's Urbanization and Industrialization. *Ecol. Econ.* **2009**, *2011*, 15–21.
39. Roser, L. Moran's I, Geary's C and Bivariate Moran's I Correlograms. Available online: https://leandroroser.github.io/EcoGenetics-documentation/reference/eco.correlog.html (accessed on 12 June 2021).
40. Aguiar, M.A.M.D.; Baranger, M.; Baptestini, E.M.; Kaufman, L.; Bar-Yam, Y. Global patterns of speciation and diversity. *Nature* **2009**, *460*, 384–387. [CrossRef]
41. Arruda, R.M.F.D.; Cardoso, D.T.; Teixeira-Neto, R.G.; Barbosa, D.S.; Ferraz, R.K.; Morais, M.H.F.; Belo, V.S.; Silva, E.S.d. Space-time analysis of the incidence of human visceral leishmaniasis (VL) and prevalence of canine VL in a municipality of southeastern Brazil: Identification of priority areas for surveillance and control. *Acta Trop. J. Biomed. Sci.* **2019**, *197*, 105052. [CrossRef]
42. Avise, M.J.C. Comparative phylogenetic analysis of male alternative reproductive tactics in ray-finned fishes. *Evolution* **2006**, *60*, 1311–1316.
43. Jayakumar, K.; Mundassery, D.A. On Moran's Bivariate Gamma and Bivariate Negative Binomial Distribution. *Calcutta Stat. Assoc. Bull.* **2007**, *59*, 15–27. [CrossRef]
44. Matkan, A.A.; Mohaymany, A.S.; Shahri, M.; Mirbagheri, B. Detecting the spatial–temporal autocorrelation among crash frequencies in urban areas. *Can. J. Civ. Eng.* **2013**, *40*, 195–203. [CrossRef]
45. Lee, S.I. Developing a bivariate spatial association measure: An integration of Pearson's r and Moran's I. *J. Geogr. Syst.* **2001**, *3*, 369–385. [CrossRef]
46. Shuang, L.; Yan-Hua, Z. The Spatial Distribution and the Convergence of Human Capital in Cities of China. *Sci. Technol. Econ.* **2019**, *32*, 71–75.
47. Anselin, L. Local Indicators of Spatial Association—LISA. *Geogr. Anal.* **1995**, *27*, 93–115. [CrossRef]
48. Shen, W.; Huang, Z.; Yin, S.; Hsu, W.-L. Temporal and Spatial Coupling Characteristics of Tourism and Urbanization with Mechanism of High-Quality Development in the Yangtze River Delta Urban Agglomeration, China. *Appl. Sci.* **2022**, *12*, 3403. [CrossRef]
49. Changfeng, L.; Xiaotian, L. Research on the development of high-tech industry and technology service industry based on industrial collaborative evolution: Case of Jiangsu province. *Jiangsu Sci. Technol. Inf.* **2018**, *35*, 1–5.
50. Souris, M.; Bichaud, L. Statistical methods for bivariate spatial analysis in marked points. *Ex. Spat. Epidemiol. Spat. Spatio-Temporal Epidemiol.* **2011**, *2*, 227–234. [CrossRef]

51. Zhao, H.; Duan, X.; Stewart, B.; You, B.; Jiang, X. Spatial correlations between urbanization and river water pollution in the heavily polluted area of Taihu Lake Basin, China. *J. Geogr. Sci.* **2013**, *23*, 735–752. [CrossRef]
52. Bellamy, S.A.H.E. The influence of hedge structure, management and landscape context on the value of hedgerows to birds: A review. *J. Environ. Manag.* **2000**, *60*, 33–49.
53. Lewis, J.A.; Zipperer, W.C.; Ernstson, H.; Bernik, B.; Hazen, R.; Elmqvist, T.; Blum, M.J. Socioecological disparities in New Orleans following Hurricane Katrina. *Ecosphere* **2017**, *8*, e01922. [CrossRef]
54. Costa, M.; Iezzi, S. Technology spillover and regional convergence process: A statistical analysis of the Italian case. *Stat. Methods Appl.* **2004**, *13*, 375–398. [CrossRef]
55. Zhang, Z.M.; Gao, J.F.; Fan, X.Y.; Lan, Y.; Zhao, M.S. Response of ecosystem services to socioeconomic development in the Yangtze River Basin, China. *Ecol. Indic.* **2017**, *72*, 481–493. [CrossRef]
56. Qiu, L.; Lindberg, S.; Nielsen, A.B. Is biodiversity attractive?—On-site perception of recreational and biodiversity values in urban green space. *Landsc. Urban Plan.* **2013**, *119*, 136–146. [CrossRef]
57. Yuan, J.; Lu, Y.; Ferrier, R.C.; Liu, Z.; Su, H.; Meng, J.; Song, S.; Jenkins, A. Urbanization, rural development and environmental health in China. *Environ. Dev.* **2018**, *28*, 101–110. [CrossRef]
58. Niu, Q.; Yu, L.; Jie, Q.; Li, X. An urban eco-environmental sensitive areas assessment method based on variable weights combination. *Environ. Dev. Sustain.* **2020**, *22*, 2069–2085. [CrossRef]
59. Guan, X.; Wei, H.; Lu, S.; Dai, Q.; Su, H. Assessment on the urbanization strategy in China: Achievements, challenges and reflections. *Habitat Int.* **2018**, *71*, 97–109. [CrossRef]
60. NDRC. Yangtze River Delta Urban Agglomeration Development Plan. National Development and Reform Commission. 2016. Available online: http://www.ndrc.gov.cn/zcfb/zcfbghwb/201606/t20160603_806390.html (accessed on 16 June 2018).
61. Ting, M.; Zhan, Y.; Alicia, Z. Delineating Spatial Patterns in Human Settlements Using VIIRS Nighttime Light Data: A Watershed-Based Partition Approach. *Remote Sens.* **2018**, *10*, 465. [CrossRef]
62. Ye, C.; Zhu, J.; Li, S.; Yang, S. Assessment and analysis of regional economic collaborative development within an urban agglomeration: Yangtze River Delta as a case study. *Habitat Int.* **2019**, *83*, 20–29. [CrossRef]
63. Jing, W.; Hong-Lian, W.; Chong, Z. Temporal and Spatial Changes and Influencing Factors of Vegetation Cover in Baoji Area Based on MODIS Data. *Acta Agric. Jiangxi* **2018**, *30*, 127–133.

land

MDPI

Article

Simulation and Attribution Analysis of Spatial–Temporal Variation in Carbon Storage in the Northern Slope Economic Belt of Tianshan Mountains, China

Kun Zhang [1,2,3,4,5], Yu Wang [2,3,4,5], Ali Mamtimin [2,3,4,5,*], Yongqiang Liu [1], Lifang Zhang [1], Jiacheng Gao [2,3,4,5], Ailiyaer Aihaiti [2,3,4,5], Cong Wen [2,3,4,5], Meiqi Song [2,3,4,5], Fan Yang [2,3,4,5], Chenglong Zhou [2,3,4,5] and Wen Huo [2,3,4,5]

1 College of Geography and Remote Sensing Sciences, Xinjiang University, Urumqi 830046, China; zkun_@stu.xju.edu.cn (K.Z.); liuyq@xju.edu.cn (Y.L.); zhanglf@xju.edu.cn (L.Z.)
2 Institute of Desert Meteorology, China Meteorological Administration, Urumqi 830002, China; wangyu@idm.cn (Y.W.); gaojiach@idm.cn (J.G.); ailiyaer@idm.cn (A.A.); wencong@idm.cn (C.W.); songmq@idm.cn (M.S.); yangfan@idm.cn (F.Y.); zhoucl@idm.cn (C.Z.); huowenpet@idm.cn (W.H.)
3 National Observation and Research Station of Desert Meteorology, Taklimakan Desert of Xinjiang, Urumqi 830002, China
4 Taklimakan Desert Meteorology Field Experiment Station of China Meteorological Administration, Urumqi 830002, China
5 Xinjiang Key Laboratory of Desert Meteorology and Sandstorm, Urumqi 830002, China
* Correspondence: ali@idm.cn

Abstract: Intensive economic and human activities present challenges to the carbon storage capacity of terrestrial ecosystems, particularly in arid regions that are sensitive to climate change and ecologically fragile. Therefore, accurately estimating and simulating future changes in carbon stocks on the northern slope economic belt of Tianshan Mountains (NSEBTM) holds great significance for maintaining ecosystem stability, achieving high-quality development of the economic belt, and realizing the goal of "carbon neutrality" by 2050. This study examines the spatiotemporal evolution characteristics of the NSEBTM carbon stocks in arid regions from 1990 to 2050, utilizing a combination of multi-source data and integrating the Patch-generating Land use Simulation (PLUS) and Integrated Valuation of Ecosystem Services and Trade-offs (InVEST) models. Additionally, an attribution analysis of carbon stock changes is conducted by leveraging land use data. The findings demonstrate that (1) the NSEBTM predominantly consists of underutilized land, accounting for more than 60% of the total land area in the NSEBTM. Unused land, grassland, and water bodies exhibit a declining trend over time, while other forms of land use demonstrate an increasing trend. (2) Grassland serves as the primary reservoir for carbon storage in the NSEBTM, with grassland degradation being the leading cause of carbon loss amounting to 102.35 t over the past three decades. (3) Under the ecological conservation scenario for 2050 compared to the natural development scenario, there was a net increase in carbon storage by 12.34 t; however, under the economic development scenario compared to the natural development scenario, there was a decrease in carbon storage by 25.88 t. By quantitatively evaluating the land use change in the NSEBTM and its impact on carbon storage in the past and projected for the next 30 years, this paper provides scientific references and precise data support for the territorial and spatial decision making of the NSEBTM, thereby facilitating the achievement of "carbon neutrality" goals.

Keywords: scenario simulation; carbon storage; spatiotemporal evolution; InVEST model; northern slope economic belt of Tianshan mountains

1. Introduction

Globally, nations have set a goal of achieving "net zero carbon emissions" in 2021. Enhancing carbon storage in terrestrial ecosystems represents a pivotal approach to mitigating carbon dioxide emissions, thus constituting one of the foremost strategies [1]. The United Nations report on the 2030 Agenda for sustainable development underscores the imperative of safeguarding, rehabilitating, and sustainably harnessing ecosystems in order to advance sustainable development goals. Human activities and the process of economic development lead to modifications in land cover, resulting in consequential alterations to factors such as the land, climate, and environment. These changes subsequently impact global terrestrial carbon storage ecosystem processes [2]. Given the escalating concerns regarding land use, urban expansion, and climate change in the foreseeable future, they pose persistent threats to the sustainable development of terrestrial ecosystems and have garnered global attention [3,4]. Carbon stocks in terrestrial ecosystems are primarily affected by climate change and human-induced alterations in land use and cover, which is widely recognized [5]. On one hand, human activities and LUCC changes are substantial contributors to the rise in global temperatures through carbon emissions [6,7]. Controlling the global carbon cycle and climate change requires carbon storage, a key measure of the health of terrestrial ecosystems [8,9]. LUCC is the second major factor contributing to the significant rise in global carbon emissions, as multiple studies have demonstrated that they have a direct impact on terrestrial ecosystems' ability to store carbon [10,11]. On the other hand, human activities play a constructive role in carbon sequestration within arid regions through their influence on land use and vegetation cover dynamics. Purposeful human intervention and the implementation of ecological restoration policies have the potential to augment vegetation coverage and enhance greenery [12]. Hence, it is evident that land use change not only represents a prominent aspect of global surface environmental alterations but also serves as a focal point for investigating variations in carbon storage within terrestrial ecosystems. Consequently, accurately evaluating the impact of land use change on carbon storage and establishing harmonious, sustainable, and high-quality human–land relationships are imperative to accomplish the mission and objectives of achieving global "carbon neutrality" by 2050.

At present, there are three main approaches to studying carbon storage in terrestrial ecosystems: conducting field surveys, utilizing remote sensing technology, and employing model simulations. For instance, utilizing data from various carbon pools along with spatial–temporal visualization capabilities enables the effective calculation of carbon storage using models like InVEST [13]. The model is simple to operate, flexible in terms of parameters, and yields accurate results [14]. The InVEST model, when combined with GIS technology, has successfully addressed the limitations of traditional methods for estimating carbon storage. These limitations include extended sampling periods and intensive labor. In addition, the InVEST model offers the benefits of simplified parameter acquisition and visually presented results. As a result, it is extensively utilized in various applications [15–18]. Research has shown that the InVEST model, combined with geographic information mapping, was used to simulate carbon storage in the Poyang Lake basin [19]. In addition, some scholars have also used land use types in conjunction with the InVEST model to quantitatively assess the impact of LUCC conversion on ecosystem carbon storage [20]. Many researchers use models such as SD, FLUS, CA-Markov, ANN-CA, and logistic-CA to predict LUCC changes and have coupled these models with the InVEST model to reveal the spatial distribution characteristics of carbon storage under different future LUCC scenarios. The utilization of the Flus model and the InVEST model has yielded favorable validation outcomes in simulating carbon storage in the arid regions of northwest China [21]. Additionally, the CA-Markov model has demonstrated strong applicability in the Sariska Tiger Reserve in India. The utilization of the future land use prediction models can effectively explore the dynamic changes in carbon storage within the research area, thereby providing valuable support for informed decision making on land management by local departments and governments [22]. Among the abovementioned models used for future LUCC predictions, the PLUS model effectively elucidates the underlying causes of the diverse LUCC changes and accurately simulates spatial transformations in various small-scale regions. It facilitates the incorporation of prospective spatial policy elements and enables a more scientifically rigorous simulation of the future LUCC alterations under distinct policy scenarios [23,24].

The north slope economic belt of Tianshan Mountains (NSEBTM) is located in a strip-shaped oasis at the foot of the northern slope of Tianshan Mountains, on the northwest border of China [25]. Due to its inland location, distance from the sea, and scarce precipitation, it belongs to China's arid region with fragile ecosystems and harsh climatic conditions [12]. The NSEBTM, endowed with abundant energy resources and strategically located, functions as a convergence zone propelled by the China–Central Asia–West Asia economic Corridor, the China–Russia–Mongolia economic Corridor, and the China–Pakistan economic Corridor. It assumes a pivotal role as an important strategic support point along the Silk Road economic belt and serves as a crucial bridge for China's western opening-up endeavors. In line with China's comprehensive opening-up strategy and modernization efforts, the NSEBTM holds significant strategic importance. In this scenario, the NSEBTM has emerged as the region with the highest level of urbanization and population density in Xinjiang [26]. However, in this resource-constrained and economically active arid region, the inherent contradiction between fragile ecosystems and high-quality economic and social development has persistently plagued the area. On one hand, there is an urgent imperative to achieve global "carbon neutrality" by 2050; on the other hand, compared to other regions in China, this particular region is situated within an arid zone where its environmental vulnerability exhibits heightened sensitivity to fluctuations in human–environment interactions [27]. As the pivotal region of the Belt and Road Initiative, the economic development and establishment of the NSEBTM as a free trade experimental zone will inevitably induce land use transformations, which in turn will have profound implications for carbon storage ecosystems [28]. Therefore, it is imperative to address these scientific inquiries. Presently, domestic and international research on the NSEBTM primarily concentrates on urbanization, urban spatial structure morphology, and landscape remote sensing monitoring, among other facets. However, further investigations are warranted to elucidate the underlying mechanisms governing the historical and future variations in carbon storage within the NSEBTM against the backdrop of achieving "carbon neutrality" by 2050. The mechanisms underlying the spatiotemporal variations in carbon storage within the NSEBTM region remain uncertain in the context of ecological engineering and urbanization.

In light of this, leveraging long-term time series remote sensing data products, this study employs the InVEST model to quantitatively analyze the spatiotemporal dynamics of carbon storage in the NSEBTM amidst land use change from 1990 to 2050. Additionally, the PLUS model is employed to construct multiple scenario simulations for natural, sustainable, and economic development by 2050. The objective is to investigate the future trends in carbon storage evolution and underlying impact mechanisms, thereby providing a decision-making foundation for well-organized urbanization and territorial spatial planning in the NSEBTM. Therefore, this study aimed to accomplish the following: (1) use LUCC data from 1990 to 2020 as well as LUCC data predicted using the PLUS model for disparate scenarios in 2050 and couple them with the InVEST model to analyze the spatiotemporal changes in LUCC and carbon storage in the NSEBTM from 1990 to 2050, (2) quantitatively assess the impact of LUCC changes on carbon storage in the NSEBTM by coupling LUCC data with the InVEST model, and (3) quantitatively assess the impact of ecological engineering construction and economic development strategies on LUCC change types, and their consequent effects on carbon storage dynamics in the NSEBTM. These studies will provide reliable data support for the NSEBTM LUCC management and decision making, filling the gap in research on the spatial and temporal changes in carbon storage under the background of LUCC change in the NSEBTM region, and providing a reference for achieving "carbon neutrality" goals in the NSEBTM and even China as a whole.

2. Materials and Methods

2.1. Study Area

The NSEBTM (40°52′ N∼47°14′ N, 79°53′ E∼96°23′ E) is situated in the Eurasian hinterland, deep within the inland region of northwest China (Figure 1). This region exhibits significant spatial heterogeneity in its natural conditions [25]. Encompassing an area of approximately 3.96 × 10^5 km^2, it constitutes 23.8% of Xinjiang's total landmass. Being distant from the ocean, this area experiences a typical temperate continental climate influenced by the westerly circulation and moisture from the Arctic Ocean, resulting in limited precipitation but high evaporation rates accompanied by ample sunshine. The elevation in this region varies from −153 to 4814 m, with the Tianshan Mountain range exhibiting relatively higher altitudes while the surrounding basins display lower elevations [29]. The natural vegetation is sparsely distributed, predominantly comprising grasslands and desert flora. As a result of the "warm-wet transformation" of climate and the implementation of ecological restoration projects, Xinjiang's vegetation has gradually exhibited a discernible trend towards increased greening. Over the past decade, there has been an incremental rise in the Normalized Difference Vegetation Index (NDVI) at a rate of 0.005/10a [12]. According to previous research, the NSEBTM range mainly includes cities such as the Urumqi City, Karamay City, Turpan City, Hami City, Changji City, Fukang City, Kuitun City, Wusu City, Tacheng City, Bole City, Alashankou City, Shihezi City, Wujiaqu City, Huyanghe City, Shuanghe City, and Xinxing City [25]. In the past few years, with the progress of China's economy and the advancement of urbanization towards a phase of superior quality expansion, constructive human endeavors and initiatives for ecological restoration have had a notable impact on the field of ecological engineering in this particular area [12,30,31].

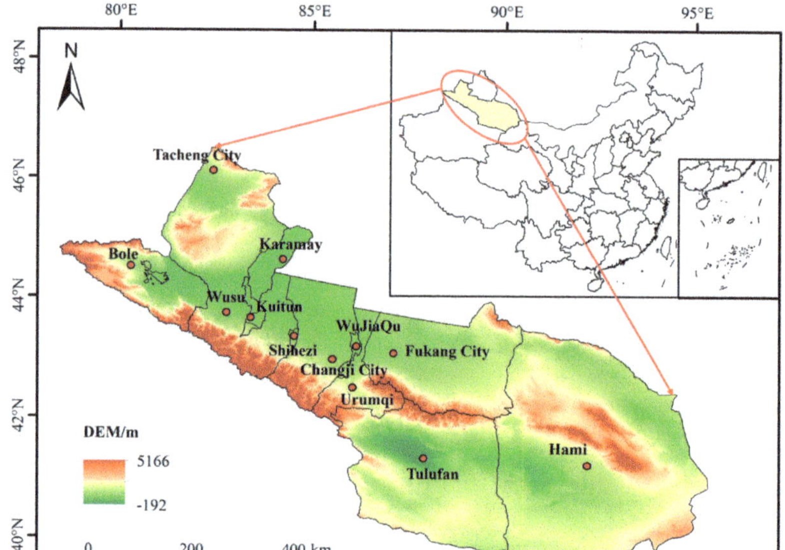

Figure 1. NSEBTM of Xinjiang, China. (Drawing review number: GS (2019) No. 1822. There is no modification to the base map, which is the same below).

2.2. Data and Methods

2.2.1. Data Source

The data utilized in this study encompassed provincial administrative boundaries, land use and land cover change (LUCC), socioeconomic factors, climate factors, and terrain data (Table 1). Among them, the provincial administrative boundary data are in vector format. The LUCC data were obtained from the first Landsat-derived annual China land cover dataset (CLCD); these datasets were derived from Landsat images available on Google Earth Engine [32]. The team first collected training samples through visual interpretation, combining China's land-use/cover datasets (CLUDs), satellite time series data, and samples extracted from Google Earth and Google Maps. They then constructed multiple temporal dimensions using Landsat images on GEE and applied a random forest classifier to obtain classification results. Finally, a spatiotemporal post-processing method was proposed to further improve the consistency of the spatiotemporal CLCD. The overall accuracy of CLCD reached 79.31%. To ensure consistency with other studies, we reclassified these LUCC types into six categories: farmland, forest, grassland, water, built-up land, and unused land. We selected LUCC data for 1990, 2000, 2010, and 2020 primarily to analyze changes in LUCC patterns as well as simulate carbon storage under different LUCC scenarios.

Table 1. Data types and sources.

Type	Data	Resolution	Data Source
Provincial administrative boundaries	Research area boundaries	Vector data	Chinese Academy of Sciences Resource and Environmental Science and Data Center
LUCC	CLCD	30 m	Wuhan University
Socioeconomic factors	GDP Population	1 km 1 km	Chinese Academy of Sciences Resource and Environmental Science and Data Center
	Distance from the city Distance from the road Distance from the water Distance from the train station	Vector data Vector data Vector data Vector data	OpenStreetMap
	Nighttime lighting data	1 km	National Oceanic and Atmospheric Administration of the United States
Climate and environmental factors	Annual precipitation Annual temperature Soil type	1 km 1 km 1 km	Chinese Academy of Sciences Resource and Environment Science and Data Center
	NDVI	1 km	National Aeronautics and Space Administration
Topographic data	DEM	30 m	Geospatial data cloud
	Slope Aspect	30 m 30 m	Based on ArcGIS

The driving factor data include socioeconomic factors and terrain data. The socioeconomic factors include the Gross Domestic Product (GDP), population data, and distance data to cities, roads, water bodies, and stations. The terrain data include digital elevation model (DEM) data, slope, and aspect data obtained through surface analysis using DEM data in ArcGIS 10.8.1 software with a resolution of 30 m. The nighttime light data are used to characterize the urbanization process in the study area, while other driving factor data are mainly used for simulating future LUCC changes.

The climate and environmental factors include annual precipitation, temperature data, the number of soil types, and Normalized Difference Vegetation Index (NDVI) data. The NDVI data are used to represent the progress achieved in ecological restoration projects in the study area, while other climate and environmental factor data are used to adjust carbon density. The soil type data will also be involved in predicting the future LUCC data.

Considering the inconsistent spatial resolutions of the above data and the size of the study area, we standardized all the data to a spatial resolution of 500 m.

2.2.2. Human Activity Data Processing

Because of the different sensors used for the 2000 and 2020 nighttime light data, it was necessary to perform a data uniformity correction. First, the relative invariant target area method was used to perform relative correction on the 1992–2013 NSEBTM DMSP-OLS images, combined with saturation and continuity correction after the radiometric calibration of the reference images. Simultaneously, noise processing and logarithmic transformation were used to correct the NPP-VIIRS images from 2012 to 2020. A significant correlation was found between the DN values of the DMSP-OLS images and the radiance values of the NPP-VIIRS images after index transformation. The BiDoseResp function model was then selected to perform a consistency correction on the DMSP-OLS and NPP-VIIRS images [33], ultimately obtaining a long-term time series of the nighttime light image datasets for the NSEBTM from 1997 to 2020.

Furthermore, because of the inconsistency in the DMSP-OLS data from different sensors for the same year, to fully utilize the image data obtained by different sensors, continuity correction was performed on the images after mutual and saturation correction. The continuity correction is divided into sensor and time-series continuity corrections. In this study, Equation (1) was used to perform sensor correction on the DMSP-OLS data, while Equation (2) was used to perform time-series continuity correction on both the DMSP-OLS andNPP-VIIRS data [34].

$$DN_{(i,j)} = \begin{cases} 0, & DN^{a}_{(i,j)} = 0 \text{ and } DN^{b}_{(i,j)} = 0 \\ (DN^{a}_{(i,j)} + DN^{b}_{(i,j)})/2, & \text{other} \end{cases} \tag{1}$$

$DN_{(i,j)}$ is the DN value of the jth pixel in the corrected image in the ith year; $DH^{a}_{(i,j)}$ and $DH^{b}_{(i,j)}$ represents the DN value of the jth pixel in the images obtained from two different sensors before correction.

$$DN_{(i,j)} = \begin{cases} DN_{(i-1,j)}, & DN_{(i-1,j)} > DN_{(i,j)} \\ DN_{(i,j)}, & \text{other} \end{cases} \tag{2}$$

$DN_{(i,j)}$ and $DN_{(i-1,j)}$ represent the DN values of the jth pixel in the images corrected for saturation and multi-sensor continuity in the ith and i-1th year, respectively.

2.2.3. Carbon Storage Estimation Method

This study utilized the carbon storage sub-module of the InVEST model to calculate the carbon storage changes corresponding to LUCC changes over time for each remote sensing pixel. It simulates ecosystem carbon storage and carbon source/sink changes under different LUCC types and future land development scenarios [35]. To calculate ecosystem carbon storage, we used the carbon stock method to simulate carbon storage, which involves using carbon density data for aboveground, belowground, soil, and dead litter, multiplying the corresponding area by the carbon density data to calculate the storage data for each carbon pool, and then adding them together to obtain the total ecosystem carbon storage for a certain area [17]. The formula for calculation is presented below:

$$C_z = C_{z-above} + C_{z-below} + C_{z-soil} + C_{z-dead} \tag{3}$$

$$C_t = \sum_{i=1}^{n} C_z \times S_z \tag{4}$$

C_z is the total carbon density ($t{\cdot}hm^{-2}$) of land-use type; $C_{z\text{-}above}$, $C_{z\text{-}below}$, $C_{z\text{-}soil}$, and $C_{z\text{-}dead}$ represent the aboveground carbon density, belowground carbon density, soil carbon density, and dead organic matter carbon density of LUCC type, respectively($t{\cdot}hm^{-2}$). C_t is the total carbon stock of the ecosystem (t); S_z is the area (hm^2) of LUCC type; n is the number of LUCC types (in this study, n is 6).

2.2.4. Selection and Correction of Carbon Density

The carbon density data for different LUCC types were obtained from the National Ecological Data Center's Resource Sharing Service Platform (http://www.nesdc.org.cn/, accessed on 10 February 2024), supplemented with relevant information from related research studies. Initially, we focused on selecting pertinent studies in the NSEBTM and then expanded our search to include studies conducted in the arid and semi-arid regions of Xinjiang. Previous investigations have demonstrated that regional climate and soil type factors can significantly influence carbon density. Temperature exhibited a positive correlation with biomass, soil organic carbon density, and precipitation, while it displayed a negative correlation with them as well. To address this issue, we applied a correction method proposed by Alam et al. [36] to adjust the relationships between annual precipitation, biomass, and soil carbon density. The relationship between annual temperature and biomass carbon density was corrected using methods described by Giardina and Ryan [37] and Chen et al. [38]. Therefore, this approach was employed to rectify the carbon density data for various LUCC change categories within the study area, thereby obtaining localized carbon density data for the investigation region (Table 2). The specific approach is outlined as follows:

$$C_{SP} = 3.3968 \times MAP + 3996.1 \tag{5}$$

$$C_{BP} = 6.798 \times e^{0.0054 \times MAP} \tag{6}$$

$$C_{BT} = 28 \times MAT + 398 \tag{7}$$

C_{SP} represents the soil carbon density corrected for annual precipitation; C_{BP} represents the biomass carbon density corrected for annual precipitation; C_{BT} represents the biomass carbon density corrected for annual temperature; MAP represents the mean annual precipitation (mm); MAT represents the mean annual temperature (°C).

$$K_{BP} = \frac{C'_{BP}}{C''_{BP}} K_{BT} = \frac{C'_{BT}}{C''_{BT}} \tag{8}$$

$$K_B = \frac{K_{BP}}{K_{BT}} K_s = \frac{C'_{SP}}{C''_{SP}} \tag{9}$$

K_{BP} is the precipitation correction factor for biomass carbon density; K_{BT} is the temperature correction factor for biomass carbon density; K_B is the correction factor for biomass carbon density; K_S represents the correction factor for soil carbon density; C' and C'' are the carbon density data for the NSEBTM and the whole country, respectively.

Table 2. Carbon density variation among different LUCC types in NSEBTM.

LUCC	C_{above}	C_{below}	C_{soil}	C_{dead}	References
Farmland	4.18	3.38	80.22	0	[20,21,39]
Forest	44.51	3.37	137.12	0	[20,21,39]
Grassland	8.49	2.61	73.93	0	[20,21,39]
Water	0.92	0	0	0	[20,21,39]
Built-up	3.26	2.09	0	0	[20,21,39]
Unused	0.65	1.25	15.99	0	[20,21,39]

2.2.5. Scenario Simulation

When simulating the future LUCC map, three scenarios were set: natural growth, ecological protection, and economic development. (1) Natural growth scenario: this scenario is considered the baseline scenario, maintaining the historical trend of LUCC development and simulating future LUCC based on this trend. (2) Ecological protection scenario: in this scenario, ecological protection is strengthened, economic growth is slowed down, and the conversion of forests and grasslands into other land types is restricted. (3) Economic development scenario: This scenario focuses mainly on economic growth and neglects the protection of resources and the environment. Therefore, it restricts the conversion of built-up land to other land types, while increasing the probability of the other LUCC types being converted to built-up land.

The PLUS model was initially used to convert the LUCC data formats for 2010 and 2020. The LEAS module incorporated various driving factors to extract the expansion of initial LUCC to final land use. To assess the development potential of the different LUCC types and investigate the relationship between these driving factors and LUCC expansion, we employed the random forest classification algorithm, which determined the contribution of each driving factor to LUCC expansion [40]; among them, the sampling rate of random forest was set to 0.01, and the running parameter was set to 5. The Markov chain method was used to predict future LUCC demand, and simulated patches were generated in the CARS module to obtain a simulated future LUCC map, among them, the default value within the domain range was set to 3, with 5 parallel threads, a decay coefficient of 0.9 for the decrement threshold, and a diffusion coefficient of 0.1. Based on adherence to the actual development situation in the research area and the transfer matrix law of land use area, three typical scenarios were set for each type of land's cost matrix (Table 3), where a value of "1" represented allowed conversion and a value of "0" represented otherwise [41–43]. The domain weights are shown in Table 4.

Table 3. Transition cost matrix for three land use scenarios.

Land Use Scenarios	Land Use Types	Farmland	Forest	Grassland	Water	Built-up	Unused
Natural growth scenario	Farmland	1	0	1	0	0	0
	Forest	1	1	0	0	0	0
	Grassland	1	1	1	1	1	1
	Water	1	0	1	1	1	1
	Built-up	0	0	0	1	1	0
	Unused	1	0	1	1	1	1
Ecological protection scenario	Farmland	1	1	1	1	0	1
	Forest	0	1	1	0	0	0
	Grassland	0	1	1	0	0	0
	Water	1	1	1	1	0	1
	Built-up	1	1	1	1	1	1
	Unused	1	1	1	1	0	1
Economic development scenario	Farmland	1	0	0	0	1	0
	Forest	0	1	0	0	1	0
	Grassland	0	0	1	0	1	0
	Water	0	0	0	1	1	1
	Built-up	0	0	0	0	1	0
	Unused	0	0	0	0	1	1

Table 4. Neighborhood weight settings.

Land Use Type	Farmland	Forest	Grassland	Water	Built-up	Unused
Neighborhood Weight	0.27	0.03	0.39	0.11	0.06	0.14

The evaluation of the PLUS model's simulation performance was conducted by measuring two metrics: overall accuracy (OA) and Kappa coefficient. The Kappa coefficient was computed according to the following formula:

$$\text{Kappa} = \frac{p_o - p_c}{p_p - p_c} \tag{10}$$

where Kappa is the simulation accuracy index, P_o is the actual simulation accuracy, P_c is the expected simulation accuracy under random conditions, and Pp is the simulation accuracy under ideal conditions. Generally, when the Kappa value is greater than 0.75, the simulation accuracy is high; a value between 0.4 and 0.75 means that the simulation accuracy is moderate; and when it is less than 0.4, the simulation accuracy is poor. The Kappa coefficient of the simulation accuracy in this study was 0.86, the OA was 0.93, and the simulation results met the research requirements.

This research was conducted in three stages: (a) historical LUCC data from 2010 and 2020 were utilized along with the PLUS model to forecast LUCC data for 2050, with separate predictions made for three scenarios; (b) carbon density data were adjusted based on temperature and precipitation information; and (c) the InVEST model was employed to estimate carbon storage by combining historical and projected LUCC data with carbon density data from 1990 to 2050. The methodological process is illustrated in Figure 2.

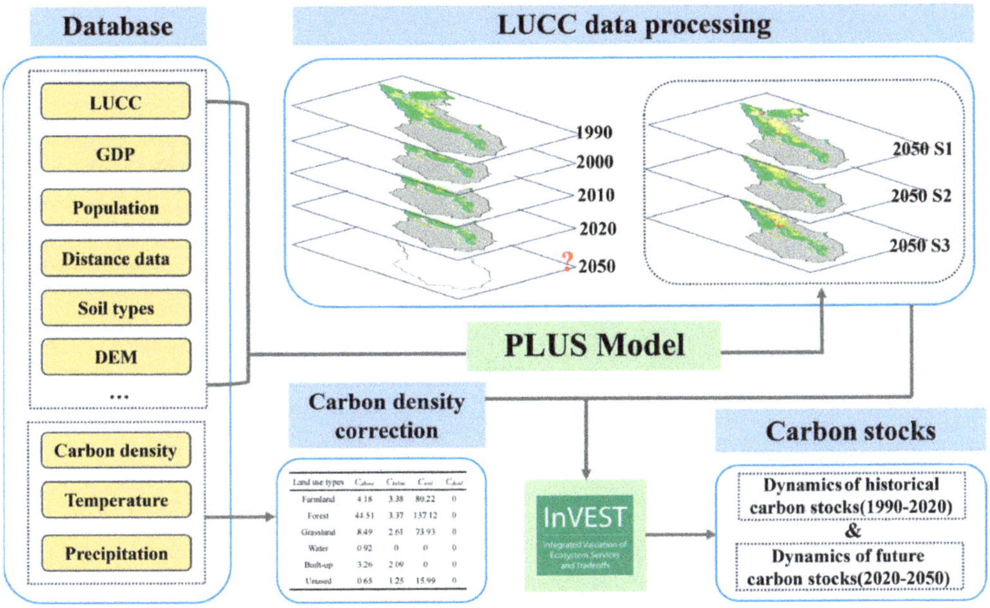

Figure 2. Methodology flowchart for this study.

3. Results

3.1. LUCC Dynamics in NSEBTM from 1990 to 2050

3.1.1. LUCC Dynamics from 1990 to 2020

In terms of spatial arrangement, owing to differences in terrain, landforms, and climatic conditions, the LUCC distribution characteristics of the NSEBTM show obvious spatial heterogeneity. As shown in Figure 3, arable land is mainly distributed in the southeastern part of the Tacheng region, Huyanghe City, and Karamay City. There are also scattered distributions in the central part of Hami City and the western part of the Turpan City. Forest land is relatively dispersed, mainly found in the sporadic areas of the Tianshan Mountains. Grassland is primarily distributed in the northwestern and southern parts of the Tacheng region, Shuanghe City, Urumqi City, as well as the central regions of the Hami City and Turpan City. Water bodies are mainly located in the western region of the Ili Kazakh Autonomous Prefecture. Construction land is predominantly distributed within urban clusters and their surrounding areas. Unused land is mostly found in large portions of the Hami City, Turpan City, and Changji Prefecture.

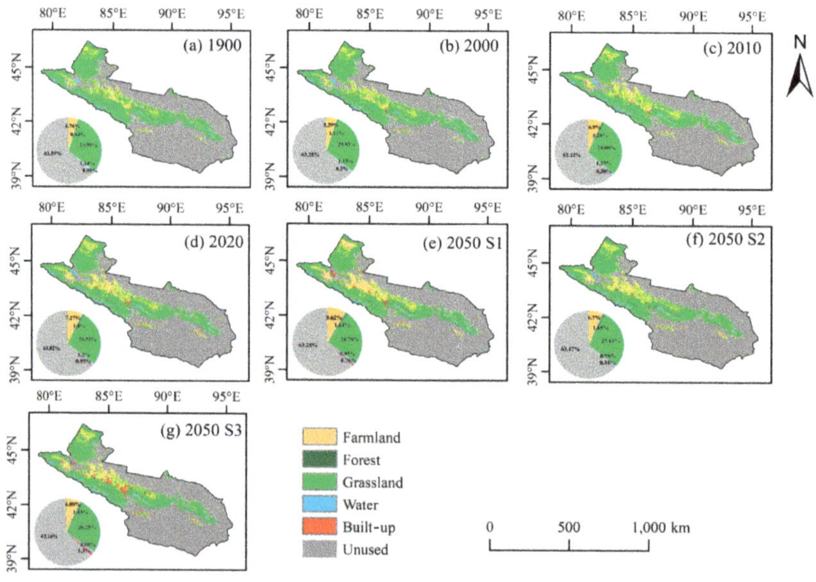

Figure 3. Spatial distribution map of LUCC in the NSEBTM from 1990 to 2050 and the proportion of LUCC types (S1 represents a natural development scenario; S2 represents an ecological protection scenario; S3 represents an economic development scenario).

The unused land area in the NSEBTM accounted for 63.59%, 63.25%, 62.12%, and 63.02% of the total area in 1990, 2000, 2010, and 2020, respectively, and was the most important LUCC type in the NSEBTM (Figure 3), with a much higher proportion than the national average (27.9%) [44]. Grassland and farmland were the next most important, accounting for 26.55–29.59% and 4.76–7.24%, respectively, with the proportion of farmland being much lower than that of the national average (14.3%) [44]. Water bodies, forest land, and built-up land had the smallest area, accounting for 1.13–1.25%, 0.84–1.40%, and 0.09–0.55% of the total area, respectively, with the proportion of built-up land being lower than the national average (4%) [44].

To clearly describe the changes in LUCC in the NSEBTM, we quantitatively expressed the conversion relationships between the different LUCC types using a Sankey diagram. The findings indicate that there have been diverse transformations in the LUCC categories in the NSEBTM during the last three decades (Figure 4). The area of unused land, which had the highest proportion, experienced an annual decrease. Specifically, 7394.17 km^2, 3744.89 km^2, 717.61 km^2, 9 km^2, and 718 km^2 were transformed into grassland, farmland, built-up land, forest land, and water bodies, respectively. Grassland also showed a decreasing trend every year, with 9558.73 km^2, 643.08 km^2, 2201.26 km^2, 342.45 km^2, and 9479.47 km^2 being converted to farmland, built-up land, forest land, water bodies, and unused land, respectively. However, it is worth noting that the area of farmland increased every year, with 2718.41 km^2, 463 km^2, 11 km^2, 98 km^2, and 100 km^2 being converted from grassland, built-up land, forest land, water bodies, and unused land, respectively. However, over the past 30 years, the areas converted from grassland, forest land, water bodies, and unused land to farmland were 9558.73 km^2, 8 km^2, 38 km^2, and 3744.89 km^2, and conversion into farmland far exceeded conversion from farmland. Forest land also showed an increasing trend, mainly being converted from grassland, farmland, water bodies, and unused land with areas of 2201.26 km^2, 11 km^2, 26 km^2, and 9 km^2, respectively. The main factor contributing to alterations in all types of LUCC was the unused land. The area of water bodies has remained stable over the past 30 years, with areas of 12,584.20 km^2 and 10,315.19 km^2 being converted from and to, respectively.

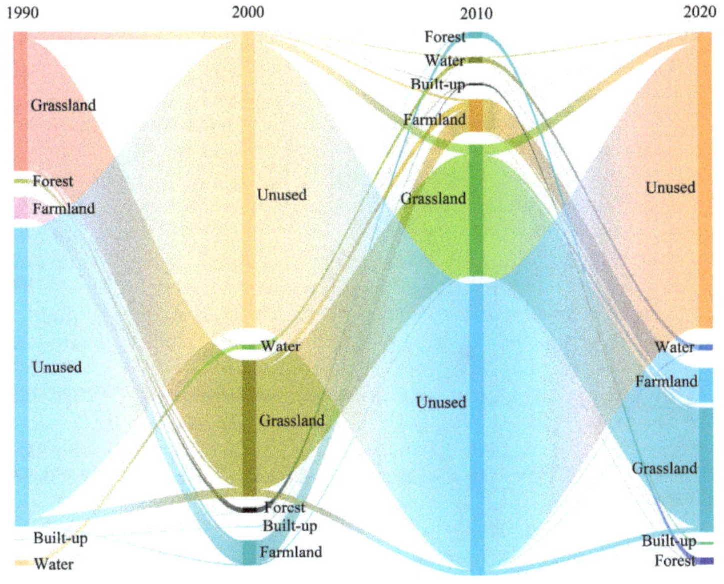

Figure 4. Dynamic changes in LUCC types in NSEBTM from 1990 to 2020.

3.1.2. LUCC Dynamics from 2020 to 2050

The validated PLUS model was employed to forecast the spatial distribution of LUCC change in 2050 under various scenarios, based on the LUCC data for the study area in 2020 (Figure 3e–g). Under the natural change scenario, the predominant LUCC type in 2050 is still unused land, accounting for 63.25% of the total area (Figure 3e), followed by grassland and farmland, accounting for 24.79% and 8.62%, respectively. Water bodies, forests, and built-up areas accounted for smaller proportions of the area (0.95%, 1.64%, and 0.76%, respectively). Grassland and water bodies continued to show decreasing trends, with cumulative areas decreasing by 6940.98 km^2 and 1010.38 km^2, respectively. Farmland, forests, built-up, and unused land are areas that showed cumulative increases in area of 5324.29 km^2, 942.03 km^2, 32.08 km^2, and 975.75 km^2, respectively. In terms of LUCC change direction (Figure 5a), all LUCC types showed trends of mutual transformation, with unused land being the main contributor to conversion to other LUCC types (8300.19 km^2, 1167.21 km^2, 378.43 km^2, 19 km^2, and 814.44 km^2 were converted to grassland, farmland, built-up areas, forests, and water bodies, respectively). Unused land remained the main contributor to LUCC change for the next 30 years.

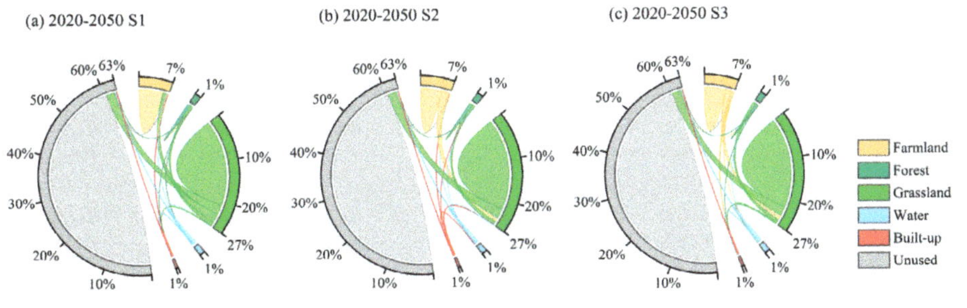

Figure 5. Dynamic changes in LUCC types in the NSEBTM in 2050 under different scenarios and the proportion of LUCC types. (S1 represents a natural development scenario; S2 represents an ecological protection scenario; S3 represents an economic development scenario).

Under the ecological protection scenario, the proportions of unused land, grassland, farmland, water bodies, forests, and built-up areas in the NSEBTM for 2050 were 63.17%, 27.15%, 6.70%, 0.98%, 1.65%, and 0.35%, respectively (Figure 3f). Compared to 2020, farmland, water bodies, and built-up land showed decreasing trends, with decreases of 2257.17 km^2, 874.93 km^2, and 792.40 km^2, respectively. Forests, grassland, and unused land showed increasing trends, with increases of 991.43 km^2, 2402.6 km^2, and 653.26 km^2, respectively. In terms of LUCC change direction (Figure 5b), all LUCC types showed similar trends of mutual transformation, with the increase in forest land mainly coming from the conversion of farmland to forests, and the increase in grassland mainly coming from the conversion of farmland, built-up land, and unused land. Overall, compared to the natural development scenario, the proportion of grassland increased from 24.79% to 27.15%, while the proportion of forests remained relatively stable, mainly because of the conversion of forests to grassland covering a larger area of 2669.87 km^2. There was relatively less conversion to other LUCC types.

In the context of economic development, it is projected that by 2050, the NSEBTM will have a distribution of unused land (63.16%), grassland (26.25%), farmland (6.89%), water bodies (0.98%), forest land (1.43%) and built-up land (1.30%) (Figure 3g). Compared with 2020, farmland, grassland, and water bodies showed significant decreases in area of 1509.66 km^2, 1151.62 km^2, and 865.42 km^2, respectively. The extent of built-up and unused land exhibited substantial increases, measuring 2938.08 km^2 and 611.74 km^2, respectively. However, there was a slight decline in the forested area. In terms of LUCC conversion (Figure 5c), all LUCC types showed a trend of mutual conversion, with conversion to built-up land being the most significant. Specifically, the areas of grassland, farmland, forest land, water bodies, and unused land that converted to built-up land were 1312.47 km^2, 1902.76 km^2, 8 km^2, 513.78 km^2, and 393.43 km^2, respectively. The primary driver of LUCC change in the economic development scenario was an increase in built-up land. In comparison to the natural development scenario, all other LUCC types experienced significant decreases, except for the proportion of built-up land which increased from 0.76% to 1.30%.

3.2. *Dynamics of Carbon Storage in NSEBTM from 1990 to 2050*

The InVEST model was employed to compute carbon storage in the NSEBTM from 1990 to 2050. However, the spatial distribution of carbon storage in the NSEBTM exhibited no significant changes (Figure 6). In general, high carbon reserves are mainly distributed in the central regions of the Tacheng City, Shuanghe City, Huyanghe City, Shihezi City, and Urumqi City. The highest values are found in scattered areas of the Tianshan Mountains. It is worth noting that the desert areas surrounding Hami City, Turpan City, and Changji Prefecture also store carbon reserves.

In order to enhance the clarity of illustrating the spatial changes in carbon storage in the NSEBTM, we conducted raster subtraction operations on the maps depicting carbon storage distribution during different time periods (Figure 7). According to the actual distribution of carbon storage, the areas were divided into carbon sink areas, balance areas, and carbon source areas, with values of 0 values and values close to 0 (−500 to 500 tons) being classified as balance areas [21]. From 1990 to 2000, 2000 to 2010, and 2010 to 2020, over 95% of the area did not show obvious changes in carbon storage, indicating that LUCC in most areas of the NSEBTM was minimally disturbed by human activities and did not undergo significant changes (Figure 7a–c). From 1990 to 2020, more than 3.60% of the total area showed an increase in carbon storage, indicating an enhancement in carbon sequestration capacity due to changes in LUCC, which led to an increase in carbon sink intensity. More than 3.18% of the area showed a decrease in carbon storage, indicating a weakening of the carbon sequestration capacity of the underlying surface, which turned into a carbon source area (Figure 7d). In general, the NSEBTM has maintained a relatively stable level of carbon storage over the past three decades, with approximately 95% of the region experiencing a balanced state between carbon storage gain and loss.

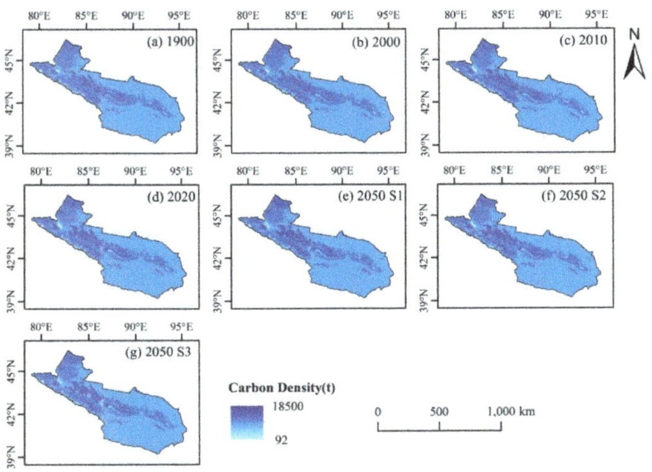

Figure 6. Spatial distribution of carbon storage in the NSEBTM from 1990 to 2050. (S1 represents a natural development scenario; S2 represents an ecological protection scenario; S3 represents an economic development scenario).

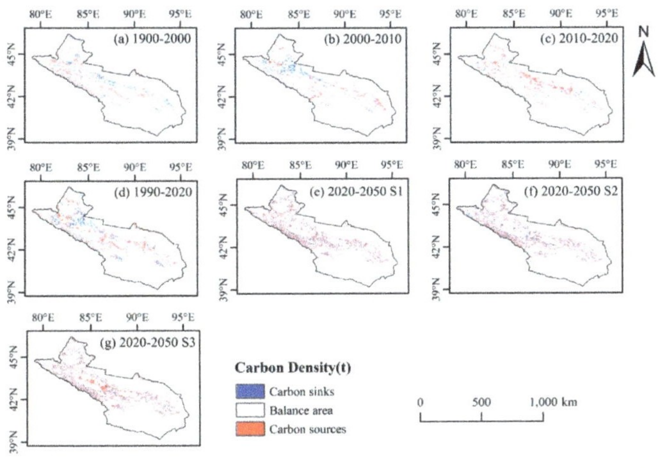

Figure 7. Spatial distribution changes in carbon storage in the NSEBTM from 1990 to 2050. (S1 represents a natural development scenario; S2 represents an ecological protection scenario; S3 represents an economic development scenario).

We also simulated carbon storage scenarios for the three development scenarios for 2050. Compared with 2020, in the natural development scenario, the areas with increased carbon storage exceeded 3.89%, the areas with decreased carbon storage exceeded 4.02%, and nearly 92.09% of the areas maintained a relatively stable carbon storage status (Figure 7e). In the ecological conservation scenario, areas with increased carbon storage exceeded 4.26%, areas with decreased carbon storage exceeded 3.60%, and over 92.14% of these areas maintained relatively stable carbon storage status (Figure 7f). Compared to the scenario of natural development, the ecological conservation scenario showed an increase in the area that absorbs carbon dioxide and a decrease in the area that emits carbon dioxide, while maintaining a relatively stable balance area. In the economic development scenario, areas with increased carbon storage exceeded 3.87%, areas with decreased carbon storage exceeded 5.20%, and nearly 91.74% of the areas were in a state of carbon balance (Figure 7g).

In the context of economic development, carbon sequestration decreases and the carbon source area expands compared to natural development, while the area of equilibrium remains relatively stable.

3.3. Revisions in Carbon Storage Resulting from LUCC Change

Based on the data presented in Table 5, the carbon storage contribution of various LUCC change types to the overall carbon storage can be ranked from highest to lowest as follows: grassland, unused land, farmland, forest land, water bodies, and built-up land. From 1990 to 2020, despite a declining trend in grassland areas (Figure 3) and a reduction of 102.35 t in carbon storage, it still constituted over 55% of the total carbon storage, thereby establishing itself as the predominant carbon sink. The area and carbon storage of unused land exhibited a declining trend, with a reduction of 4.05 t in carbon storage, constituting over 25% of the total carbon storage and positioning it as the second largest carbon sink within the region. The farmland area carbon storage exhibited a consistent annual increase, contributing to approximately 15% of the overall carbon storage. Over the past three decades, forest land witnessed a continuous growth in carbon storage, with an increment of 41.44 t. Although the forest land area constituted a relatively small proportion of the NSEBTM's total area, its share of carbon storage increased from 3.66% to 6.05%. Overall, the degradation of grasslands over the past three decades has been a significant contributing factor to carbon loss.

Table 5. Changes in carbon storage of LUCC types in NSEBTM from 1990 to 2050.

LUCC	Carbon Storage/t						
	1990	2000	2010	2020	2050 S1	2050 S2	2050 S3
Farmland	165.33	183.65	239.69	252.75	299.49	232.93	239.50
Forest	61.20	81.20	93.55	102.64	120.07	120.95	104.47
Grassland	995.93	976.99	945.24	893.58	834.65	914.10	883.87
Water	0.41	0.41	0.46	0.44	0.35	0.36	0.36
Built-up	0.18	0.43	0.80	1.17	1.62	0.75	2.75
Unused	450.40	447.95	439.94	446.35	448.09	447.52	447.44

We also calculated the carbon storage of each LUCC type in 2050 under the three development scenarios. Under the scenario of natural development, grassland and water bodies experienced a decrease in carbon storage by 58.93 t and 0.09 t, respectively, compared to 2020. Meanwhile, unused land, farmland, forest land, and built-up land saw an increase in carbon storage by 1.74 t, 46.74 t, 17.43 t, and 0.45 t, respectively. Under the ecological protection scenario, grassland and forest land exhibited an increase in carbon storage by 79.45 t and 18.31 t, respectively, compared to the levels observed in 2020. Conversely, farmland, built-up land, and water bodies experienced a decrease in carbon storage by 19.82 t, 0.42 t, and 0.08 t correspondingly. In the context of economic development, a significant increase in carbon storage was observed on developed land compared to the levels recorded in 2020, with an increment of 1.58 t, while the changes in the carbon storage of the other LUCC types were relatively inconspicuous. Overall, in the three development scenarios for 2050, grassland remained the most important carbon sink in the NSEBTM, and the increase or decrease in the grassland area mainly affected the carbon balance in the NSEBTM.

Changes in LUCC type significantly affected changes in carbon storage. Based on the changes in LUCC transformation as well as differences in soil and vegetation density from 1990 to 2050, the impact of LUCC changes on carbon storage in the NSEBTM was calculated. As shown in Figure 8, the changes in carbon storage from 1990 to 2020 were mainly caused by the conversion between grassland and unused land. From 1990 to 2020, the conversion from grassland resulted in a decrease of 7.17×10^7 t in the region's carbon storage, and there was an increase of 2.20×10^7 t in carbon storage from conversion to forest land, a decrease of 2.88×10^6 t from conversion to water bodies, a decrease of 5.12×10^6 t from conversion to built-up land, and a decrease of 6.37×10^7 t from conversion to unused land. This decrease was much greater than the increase, indicating that conversion from grassland was not conducive to increasing regional carbon storage. The conversion of a substantial amount of unused land into farmland and grassland resulted in respective increases of 2.62×10^7 t and 4.97×10^7 t in carbon storage. A minor conversion between unused land and water bodies resulted in a basic balance between the increase or decrease in carbon storage. Similarly, the bidirectional conversion between unused land and grassland resulted in a "balance between gains and losses" in the regional carbon storage. In the natural progression from 2020 to 2050, there was a net decrease of 6.15×10^7 t in carbon storage due to the transformation of grassland into unused land. Conversely, the conversion of this unused land into forest resulted in an increase of 3.59×10^7 t in carbon storage. Additionally, there was a significant boost of 5.57×10^7 t in carbon storage when converting unused land back into grassland, and a further increase of 8.16×10^6 t when transforming it into farmland. In the context of ecological conservation, between 2020 and 2050, the transition from grassland to forest land resulted in a carbon storage increase of 3.65×10^7 t. Similarly, converting water bodies to grassland and unused land led to respective increments of 3.73×10^6 t and 2.02×10^6 t in carbon storage. Furthermore, transforming built-up land into farmland and grassland contributed to an additional carbon storage of 4.29×10^6 t and 5.66×10^6 t, respectively. Notably, there were significant increases in carbon storage by converting unused land into farmland (6.99×10^6 t) and grassland (5.85×10^7 t). In terms of carbon storage decrease, the conversion from farmland to unused land resulted in a reduction of 6.53×10^6 t in carbon storage. The conversion from forest land to grassland led to a substantial decline of 2.67×10^7 t in carbon storage. Similarly, the transformation from grassland to water bodies and unused land caused significant decreases of 1.51×10^6 t and 5.91×10^7 t in carbon storage, respectively. Overall, in the ecological protection scenario, the increase in carbon storage resulting from land type conversion was much greater than the decrease, the main manifestation was the conversion of other LUCC types to grassland, resulting in a stronger regional carbon storage capacity. Under the economic development scenario, the main conversion was between grassland and unused land, and the increases and decreases in carbon storage were generally balanced. The conversion of farmland and grassland into built-up land resulted in reductions of 1.57×10^7 t and 1.05×10^7 t in carbon storage, respectively. Overall, under the economic development scenario, the decrease in carbon storage exceeded the increase, mainly due to the conversion of other LUCC types to built-up land, leading to a net loss of approximately 1.79×10^7 t.

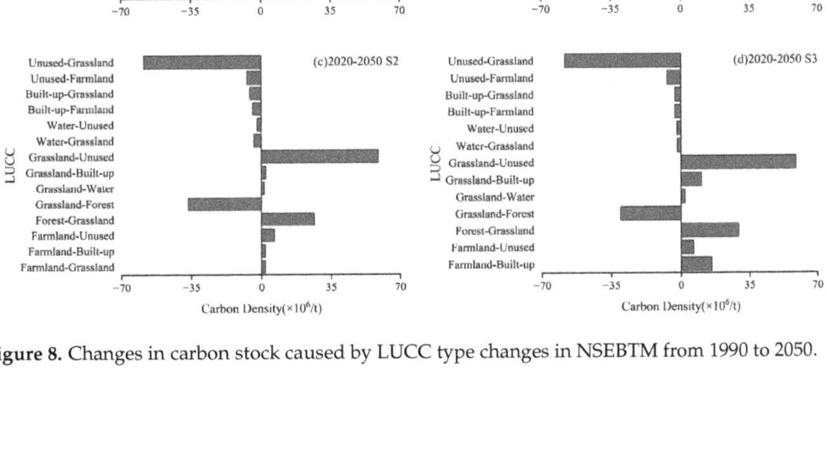

Figure 8. Changes in carbon stock caused by LUCC type changes in NSEBTM from 1990 to 2050.

4. Discussion

4.1. Implications of Human Actions on Changes in LUCC

The NSEBTM, located on the western frontier of China, has a large regional area and is one of the most sparsely populated areas in China's geographical space [12]. The NSEBTM is deep inland and far from the sea with a closed terrain, which makes it difficult for marine moisture to reach the region. It is controlled by continental air masses throughout the year, with scarce precipitation and an arid climate [45], which pose significant challenges to vegetation growth in the region. The main LUCC types in the NSEBTM are unused land, grassland, and farmland. Although the forest land area is relatively sparse, it is particularly important in LUCC planning, especially in the ecologically fragile areas of the NSEBTM. In 1979, China planned to implement the "Three-North Shelterbelt Project", which has a planned duration of 73 years, is divided into eight phases, and aims to establish large-scale artificial forestry. As an important part of northwest China, the NSEBTM has made substantial progress in ecological engineering construction under the Three-North Shelterbelt Policy [46]. By 2020, the fifth phase of the "Three-North Shelterbelt Project" in China was almost completed, and the forest area in the NSEBTM increased from 3307.17 km^2 in 1990 to 5546.42 km^2 in 2020 (Figure 3). NDVI data can characterize vegetation coverage; therefore, we established a spatial correlation between the NSEBTM NDVI data and the carbon storage data simulated using the InVEST model (Figure 9a). The graph reveals a distinct correlation between NDVI and carbon storage, particularly in the Huyanghe City, Karamay City, and Yili Prefecture. Taking the central region of the Yili Prefecture as an illustrative example from 2000 to 2020, there has been a notable augmentation in vegetation coverage within this area. Consequently, the expansion of vegetation coverage has led to a concurrent increase in carbon storage (Figure 9b,c).

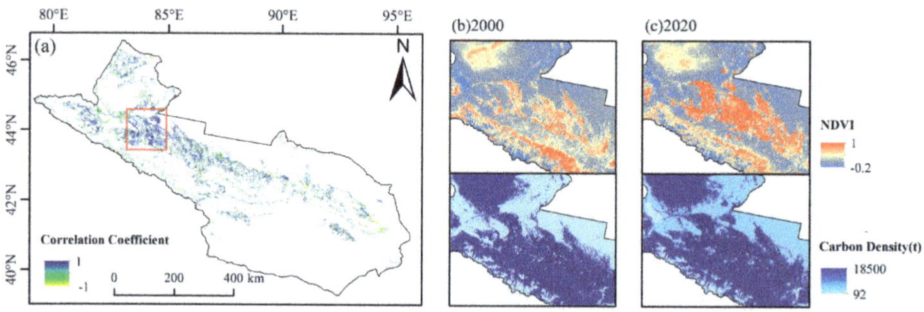

Figure 9. Changes in NDVI and corresponding variations in carbon storage. (**a**) Correlation between carbon stock and NDVI; (**b**) spatial distribution of NDVI and carbon stock in local areas in 2000; (**c**) spatial distribution of NDVI and carbon stock in local areas in 2020.

From 1990 to 2020, the farmland area in the NSEBTM increased from 18,832.5 km^2 in 1990 to 28,791.71 km^2. In 2021, the Xinjiang Uygur Autonomous Region issued the "14th Five-Year Plan for the Protection and Development of Land Resources in Xinjiang Uygur Autonomous Region", which mentioned the management policy for farmland: strengthening the protection of farmland, focusing on curbing the "non-agriculturalization" and preventing the "non-foodization" of farmland, and strictly observing the red line and bottom line of farmland. The overall LUCC plan issued by the Xinjiang Uygur Autonomous Region (2006–2020) also emphasized the effective protection of farmland, which plays a very important role as the third largest carbon sink in the NSEBTM. Over the past 30 years, the grassland area in the NSEBTM has shown a decreasing trend, and under our predicted natural development scenario for 2050, the grassland area will continue to decrease without human intervention. Conversely, under the ecological protection scenario, which adheres to the policy of protecting forests and grasslands, the grassland area will continue to increase. Overall, LUCC types in the NSEBTM are strongly influenced by human activities, which means that in the future, it will be necessary to reasonably restrict human activities to protect the ecological land in the NSEBTM.

4.2. Impact of Urbanization on Carbon Storage in NSEBTM

The findings of the research indicate that the process of rapid urbanization and economic growth has significantly expedited land degradation [47]. The expansion of urbanized land area signifies the process of urbanization, and it is crucial not to underestimate the conversion from non-built-up land to built-up land, as it results in a reduction in carbon storage. Therefore, we used nighttime light remote sensing data to characterize the changes in built-up land types in the NSEBTM. According to the statistics of the NSEBTM's illuminated area (Figure 10(b-3,b-4,c-3,c-4)), the illuminated area expanded four-fold from 2000 to 2020, increasing from 2.42% in 2000 to 7.20% in 2020. The spatial distribution of nighttime light data shows that built-up land has expanded to surrounding areas based on the original area, unlike the increase in light intensity in coastal areas. Urbanization in the inland areas of the NSEBTM is manifested by the expansion of nighttime lighting areas. The increase in nighttime lighting areas contributed to a reduction in carbon storage. With the development of urbanization in the NSEBTM, three central cities will be built, driving the joint development of the Urumqi metropolitan area, northern Xinjiang urban belt, and southern Xinjiang urban cluster. It is evident that, under the guidance of urbanization policies in Xinjiang, from 2000 to 2020, there has been an obvious increase in built-up land in the NSEBTM (Figure 10b,c), corresponding to a decrease in carbon storage.

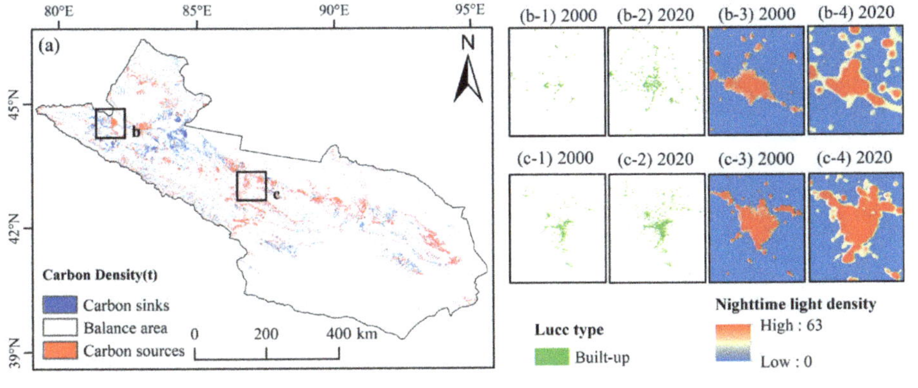

Figure 10. The relationship between built-up land change and carbon storage. (**a**) Spatial changes in carbon storage in the NSEBTM from 2000 to 2020, The black boxes in Figure (**a**) correspond to Figure (**b**) and Figure (**c**), respectively; (**b**,**c**) the distribution map of urban built-up land, and the spatial distribution map of light remote sensing data.

Research has shown that the level of urbanization in China has increased obviously from 2000 to 2019, and industrialization has accelerated obviously. Urban development requires more built-up land, and the continuous expansion of built-up land areas means that more energy and industrial production activities will occur, leading to a loss of carbon storage [48]. Numerous studies have quantitatively demonstrated that an increase in urban built-up land leads to a loss of carbon storage, such as the 8.64 Tg loss of carbon storage in the Chang-Zhu-Tan urban agglomeration due to the urban expansion from 1995 to 2009 [49]. In addition, numerous studies have shown that an increase in urban land will lead to a large amount of artificial greening and landscaping activities, which, although helpful in increasing urban green coverage, will also have adverse effects on soil quality and soil carbon storage capacity [50,51]. Furthermore, an increase in urban land will lead to an increase in traffic and industrial pollution, which will adversely affect soil and vegetation, further reducing carbon storage [52].

4.3. Implications of Different Scenarios of Carbon Storage Results for Future Planning

Our findings demonstrate that, under the ecological protection scenario for 2050, there was a significant increase in carbon storage by 12.34 t compared to the natural development scenario (Table 5). Notably, grasslands emerged as the primary carbon reservoirs in the NSEBTM, contributing over 55% of the total carbon storage (Figure 6). The degradation of grassland in the NSEBTM has been identified as the primary factor contributing to the loss of carbon storage, resulting in a significant decline of 102.35 t over the past three decades. The NSEBTM has abundant grassland resources and a variety of grassland types and species [53]; therefore, in future planning, it is necessary to strengthen the protection and rational utilization of grassland and steppe resources to increase carbon storage. Unused land was the main LUCC type in the NSEBTM, with deserts being the predominant LUCC type. Deserts are an important part of terrestrial ecosystems, and their contribution to carbon storage cannot be ignored. Gulnur et al. [54] explained that the Gurbantunggut Desert is a carbon sink during the vegetation growing season. Similarly, the Taklimakan Desert in southern Xinjiang is also an important carbon sink, sequestering 148.85×10^4 tons a^{-1} of carbon annually [55]. Our study confirms this conclusion, with unused land accounting for over 25% of the total carbon storage in the NSEBTM, making it the second largest carbon reservoir among the NSEBTM's land types. Therefore, given the NSEBTM's unique geographical location and climatic conditions, it is important to develop desert ecological projects that effectively increase land carbon storage. In future plans, investments in desert ecological projects should be increased to enhance carbon storage. Additionally, it is worth noting that forests and farmland are important sources of carbon storage in terrestrial ecosystems. Although the forests and farmlands in the NSEBTM occupy relatively small areas and are dispersed, their contribution to carbon storage should not be overlooked. Studies have shown that extensive logging and grazing activities reduce the carbon stored in shrubs, trees, roots, litter, and dead plants [56]. In related studies on farmland carbon storage, researchers have shown through 40 years of field cultivation data that fallow farming can increase soil organic carbon storage in the Yellow River Delta farmland [57]. Therefore, it is necessary to plan the use of forest and farmland resources in a rational manner to increase carbon storage.

Our study also shows that, compared to natural development scenarios, economic development scenarios resulted in a loss of 25.88 t of carbon storage (Table 5). Urbanization leads to an expansion of built-up land, resulting in a reduction in vegetation cover and subsequently diminishing carbon storage capacity. Furthermore, urbanization induces alterations in land use patterns, such as the conversion of farmland or forests into developed areas, leading to a decline in carbon storage. In the future process of economic development, it will be necessary to strengthen urban greening and ecological construction, protect and restore vegetation cover around cities, and increase urban green space and forest cover to increase urban carbon storage. In addition, it is necessary to strengthen urban planning and management, rationally use land resources, protect farmland and forest resources, and reduce the impact of land development on carbon storage.

In conclusion, based on the impact of LUCC changes on carbon storage in different scenarios, our findings provide a scientific basis and reference for LUCC planning and can help the government and relevant departments better formulate future LUCC policies and plans, promote ecological protection and sustainable development, and achieve a virtuous cycle of ecological environment and economic development.

4.4. Potential Applications and Limitations

The novelty of this study lies in the comprehensive analysis of the spatiotemporal variations in the NSEBTM carbon storage under past and future 30-year land use change scenarios using the PLUS model and InVEST model. By combining NDVI data and nighttime light remote sensing data, it reflects the changes in the NSEBTM carbon storage under the background of ecological engineering and urbanization. The advantage of the PLUS model over other land use prediction models is its application of a new analytical strategy that includes a new multi-seed growth mechanism coupled with multi-objective optimization algorithms, which can better support planning policies for sustainable development. The high accuracy of the future LUCC simulation results conforms to the development patterns under different scenarios for the NSEBTM, thus providing an approach to simulate the regional LUCC and carbon storage, serving as an example and reference for carbon storage research in other regions, and promoting the sustainable development of the NSEBTM.

This study also has certain limitations. The future LUCC patterns of the NSEBTM will change due to factors such as climate change, natural disasters, and policies. As the "warm-wetting" trend intensifies in northwest China [58], the LUCC patterns will also undergo changes, which will increase the uncertainty of the LUCC predictions. Additionally, carbon density will also change over time. Future research will establish models that will capture the relationship between carbon density and time to predict data that will align with future periods.

5. Conclusions

We conducted an analysis using the PLUS and InVEST models to examine how LUCC and carbon storage in the NSEBTM have changed over the past three decades, as well as projected changes for the next 30 years. Additionally, we performed a quantitative assessment to determine the factors contributing to variations in carbon storage under different scenarios. Our findings indicate that between 1990 and 2020, unused land was the primary type of LUCC change observed in the NSEBTM, followed by grassland. By 2050, different development scenarios will induce alterations in land use; nevertheless, unused land will remain the dominant category. Over the past three decades, grasslands have served as a crucial carbon sink in the NSEBTM with substantial carbon sequestration capacity compared to other types of land such as unused land. Notably, the degradation of grasslands has been identified as a key driver behind the declining carbon storage levels within this region. Under diverse development scenarios for 2050, ecological conservation initiatives are expected to contribute towards energy savings along with reduced emissions while simultaneously enhancing terrestrial carbon stocks; however, urban expansion particularly under economic development scenarios may result in diminished carbon storage capacities. Therefore, it is imperative to strategically plan future LUCC considering economic development objectives alongside tailored management strategies for distinct LUCC types aiming at mitigating the potential losses of carbon storage within the NSEBTM.

Author Contributions: Conceptualization, K.Z. and A.M.; methodology, K.Z. and Y.L.; software, Y.W.; validation, C.W. and M.S.; formal analysis, F.Y., C.Z. and W.H.; data curation, J.G. and A.A.; resources, A.M.; writing—original draft preparation, K.Z.; writing—review and editing, K.Z., A.M. and Y.L.; visualization, K.Z. and L.Z. All authors have read and agreed to the published version of the manuscript.

Funding: This research was funded by the National Natural Science Foundation of China, grant number: 42305132 and 42375084, the Scientific and Technological Innovation Team (Tien Shan Innovation Team) project, grant number: 2022TSYCTD0007, the Third Xinjiang Scientific Expedition, grant number: 2022xjkk0903, the Special Funds for Basic Scientific Research Business Expenses of Central-level Public Welfare Scientific Research Institutes, grant number: IDM2021005, the Special Project for the Construction of Innovation Environment in the Autonomous Region, grant number: PT2203, and the Graduate Education Innovation Program of the Autonomous Region, grant number: XJ2023G032.

Data Availability Statement: The data are available on request.

Conflicts of Interest: The authors declare no conflicts of interest.

References

1. Zhou, W.; Cao, X.; Dong, X.; Zhen, X. The effects of carbon-related news on carbon emissions and carbon transfer from a global perspective: Evidence from an extended STIRPAT model. *J. Clean. Prod.* **2023**, *425*, 138974. [CrossRef]
2. IPCC. AR6 Synthesis Report: Climate Change 2023. In Proceedings of the Sixth Assessment Report during the Panel's 58th Session, Interlaken, Switzerland, 13–19 March 2023.

3. Hong, X.; Liu, C.; Zhang, C.; Tian, Y.; Wu, H.; Yin, H.; Zhu, Y.; Cheng, Y. Vast ecosystem disturbance in a warming climate may jeopardize our climate goal of reducing CO_2: A case study for megafires in the Australian 'black summer'. *Sci. Total Environ.* **2023**, *866*, 161387. [CrossRef] [PubMed]
4. Wang, X.; Hu, H.; Zheng, X.; Deng, W.; Chen, J.; Zhang, S.; Cheng, C. Will climate warming of terrestrial ecosystem contribute to increase soil greenhouse gas fluxes in plot experiment? A global meta-analysis. *Sci. Total Environ.* **2022**, *827*, 154114. [CrossRef] [PubMed]
5. Findell, K.L.; Berg, A.; Gentine, P.; Krasting, J.P.; Lintner, B.R.; Malyshev, S.; Santanello, J.A.; Shevliakova, E. The impact of anthropogenic land use and land cover change on regional climate extremes. *Nat. Commun.* **2017**, *8*, 989. [CrossRef] [PubMed]
6. Li, Z.; Chen, J.; Wang, P.; Zhou, Z.; Chen, X. The synergy between temporal and spatial effects of human activities on CO_2 emissions in Chinese cities. *Environ. Impact Asses* **2023**, *103*, 107264. [CrossRef]
7. Wang, Q.; Yang, C.; Wang, M.; Zhao, L.; Zhao, Y.; Zhang, Q.; Zhang, C. Decoupling analysis to assess the impact of land use patterns on carbon emissions: A case study in the Yellow River Delta efficient eco-economic zone, China. *J. Clean. Prod.* **2023**, *412*, 137415. [CrossRef]
8. Aneseyee, A.B.; Soromessa, T.; Elias, E.; Noszczyk, T.; Hernik, J.; Benti, N.E. Expressing carbon storage in economic terms: The case of the upper Omo Gibe Basin in Ethiopia. *Sci. Total Environ.* **2022**, *808*, 152166. [CrossRef] [PubMed]
9. Ito, A.; Nishina, K.; Noda, H.M. Impacts of future climate change on the carbon budget of northern high-latitude terrestrial ecosystems: An analysis using ISI-MIP data. *Polar Sci.* **2016**, *10*, 346–355. [CrossRef]
10. Gui, D.; He, H.; Liu, C.; Han, S. Spatio-temporal dynamic evolution of carbon emissions from land use change in Guangdong Province, China, 2000–2020. *Ecol. Indic.* **2023**, *156*, 111131. [CrossRef]
11. Houghton, R.A.; Skole, D.L.; Nobre, C.A.; Hackler, J.L.; Lawrence, K.T.; Chomentowski, W.H. Annual fluxes of carbon from deforestation and regrowth in the Brazilian Amazon. *Nature* **2000**, *403*, 301–304. [CrossRef]
12. Guan, J.; Yao, J.; Li, M.; Zheng, J. Assessing the Spatiotemporal Evolution of Anthropogenic Impacts on Remotely Sensed Vegetation Dynamics in Xinjiang, China. *Remote Sens.* **2021**, *13*, 4651. [CrossRef]
13. He, Y.; Ma, J.; Zhang, C.; Yang, H. Spatio-Temporal Evolution and Prediction of Carbon Storage in Guilin Based on FLUS and InVEST Models. *Remote Sens.* **2023**, *15*, 1445. [CrossRef]
14. Li, G.; Cheng, G.; Liu, G.; Chen, C.; He, Y. Simulating the Land Use and Carbon Storage for Nature-Based Solutions (NbS) under Multi-Scenarios in the Three Gorges Reservoir Area: Integration of Remote Sensing Data and the RF–Markov–CA–InVEST Model. *Remote Sens.* **2023**, *15*, 5100. [CrossRef]
15. Xiang, S.; Wang, Y.; Deng, H.; Yang, C.; Wang, Z.; Gao, M. Response and multi-scenario prediction of carbon storage to land use/cover change in the main urban area of Chongqing, China. *Ecol. Indic.* **2022**, *142*, 109205. [CrossRef]
16. Xu, C.; Zhang, Q.; Yu, Q.; Wang, J.; Wang, F.; Qiu, S.; Ai, M.; Zhao, J. Effects of land use/cover change on carbon storage between 2000 and 2040 in the Yellow River Basin, China. *Ecol. Indic.* **2023**, *151*, 110345. [CrossRef]
17. Zhu, L.; Song, R.; Sun, S.; Li, Y.; Hu, K. Land use/land cover change and its impact on ecosystem carbon storage in coastal areas of China from 1980 to 2050. *Ecol. Indic.* **2022**, *142*, 109178. [CrossRef]
18. Li, J.; Shao, Z. Spatiotemporal evolution and prediction of carbon stock in Urumqi City based on PLUS and InVEST models. *Arid. Zone Res.* **2024**, *41*, 499–508.
19. Wang, Y.; Zhang, Z.; Chen, X. Land Use Transitions and the Associated Impacts on Carbon Storage in the Poyang Lake Basin, China. *Remote Sens.* **2023**, *15*, 2703. [CrossRef]
20. Lu, Y.; Xu, X.; Li, J.; Feng, X.; Liu, L. Research on the spatio-temporal variation of carbon storage in the Xinjiang Tianshan Mountains based on the InVEST model. *Arid. Zone Res.* **2022**, *39*, 1896–1906.
21. Han, M.; Xu, C.; Long, Y.; Liu, F. Simulation and Prediction of Changes in Carbon Storage and Carbon Source/Sink under Different Land Use Scenarios in Arid Region of Northwest China. *Bull. Soil. Water Conserv.* **2022**, *42*, 335–344.
22. Babbar, D.; Areendran, G.; Sahana, M.; Sarma, K.; Raj, K.; Sivadas, A. Assessment and prediction of carbon sequestration using Markov chain and InVEST model in Sariska Tiger Reserve, India. *J. Clean. Prod.* **2021**, *278*, 123333. [CrossRef]
23. Li, C.; Wu, Y.; Gao, B.; Zheng, K.; Wu, Y.; Li, C. Multi-scenario simulation of ecosystem service value for optimization of land use in the Sichuan-Yunnan ecological barrier, China. *Ecol. Indic.* **2021**, *132*, 108328. [CrossRef]
24. Wei, Q.; Abudureheman, M.; Halike, A.; Yao, K.; Yao, L.; Tang, H.; Tuheti, B. Temporal and spatial variation analysis of habitat quality on the PLUS-InVEST model for Ebinur Lake Basin, China. *Ecol. Indic.* **2022**, *145*, 109632. [CrossRef]
25. Wang, K.; Ma, H.; Fang, C. The relationship evolution between urbanization and urban ecological resilience in the Northern Slope Economic Belt of Tianshan Mountains, China. *Sustain. Cities Soc.* **2023**, *97*, 104783. [CrossRef]
26. Yang, Z.; Zhang, S.; Lei, J.; Zhang, X.; Tong, Y.; Duan, Z.; Fan, L. Evolution of economic linkage network of the cities and counties on the northern slope of the Tianshan Mountains, China. *Reg. Sustain.* **2023**, *4*, 173–184. [CrossRef]
27. Jin-yan, L.; Lan-bo, C.; Miao, D.; Ali, A. Water resources allocation model based on ecological priority in the arid region. *Environ. Res.* **2021**, *199*, 111201. [CrossRef]
28. Joshua, J. Economic Development and the Belt and Road Initiative. In *The Belt and Road Initiative and the Global Economy: Volume I—Trade and Economic Development*; Joshua, J., Ed.; Springer International Publishing: Cham, Switzerland, 2019; pp. 47–81.
29. Zhu, C.; Fang, C.; Zhang, L. Analysis of the coupling coordinated development of the Population–Water–Ecology–Economy system in urban agglomerations and obstacle factors discrimination: A case study of the Tianshan North Slope Urban Agglomeration, China. *Sustain. Cities Soc.* **2023**, *90*, 104359. [CrossRef]

30. Zhang, L.; Fang, C.; Zhu, C.; Gao, Q. Ecosystem service trade-offs and identification of eco-optimal regions in urban agglomerations in arid regions of China. *J. Clean. Prod.* **2022**, *373*, 133823. [CrossRef]
31. Zhu, C.; Fang, C.; Zhang, L.; Wang, X. Simulating the interrelationships among population, water, ecology, and economy in urban agglomerations based on a system dynamics approach. *J. Clean. Prod.* **2024**, *439*, 140813. [CrossRef]
32. Yang, J.; Huang, X. The 30 m annual land cover dataset and its dynamics in China from 1990 to 2019. *Earth Syst. Sci. Data* **2021**, *13*, 3907–3925. [CrossRef]
33. Lu, X.; Li, J.; Duan, P.; Zhang, B.; Li, C. Correction of nighttimelight images of DMSP/OLS in China. *Bull. Surv. Mapp.* **2019**, *7*, 127–131. [CrossRef]
34. Guan, J.; Dong, L.; Wang, Y.; Wang, X. DMSP-OLS and NPP-VIIRS night light image correction in China. *Bull. Surv. Mapp.* **2021**, *9*, 1–8.
35. Maanan, M.; Maanan, M.; Karim, M.; Ait Kacem, H.; Ajrhough, S.; Rueff, H.; Snoussi, M.; Rhinane, H. Modelling the potential impacts of land use/cover change on terrestrial carbon stocks in north-west Morocco. *Int. J. Sust. Dev. World* **2019**, *26*, 560–570. [CrossRef]
36. Alam, S.A.; Starr, M.; Clark, B.J.F. Tree biomass and soil organic carbon densities across the Sudanese woodland savannah: A regional carbon sequestration study. *J. Arid. Environ.* **2013**, *89*, 67–76. [CrossRef]
37. Giardina, C.P.; Ryan, M.G. Evidence that decomposition rates of organic carbon in mineral soil do not vary with temperature. *Nature* **2000**, *404*, 858–861. [CrossRef]
38. Chen, G.; Yang, Y.; Liu, L.; Li, X.; Zhao, C.; Yuan, Y. Research Review on Total Below ground C arbon Allocation in Forest Ecosystems. *J. Subtrop. Resour. Environ.* **2007**, *2*, 34–42.
39. Han, C.; Zheng, J.; Wang, Z.; Yu, W. Spatio-temporal variation and multi-scenario simulation of carbon storage in terrestrial ecosystems in Turpan Hami Basin based on PLUS-InVEST model. *Arid. Land. Geogr.* **2024**, *47*, 260–269.
40. Li, P.; Chen, J.; Li, Y.; Wu, W. Using the InVEST-PLUS Model to Predict and Analyze the Pattern of Ecosystem Carbon storage in Liaoning Province, China. *Remote Sens.* **2023**, *15*, 4050. [CrossRef]
41. Liang, X.; Guan, Q.; Clarke, K.C.; Liu, S.; Wang, B.; Yao, Y. Understanding the drivers of sustainable land expansion using a patch-generating land use simulation (PLUS) model: A case study in Wuhan, China. *Comput. Environ. Urban Syst.* **2021**, *85*, 101569. [CrossRef]
42. Zeng, S.; Liu, X.; Tian, J.; Zeng, J. Spatial–Temporal Pattern Analysis and Development Forecasting of Carbon Stock Based on Land Use Change Simulation: A Case Study of the Xiamen–Zhangzhou–Quanzhou Urban Agglomeration, China. *Land* **2024**, *13*, 476. [CrossRef]
43. Zhang, Y.; Liao, X.; Sun, D. A Coupled InVEST-PLUS Model for the Spatiotemporal Evolution of Ecosystem Carbon Storage and Multi-Scenario Prediction Analysis. *Land* **2024**, *13*, 509. [CrossRef]
44. Zong, S.; Hu, Y.; Zhang, Y.; Wang, W. Identification of land use conflicts in China's coastal zones: From the perspective of ecological security. *Ocean. Coast. Manag.* **2021**, *213*, 105841. [CrossRef]
45. Zhu, Y.; Sun, L.; Luo, Q.; Chen, H.; Yang, Y. Spatial optimization of cotton cultivation in Xinjiang: A climate change perspective. *Int. J. Appl. Earth Obs.* **2023**, *124*, 103523. [CrossRef]
46. Zhang, D.; Zuo, X.; Zang, C. Assessment of future potential carbon sequestration and water consumption in the construction area of the Three-North Shelterbelt Programme in China. *Agr. For. Meteorol.* **2021**, *303*, 108377. [CrossRef]
47. Wang, C.; Wang, Q.; Liu, N.; Sun, Y.; Guo, H.; Song, X. The impact of LUCC on the spatial pattern of ecological network during urbanization: A case study of Jinan City. *Ecol. Indic.* **2023**, *155*, 111004. [CrossRef]
48. Pan, Z.; Gao, G.; Fu, B.; Liu, S.; Wang, J.; He, J.; Liu, D. Exploring the historical and future spatial interaction relationship between urbanization and ecosystem services in the Yangtze River Basin, China. *J. Clean. Prod.* **2023**, *428*, 139401. [CrossRef]
49. Jiang, W.; Deng, Y.; Tang, Z.; Lei, X.; Chen, Z. Modelling the potential impacts of urban ecosystem changes on carbon storage under different scenarios by linking the CLUE-S and the InVEST models. *Ecol. Model.* **2017**, *345*, 30–40. [CrossRef]
50. Mishra, A.; Humpenöder, F.; Churkina, G.; Reyer, C.P.O.; Beier, F.; Bodirsky, B.L.; Schellnhuber, H.J.; Lotze-Campen, H.; Popp, A. Land use change and carbon emissions of a transformation to timber cities. *Nat. Commun.* **2022**, *13*, 4889. [CrossRef]
51. Zhang, Z.; Gao, X.; Zhang, S.; Gao, H.; Huang, J.; Sun, S.; Song, X.; Fry, E.; Tian, H.; Xia, X. Urban development enhances soil organic carbon storage through increasing urban vegetation. *J. Environ. Manag.* **2022**, *312*, 114922. [CrossRef]
52. Wang, F.; Dong, M.; Ren, J.; Luo, S.; Zhao, H.; Liu, J. The impact of urban spatial structure on air pollution: Empirical evidence from China. *Environ. Dev. Sustain.* **2022**, *24*, 5531–5550. [CrossRef]
53. Zhang, R.; Liang, T.; Guo, J.; Xie, H.; Feng, Q.; Aimaiti, Y. Grassland dynamics in response to climate change and human activities in Xinjiang from 2000 to 2014. *Sci. Rep.* **2018**, *8*, 2888. [CrossRef]
54. Amar, G.; Mamtimin, A.; Wang, Y.; Wang, Y.; Gao, J.; Yang, F.; Song, M.; Aihaiti, A.; Wen, C.; Liu, J. Factors controlling and variations of CO_2 fluxes during the growing season in Gurbantunggut Desert. *Ecol. Indic.* **2023**, *154*, 110708. [CrossRef]
55. Yang, F.; Huang, J.; Zheng, X.; Huo, W.; Zhou, C.; Wang, Y.; Han, D.; Gao, J.; Ali, M.; Yang, X.; et al. Evaluation of carbon sink in the Taklimakan Desert based on correction of abnormal negative CO_2 flux of IRGASON. *Sci. Total Environ.* **2022**, *838*, 155988. [CrossRef] [PubMed]
56. Magnano, A.L.; Meglioli, P.A.; Vazquez Novoa, E.; Chillo, V.; Alvarez, J.A.; Alvarez, L.M.; Sartor, C.E.; Vázquez, D.P.; Vega Riveros, C.C.; Villagra, P.E. Relationships between land-use intensity, woody species diversity, and carbon storage in an arid woodland ecosystem. *Forest Ecol. Manag.* **2023**, *529*, 120747. [CrossRef]

57. Liu, D.; Gong, H.; Li, J.; Liu, Z.; Wang, L.; Ouyang, Z.; Xu, L.; Wang, T. Continuous crop rotation increases soil organic carbon stocks in river deltas: A 40-year field evidence. *Sci. Total Environ.* **2024**, *906*, 167749. [CrossRef]
58. Ding, Y.; Liu, Y.; Xu, Y.; Wu, P.; Xue, T.; Wang, J.; Shi, Y.; Zhang, Y.; Song, Y.; Wang, P. Regional responses to global climate change: Progress and prospects for trend, causes, and projection of climatic warming-wetting in Northwest China. *Adv. Earth Sci.* **2023**, *38*, 551–562.

land

MDPI

Article

Study on the Evolution of Spatial and Temporal Patterns of Carbon Emissions and Influencing Factors in China

Maowen Sun, Boyi Liang *, Xuebin Meng, Yunfei Zhang, Zong Wang and Jia Wang

Precision Forestry Key Laboratory of Beijing, College of Forestry, Beijing Forestry University, Beijing 100083, China; sunmaowen@bjfu.edu.cn (M.S.); mengxuebin@bjfu.edu.cn (X.M.); zhangyunnfei@bjfu.edu.cn (Y.Z.); wangzong@bjfu.edu.cn (Z.W.); wangjia2009@bjfu.edu.cn (J.W.)

* Correspondence: liangboyi@bjfu.edu.cn

Abstract: Industrialization has increased global carbon emissions, necessitating effective climate change mitigation measures. China, the most populous developing nation, faces the challenge of strategizing emissions to meet national carbon neutrality objectives. However, research on specific regions' carbon emissions drivers and causal factors is limited, particularly across prefectural-level cities. This study estimates the spatial and temporal patterns of carbon emissions across China's prefectural cities and utilizes both OLS regression and stepwise regression models to analyze the impact of various factors influencing carbon emissions in these cities. Results reveal the following: (1) The country's overall 20-year carbon emissions continue to grow from 3020.29 Mt in 2001 to 9169.74 Mt in 2020, with an average annual growth rate of 5.71%; the eastern region has seen a gradual deceleration in emissions, whereas the western region continues to experience an increase. Carbon emissions in cities within each subregion consistently rise. (2) Carbon emissions in Chinese prefectural-level cities exhibit strong spatial autocorrelation and clustering ($Z > 1.96$, $p < 0.05$), with hot spots primarily in the eastern coastal areas and cold spots in the northwest to southwest regions. (3) Economic and demographic factors significantly increase carbon emissions, while climate and urbanization effects are more complex and variable. Economic growth and population increase are the most significant influencing factors, but regional variances exist in carbon emissions determinants in subregional prefectural cities. These insights provide valuable insights into national emission dynamics at the prefectural level, providing a theoretical basis for enhancing carbon emission strategies across various jurisdictions.

Keywords: carbon neutrality; carbon emissions; spatiotemporal pattern; attribution analysis; urbanization

1. Introduction

Global climate change is attracting increasing international attention as it is causing a range of environmental problems [1–3]. According to the Sixth Assessment Report of the Intergovernmental Panel on Climate Change (IPCC), the burning of fossil fuels and inequitable and unsustainable energy and land use have led to a steady increase in global temperatures over the past century, resulting in an increase in the frequency and intensity of extreme weather events, putting nature and people at increasing risk in all regions of the world [4]. It has also been pointed out that, globally, the increase in carbon emissions is mainly driven by industry, transport and energy supply, while residential and commercial buildings, forestry/deforestation, and agriculture also contribute significant amounts of carbon dioxide, methane, and other greenhouse gases [5–7]. In terms of the impact of economic growth on carbon emissions, China is one of the world's fastest-growing economies, and its energy consumption and carbon emissions have increased prominently in recent decades [8,9]. From 1980 to 2007, GDP of China grew at an average annual rate of over 9%, primary energy consumption increased by about 340%, and carbon dioxide emissions increased dramatically by about 352% [10]. As the world's largest developing country, China must assume the role of a major power, and the Chinese government has made its dual carbon target, i.e., to achieve carbon peak by 2030 and carbon neutrality by 2060 [11–13]. Achieving carbon neutrality in China means absorbing the CO_2 emitted directly and indirectly by human activities in a given period (usually one year) through carbon capture and storage or sequestration techniques, such as planting trees and forests, to achieve "zero emissions" of CO_2 [14,15]. Compared with the historical process of Europe, the United States and other developed countries, China is facing the severe challenge of time constraints and heavy tasks to achieve the goal of carbon neutrality and needs to implement a larger amount of carbon neutrality in a shorter period of time than developed countries [16].

The accelerated growth of urbanization has made cities a crucial element in the reduction of carbon emissions [17–19]. City clusters, as a pivotal area for carbon emissions and regional economic development, are connected through close economic ties, creating a spatial connection between diverse urban areas [20–23]. China's CO_2 emissions exhibit a typical pattern of spatial intensity and high emission levels in the prefectures. While there was no significant change in this pattern from 2007 to 2012, the results indicate that there was a 3% relaxation in intensity during this period. Furthermore, the results indicated that the total CO_2 emissions had increased by 33.5% during the same period. This emission pattern also reflected the impact of the typical urbanization process in China [24].

In recent years, numerous studies have investigated carbon emissions within specific urban regions, including the Beijing–Tianjin–Hebei area, the Yangtze River Delta, and the Chengdu–Chongqing area, employing various methodologies. For instance, Zeng et al. [25] selected the Chengdu–Chongqing urban agglomeration to analyze the spatial and temporal evolution pattern of carbon emissions. They employed the ridge regression model and the STIRPAT (Stochastic Impacts by Regression on Population, Affluence, and Technology) model to explore the influence of key factors on carbon emissions in the Chengdu–Chongqing urban agglomeration. Luo et al. [26] used data from Xi'an as an example to establish a spatial simulation and prediction model of carbon emissions, with the aim of providing references for the regional planning of carbon emission reduction and the implementation of carbon emission reduction technologies. Some scholars [27] chose to start from the land use to assess the impact of land use patterns on carbon emissions under the Yellow River Delta region, providing a theoretical framework for sustainable land use. Additionally, other studies on national and regional carbon emissions are predominantly based on estimation and analysis of carbon emissions based on data such as the nighttime lighting index [28–30] or focus on the relationship between carbon emissions and the economic level [31]. Despite existing research on urban carbon emissions, there is a notable lack of detailed studies on the spatial correlation and key factors influencing carbon emissions across prefecture-level cities nationwide. Understanding the distribution and determinants of these emissions is crucial for aligning regional economic development with high-quality, sustainable growth in China's new era [32,33]. Such knowledge will aid in crafting more targeted and effective carbon reduction policies. In the current research in this direction, the analysis of spatial and temporal patterns focuses on methods such as spatial autocorrelation analysis [34–36], kernel density estimation [36], and center of gravity transfer trajectory [37], among others. The methods of attribution analysis have also gone through the process of developing from simple linear [38] to non-linear machine learning methods [39,40], and the process models have gradually developed from single-factor to multi-factor [41,42] and multi-modal large model simulations [43,44], etc.

Therefore, this paper tries to contribute to achieving China's carbon peak and carbon neutrality goals by selecting the association between prefecture-level cities and carbon emissions for analysis. First, it aims to analyze the spatial and temporal distribution patterns of carbon emissions across prefecture-level cities in various subregions from 2001 to 2020. Second, it seeks to identify the factors influencing carbon emissions using stepwise and OLS regression models, exploring the extent of each factor's impact and their variations across space and time (Figure 1). The results of this study can provide a foundational and scientific reference for China's future strategies on carbon emission reduction and for the planning and development of urban economies.

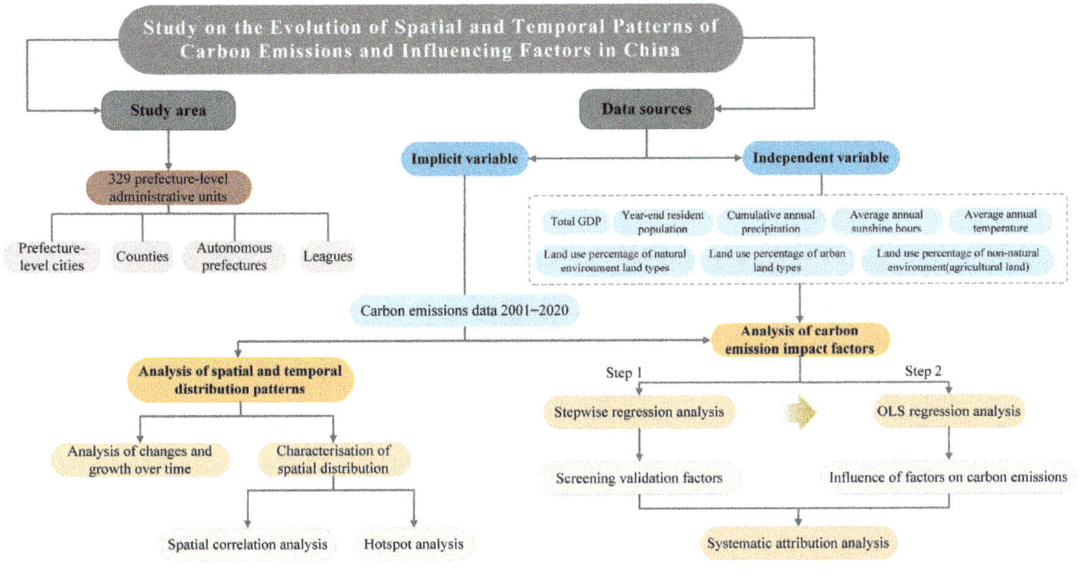

Figure 1. Flowchart of the study route (Note: Stepwise regression analysis is firstly used as an exploratory regression to initially screen the factors and obtain some of the results to show the characteristics of each factor and to be able to verify the feasibility of the impact of the selected factors on carbon emissions, and the OLS regression presents a more detailed picture of the changes in the impact on carbon emissions among the influencing factors. The two regression methods work together to serve the final result and ascertain the systematic conclusion of the attribution analysis).

2. Materials and Methods

2.1. Study Area

This study describes and analyses the spatial and temporal evolution of carbon emissions in 329 prefectural administrative units (including prefectural cities, counties, autonomous prefectures, and leagues) across China. The temporal evolution of carbon emissions is expressed at the scale of provincial administrative regions, whereas in analyzing changes in the spatial distribution pattern and spatial correlation of carbon emissions, it is expressed at the scale of municipal administrative regions. Data from the Tibet Autonomous Region (TAR), Taiwan Province, Hong Kong Special Administrative Region (HKSAR), and Macau Special Administrative Region (MSAR) are excluded from the calculations to ensure data consistency, completeness, and accuracy. The provinces in China have been reorganized into subregions as follows: The South Central Region (CS) includes Henan, Hubei, Hunan, Guangxi Zhuang Autonomous Region, Guangdong, and Hainan. The Eastern Region (E) comprises Shanghai, Shandong, Jiangsu, Anhui, Zhejiang, Fujian, and Jiangxi. In the Northern Region (N), Beijing, Tianjin, Hebei, and Shanxi are grouped together. The Northeastern Region (NE) encompasses Heilongjiang, Jilin, Liaoning, and the Inner Mongolia Autonomous Region. The Northwest Region (NW) consists of Shaanxi, Gansu, Qinghai, Ningxia Hui Autonomous Region, and Xinjiang Uygur Autonomous Region. Lastly, the Southwest Region (SW) is made up of Chongqing, Yunnan, Sichuan, and Guizhou (Figure 2).

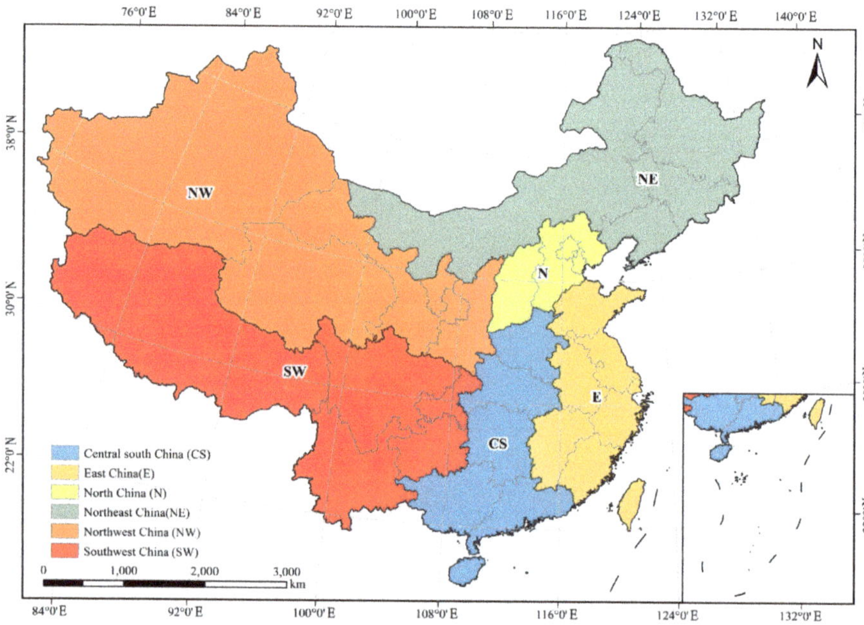

Figure 2. Regional Divisions of China.

2.2. *Data Sources*

The 2001–2020 carbon emission data used in this paper are collected from the China Carbon Accounting Database [45], and the decomposition of the carbon emission influencing factors includes the details of the year-end total GDP, year-end resident population, cumulative precipitation, average temperature, sunshine hours, the proportion of use of the natural environment land type, the proportion of use of the urban land type, and the proportion of use of the non-natural environment land type (agricultural land). The source is indicated in Table 1.

Table 1. Sources of data on impact factors.

Category	Name	Source
Economic factors	Total GDP	GDP, population and other socioeconomic data were compiled using the China Urban Statistical Yearbook [46], related urban statistical yearbooks, and the China Energy Statistical Yearbook [47] (2001–2020).
Demographic factors	Year-end resident population	
Climatic factors	Cumulative annual precipitation, Average annual temperature, Average annual sunshine hours	China National Meteorological Center (CNMC)
Land use factors	Land use percentage of natural environment land types, Land use percentage of urban land types, Land use percentage of non-natural environment (agricultural land)	Co-authored by Professors Jie Yang and Xin Huang of Wuhan University, "30 m annual landcover and its dynamics China from 1990 to 2019" (now updated to 2020)

2.3. *Research Methodology*

2.3.1. Analysis of Spatial Distribution Pattern

We use Global Moran's I and Local Moran's I. The former is a measure of spatial autocorrelation developed by Patrick Alfred Pierce Moran; as well as a global autocorrelation, followed by a local autocorrelation, the local Moran's I is the one that will show where the outliers or where the agglomeration occurs. Global Moran's I > 0 indicates positive spatial correlation; the larger the value, the more spatially clustered the regional carbon emissions are; Moran's I < 0 indicates negative spatial correlation; the smaller the value, the more significant the difference between the regional carbon emissions and the surrounding region; Moran's I = 0 indicates that there is no spatial correlation and the regional carbon emissions are spatially randomly distributed; the absolute value represents the strength of the autocorrelation. The absolute value reflects the strength of the autocorrelation. The significance of I was tested by the standardized Z-value, Z > 1.96, and passed the 5% significance level test ($p < 0.05$), indicating the existence of significant spatial correlation; the correlation formula was referenced from the literature [48]. As for the local autocorrelation, the local spatial distribution characteristics of carbon emissions in urban agglomerations under different zones are analyzed by ArcGIS using the spatial linkage local indicator LISA [49]. The LISA graph is used for visualization, and the spatial agglomeration units with significance are classified into high and high agglomeration type areas (H-H), low and low agglomeration type areas (L-L), high and low agglomeration type areas (H-L), low and high agglomeration type areas (L-H), and the rest are insignificant areas [50]. Based on the previous high/low clustering analysis, the obtained Z-score and p-value were used to calculate the Getis–Ord Gi* statistic for each element in the dataset to obtain the location of hot and cold spots where spatial clustering occurs [51], which is used as a result to further explain the relationship between the spatial distribution pattern and carbon emissions of prefecture-level cities.

2.3.2. Attribution Analysis

Two regression modelling methods, stepwise regression and OLS regression, are used to analyze the changes in the impact of different factors on carbon emissions. The idea is that stepwise regression is first used as an exploratory regression, which can first demonstrate the characteristics of each factor, initially screen the factors, and verify the feasibility of research and analysis of the impact of its selected factors on carbon emissions. Then OLS regression is selected to further analyze the change in influence on carbon emissions between each influence factor and the systematic attribution analysis of carbon emissions, and to obtain a comprehensive conclusion in time and spatial area.

Stepwise regression is a process of screening variables in regression analysis. It allows for the construction of a regression model from a set of candidate variables, with the system automatically identifying influential variables [52]. The stepwise approach to regression was selected, whereby, following the introduction of a variable, the first step is to ascertain whether this variable significantly alters the model (F-test). If this is the case, a t-test is then performed on all variables. The original variable is retained if it is not deemed to be significant due to the introduction of variables at a later stage. If the variable is no longer significant due to the introduction of variables that are added later, it is excluded to ensure that the regression equations contain only significant variables before the introduction of new variables. This process is repeated until there are neither significant explanatory variables selected nor significant explanatory variables in the regression equations. Until the regression equation contains neither significant explanatory variables nor insignificant explanatory variables, an optimal set of variables is finally obtained [53].

The application of stepwise multiple regression necessitates the selection of a model (i.e., the determination of which regressor variables should be included in the final MAM) through the use of parametric inference (i.e., the testing of whether the parameters are significantly different from zero). This process, however, can potentially lead to the introduction of bias in the parameters, the occurrence of overfitting, and the execution of incorrect significance tests. To illustrate this point, a simple example involving a single parameter is presented. A linear model of the observations y_i as a function of the parameters α and β, the predicted values x_i, and the error ε:

$$y_i = \alpha + \beta_{x_i} + \varepsilon_i \quad (1)$$

A stepwise approach can be employed to ascertain whether the model represented by Equation (1) is more efficacious than a simpler alternative:

$$y_i = \alpha + \varepsilon_i \tag{2}$$

A straightforward approach is to calculate an estimation of β (referred to as b) prior to determining whether b is demonstrably different from zero.

In this study, carbon emission was employed as the dependent variable, with economic, demographic, climatic, and land use factors selected as independent variables. Regression analyses were conducted from 2001, 2010, and 2020 to identify significant changes between the three years under different subregions. Additionally, more calculation formulas were referenced from the literature [54–56].

OLS (ordinary least squares) regression is employed primarily for parameter estimation in linear regression [57]. The underlying principle is to identify a value that minimizes the sum of squares of the difference between the actual value and the model valuation, which will be used as the parameter estimate. The optimal function match for the data is determined by minimizing the sum of squares of the error. The least squares method facilitates the identification of unknown data and the minimization of the sum of squares of the errors between these data and the actual data. The least squares method can be employed for curve fitting, and it can also be used to express optimization problems in terms of minimizing energy or maximizing entropy.

Influence factor analysis is conducted based on the least squares linear regression model (OLS), as illustrated in Equation (3):

$$y_i = \beta_0 + \sum_{k=1}^{p} \beta_k x_k + \sum_{j=1}^{p} \beta_j x_j + \varepsilon_{k,j} \tag{3}$$

In Equation (3), y_i represents the carbon emission evaluation index; i denotes a specific evaluation index; x_k denotes an explanatory variable; x_j denotes a control variable; β_0, β_k, and β_j denote the linear regression parameters; and $\varepsilon_{k,j}$ denotes the value of the random error of linear regression.

It has been shown that multicollinearity does not affect the effect before using the OLS regression model. The model was employed with carbon emissions as the dependent variable and economic, demographic, climatic, and land use factors as independent variables. These variables collectively constructed OLS models for urban clusters in different regions at three time points: 2001, 2010, and 2020. The objective of this approach was to illustrate the specific extent of the influence of each factor on regional carbon emissions, both temporally and spatially. Further details on the calculation formulas can be found in References [58,59].

3. Results

3.1. Temporal Evolution of Carbon Emissions in Prefectural-Level Cities across the Country

Overall linear regression and segmented linear regression of trends were used separately for each subregion, with 2010 as the node, and divided into two segments over a 20-year period (2001–2010, 2010–2020) to analyze whether there was significant trend variability in different regions. As illustrated in Figure 3, the changes in carbon emissions across each subregion (Figure 3a–f) demonstrate an upward trajectory from 2001 to 2020. The growth rates of carbon emissions in each subregion were found to be $2.37a^{-1}$ (Figure 3a), $1.25a^{-1}$ (Figure 3b), $1.52a^{-1}$ (Figure 3c), $1.19a^{-1}$ (Figure 3d), and $2.01a^{-1}$ (Figure 3e), respectively, with a flat growth rate of $0.88a^{-1}$ (Figure 3f). The overall growth of carbon emissions in all regions was found to be flat, with an inflection point occurring in the period of 2010–2014. Following this, there were different decreasing trends. The carbon emissions of the Northwest, Southwest, and South Central Regions increased after 2016; also, all three regions have achieved 20 years of sustained increases in average annual carbon emissions by 8.54%, 5.51%, and 6.05%. In contrast, the carbon emissions of the

North, Northeast, and East Regions slowed down or showed a weak slowing down trend. The three regions have declined from their historical highs to 2020 by 3.63%, 6.95%, and 9.92%.

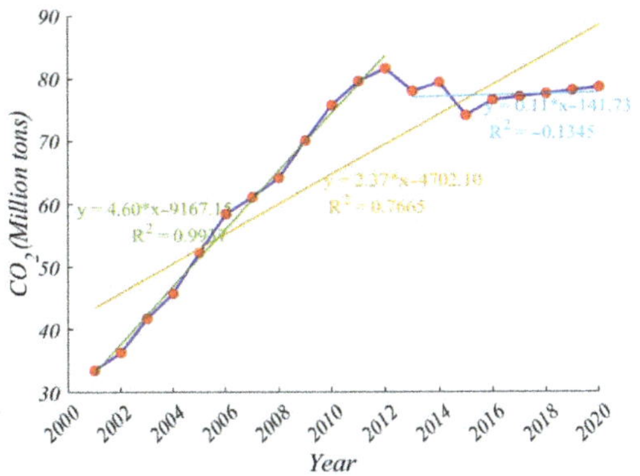

(**a**) Northern Region

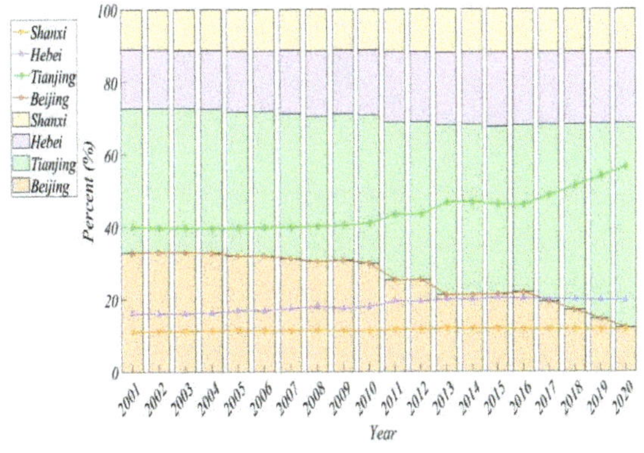

(**g**) Northern Region

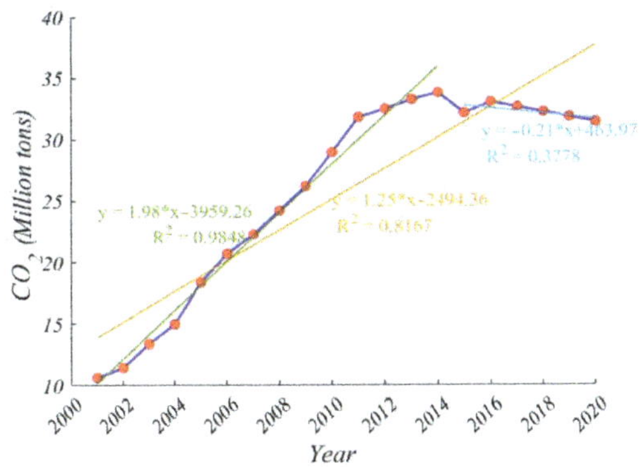

(**b**) Northeast Region

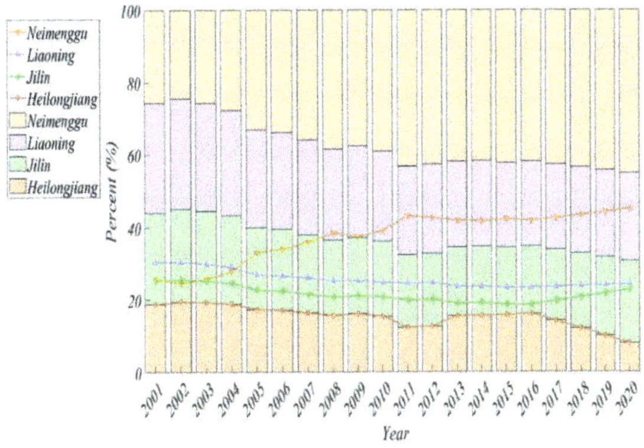

(**h**) Northeast Region

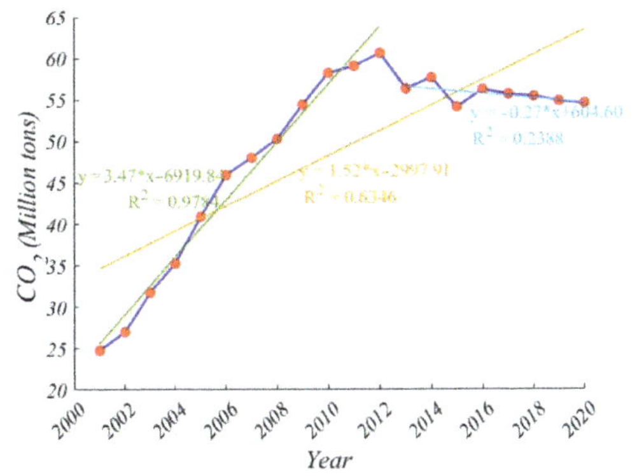

(**c**) Eastern Region

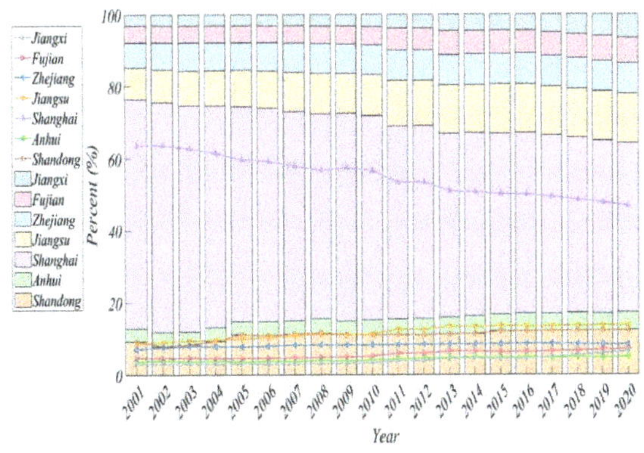

(**i**) Eastern Region

Figure 3. *Cont.*

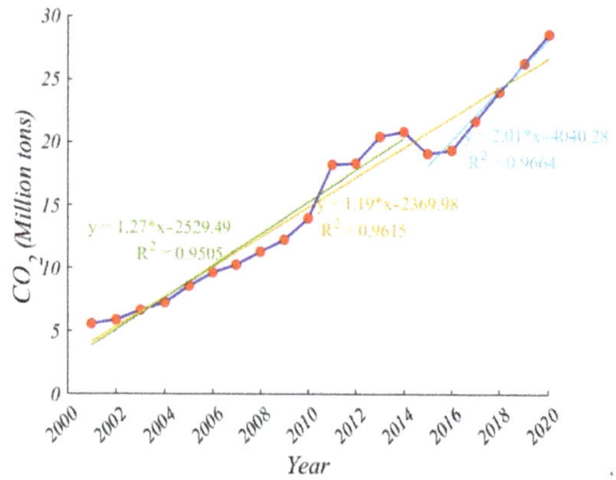

(**d**) Northwest Region

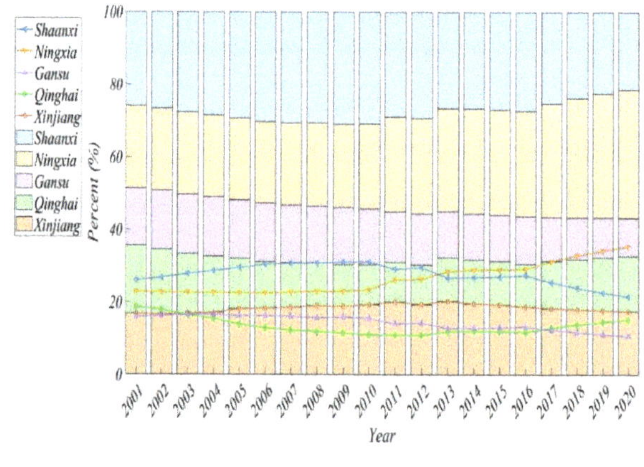

(**j**) Northwest Region

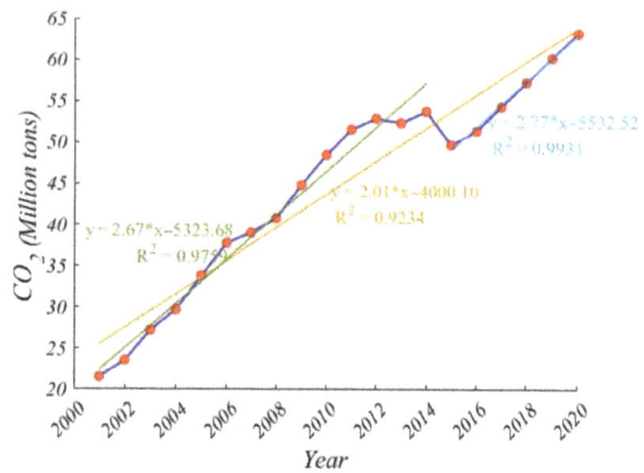

(**e**) Southwest Region

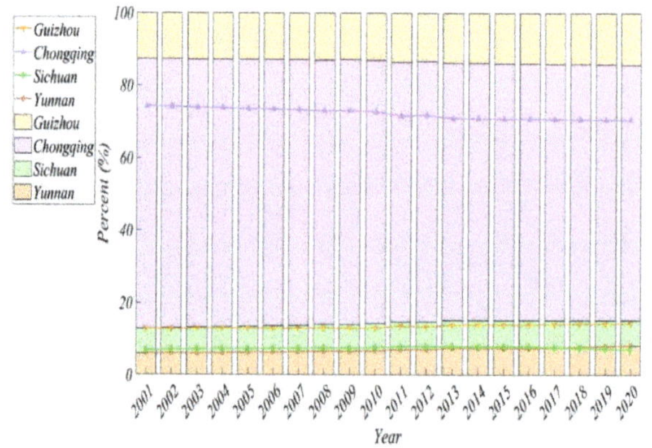

(**k**) Southwest Region

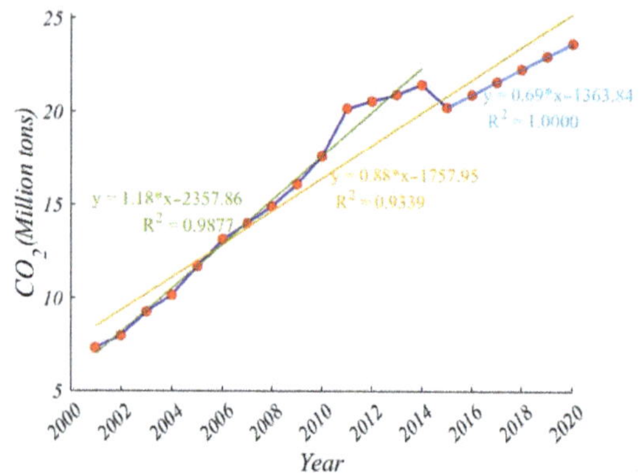

(**f**) South Central Region

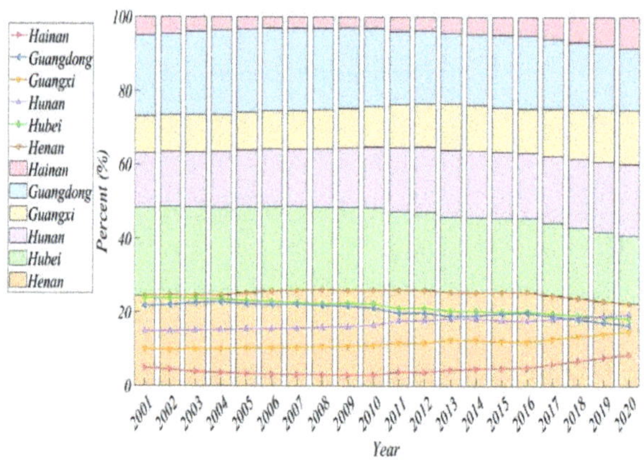

(**l**) South Central Region

Figure 3. Evolutionary trend of carbon emission changes in prefectural cities in China by subregion, 2001–2020. (Note: (**a–f**) show the changes in carbon emissions by subregion from 2001 to 2020; (**g–l**) shows the share of carbon emissions within each subregion from 2001 to 2020).

At the national level, carbon emissions grow from 3020.29 Mt in 2001 to 9169.74 Mt in 2020, with an average annual growth rate of 5.71%. From the perspective of the evolution of carbon emissions' total proportion in prefectural-level cities across different regions, from 2001 to 2020, in the Northern Region, when viewed from the provincial level, Tianjin Municipality (a directly administered municipality) had the highest proportion. Its share of carbon emissions increased from 39.93% to 56.47%. Conversely, Beijing Municipality (a directly administered municipality) exhibited a decline in its proportion, with a gradual decrease from 32.91% to 11.12%. The proportion of carbon emissions in prefectural-level cities within Shanxi Province and Hebei Province remained relatively stable over the 20-year period (Figure 3g). In the Northeast Region, only prefecture-level cities in the Inner Mongolia Autonomous Region continue to increase their share of carbon emissions over the 20-year period, from 25.69% to 45.08%. In contrast, the share of carbon emissions of Heilongjiang, Liaoning, and Jilin Provinces decreases slowly, which is probably related to the national policies and reforms (Figure 3h). In the eastern region, Shanghai (municipality directly under the central government) is the leading economic hub and has the highest share of carbon emissions. However, it has experienced a decline over the past two decades, from 63.56% to 49.60%. In contrast, other prefectural-level cities in other provinces have experienced an increase in their share of carbon emissions. For instance, Jiangsu province has risen from 8.70% to 13.72%, while Shandong province has risen steadily from 9.30% to 12.35% (Figure 3i). In the northwestern region, the carbon emission share of Gansu Province and Qinghai Province exhibits relatively stable fluctuations, while that of Ningxia Province rises from 22.77% to 35.27%, and that of Shaanxi Province and Xinjiang Uygur Autonomous Region rises at a gradual pace before declining at a similar rate (Figure 3j). In the Southwestern Region, the share of carbon emissions in Chongqing Municipality (a directly administered municipality) is the highest, but the change is only 3.84%. This situation is also observed in other provinces, such as within Yunnan Province, which is the highest. A similar situation exists in other provinces, such as Yunnan Province, where the share of carbon emissions of prefecture-level cities only changed by 2.11% (Figure 3k). In the South Central Region, Guangdong Province, the share of carbon emissions of prefecture-level cities decreased from 21.85% to 16.53%, with a smaller downward trend, and a similar change was observed in Hubei Province and Henan Province. Conversely, the proportion of carbon emissions in Hainan Province, Guangxi Zhuang Autonomous Region, and Hunan Province is gradually increasing. For instance, in Hunan Province, the figure has risen from 14.83% to 19.34% (Figure 3l).

3.2. Changes in the Spatial Distribution Pattern and Spatial Correlation of Carbon Emissions in Prefecture-Level Cities across the Country

In order to investigate the spatial dynamics of carbon emissions within the context of national prefecture-level cities, data on carbon emissions for these cities across four time periods (2001–2005, 2006–2010, 2011–2015, and 2016–2020) were processed. The first step is to analyze the spatial autocorrelation of carbon emissions of the national prefecture-level city group. This allows for us to judge whether there is aggregation or isolation of carbon emissions of prefecture-level city groups under the national division in space, and to make a judgement on the aggregation trend in space. The results of the analysis are shown in Table 2.

Table 2. Results of global spatial autocorrelation analysis of carbon emissions of prefecture-level cities in China.

Year	I	Z	*p*
2001–2005	0.207	12.539	<0.05
2006–2010	0.250	14.821	<0.05
2011–2015	0.234	13.685	<0.05
2016–2020	0.202	11.802	<0.05

As illustrated in Table 2, the global spatial autocorrelation analyses of carbon emissions of the national urban agglomerations in 2001–2005, 2006–2010, 2011–2015, and 2016–2020 passed the significance test ($Z > 1.\ 96$, $p < 0.05$), indicating that there is a significant spatial autocorrelation of carbon emissions of the national prefectural agglomerations in the spatial context. Over the past two decades, the I and Z values have exhibited a general trend of increasing and then decreasing, indicating an overall spatial aggregation of carbon emissions among prefecture-level cities across the country. However, over the 20-year period, the highest level of agglomeration is observed in the 2006–2015 period, with the level of agglomeration in the 2016–2020 period being approximately equivalent to that observed in the 2001–2005 period.

In order to gain further insight into the spatial distribution of carbon emissions within the national prefecture-level city cluster, this study selects the carbon emissions of the national prefecture-level city cluster in the four time periods of 2001–2005, 2006–2010, 2011–2015, and 2016–2020 to further analyze the local spatial correlation, as illustrated in Figure 4. On a national scale, the spatial clustering of China's overall carbon emissions over the past 20 years has been relatively stable (Figure 4). The (H-H) high-concentration area refers to the region with high carbon emissions and its neighboring municipalities, which are mainly concentrated in the Eastern and Northern Regions, including Beijing, Tianjin, and Hebei, and Shandong, Jiangsu, Zhejiang, and Shanghai, while the (H-L) high- and low-concentration area refers to the region having high carbon emissions, but its neighboring municipalities having low carbon emissions, which is consistently reflected in Chongqing, Chengdu, and Kunming in the Southwestern Region, Lanzhou in the Northwestern Region, Nanning in the South Central Region, and Nanchang in the Eastern Region; Kashgar changes from a low–low to a high–low agglomeration in the period 2016–2020, Xi'an is of this type in the period 2006–2010, and the rest are non-significant areas, as is Panzhihua in the period 2001–2005; Wuhan changes from a non-significant area at the beginning to a persistent high–low agglomeration, and Changsha, on the other hand. (L-H) Low–high agglomerations, where carbon emissions are low in the region but high in the neighboring prefectures, are embedded in high-value agglomerations with consistently low emissions compared to neighboring high-value cities, such as Yangquan, Shuozhou, and Chengde in the north; Chaoyang City in the northeast; Rizhao and Huzhou City in the east; and Puyang City in the South Central Region. In this 20-year period, there are cities that have changed from insignificant regions to this type, such as Daxinganling and Jiaozuo; there are also prefecture-level cities that have changed in the opposite direction, such as Maanshan and Xuancheng; and there are also cities that have changed to high-value agglomerations, such as Ulanqab and Chuzhou, etc. The (L-L) low-value agglomeration area is defined as an area with low carbon emissions and its neighboring prefectures. It is primarily located in the Western Region of China in a linear shape around several prefectures and cities, specifically in the Southwestern Region, around Kunming City, Chengdu City distribution; in the South Central Region is the distribution of Nanning City as the center; lastly, it is dispersed with Lanzhou City serving as the center in the northwest. During the period 2016–2020, there was a proliferation of low-value agglomerations of prefecture-level cities in the northeast of the Northeast Region (the east of Heilongjiang Province). These included Jixi, Jiamusi, and Shuangyashan. In addition, the low-value prefecture-level city area that used to be centered on Nanchang City is also diminishing over time, becoming a non-significant area.

In light of the findings of the local autocorrelation analysis presented in the above section, the carbon emission data of the national prefecture-level cities in the four time periods of integrated 2001–2005, 2006–2010, 2011–2015, and 2016–2020 are selected to further calculate the Getis–Ord Gi* statistic. This allows for the identification of high and low agglomerations of values within each geographic unit and its domains, which in turn enables the determination of spatiotemporal hotspots (Table 3 and Figure 5). The spatial distribution of carbon emission cold and hotspots exhibits aggregation characteristics, with the majority of the hotspots concentrated at the municipal scale. To analyze the national carbon emissions in four segments from 2001 to 2020, the hotspot area spreads from Beijing,

Tianjin, and Hebei to coastal cities such as Jiangsu, Zhejiang, and Shanghai. These cities are considered the overall core, with the surrounding prefectures radiating outward. In contrast, the hot spot range in the Eastern Region has been reduced in the period from 2016 to 2020. Conversely, the cold spot area is centered on the prefectures in the Northwest Region of Qinghai and Gansu Province, and gradually spreads to the Southwest Region.

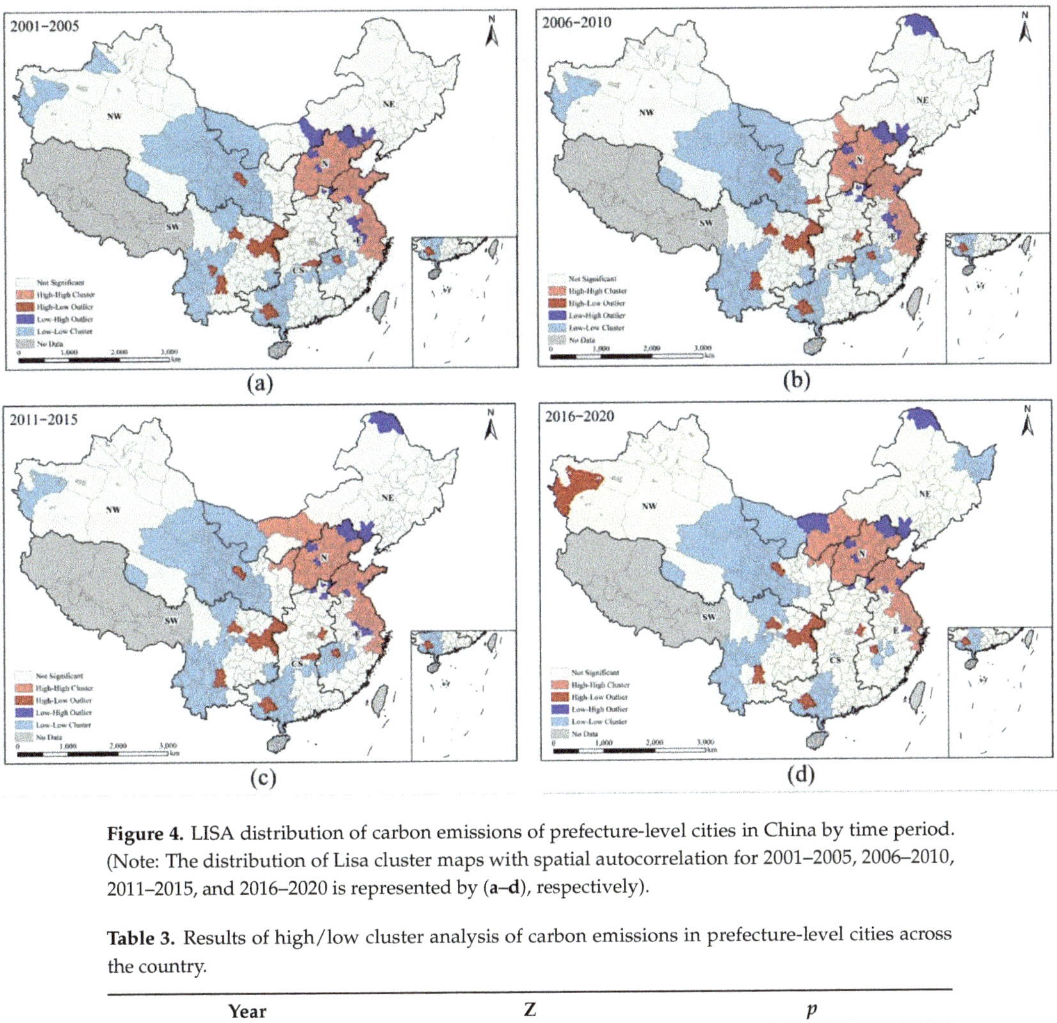

Figure 4. LISA distribution of carbon emissions of prefecture-level cities in China by time period. (Note: The distribution of Lisa cluster maps with spatial autocorrelation for 2001–2005, 2006–2010, 2011–2015, and 2016–2020 is represented by (**a**–**d**), respectively).

Table 3. Results of high/low cluster analysis of carbon emissions in prefecture-level cities across the country.

Year	Z	p
2001–2005	7.470	<0.05
2006–2010	7.911	<0.05
2011–2015	6.851	<0.05
2016–2020	5.908	<0.05

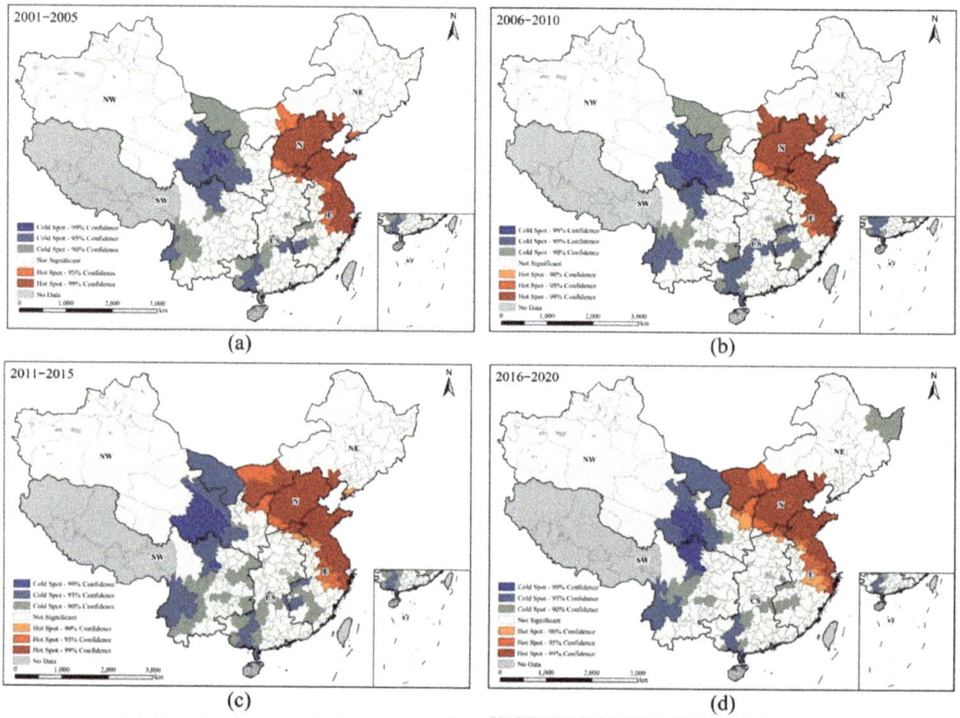

Figure 5. Distribution of carbon emission cold/hot spots in prefecture-level cities across the country by time period. (Note: The distribution of cold hotspots for 2001–2005, 2006–2010, 2011–2015, and 2016–2020 and the degree of rank are represented by (**a**–**d**), respectively).

3.3. Analysis of the Influential Factors of Carbon Emission in Prefecture-Level Cities in China

3.3.1. Stepwise Regression Exploratory Analysis with Regression Analysis of Carbon Emission Impact Factors and Model Fitting Analysis

Stepwise regression can assist in determining the significance of the factors influencing carbon emissions in prefecture-level cities within each region, while also evaluating the feasibility of analyzing the impact factors. The results are presented in Figure 6, which compares the significance changes of the factors in 2001, 2010, and 2020.

The analysis of CO_2 emissions, as detailed in Figure 6, employs stepwise regression to discern the impacts of various influencing factors across different regions over the years 2001, 2010, and 2020. The findings consistently show that GDP and population have a significant positive impact on CO_2 emissions across all regions. In contrast, the effects of climatic factors and land use ratios are marked by complexity and instability, reflecting their less predictable influence on emissions.

The OLS regression model was employed on three occasions, in 2001, 2010, and 2020, in order to further ascertain the impact of economic, demographic, climatic, and land use factors on carbon emissions. Moreover, the absence of multicollinearity between the influencing factors, as indicated by a VIF of less than 10, was demonstrated prior to the application of the OLS analysis (see Appendix A, Table A1). The results of the calculations are presented in Table 4.

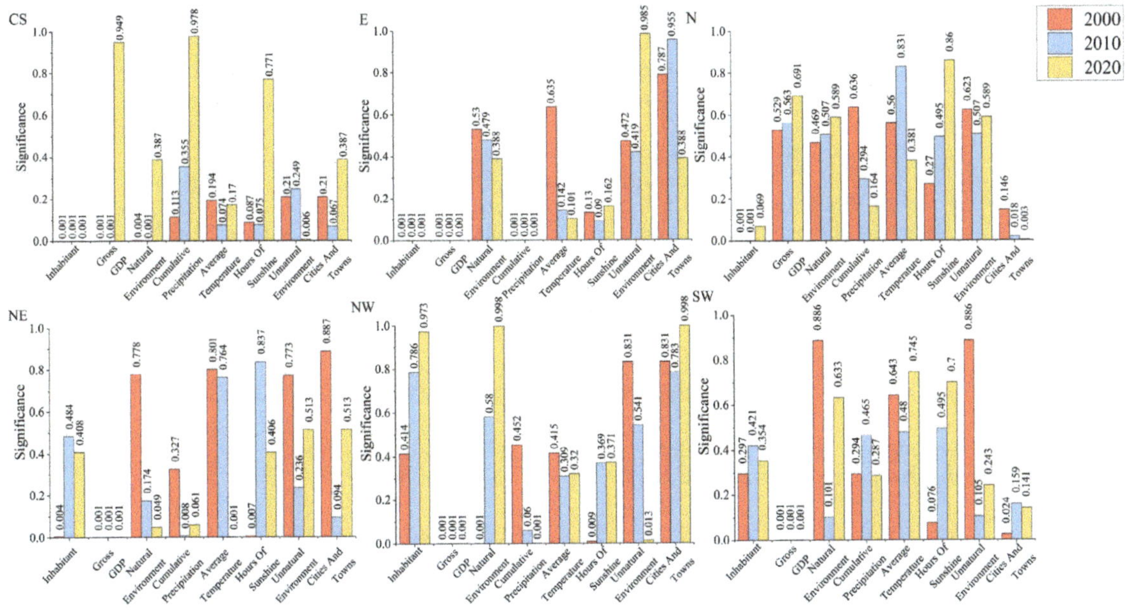

Figure 6. Stepwise regression results of carbon emission influencing factors in sub-districts of prefecture-level cities nationwide.

Table 4. Overall OLS regression results of carbon emission influencing factors in prefecture-level cities across the country (Note: the coefficients and *p*-values (*replaced) explain the degree of influence and significance of each independent variable on the dependent variable; the t-values in parentheses reflect the difference between the coefficient estimates and zero, divided by the standard error, the absolute value of which also reflects the degree of significance of the coefficients; the R^2 measures the degree of fit of the model, ranging from 0 to 1, and indicates the percentage of variation in the dependent variable that the model explains; and the F-statistic is used to test the significance of the entire model, * $p < 0.1$, ** $p < 0.05$, *** $p < 0.01$).

	2001 CO_2	**2010** CO_2	**2020** CO_2
lnGDP	3.020 ***	8.797 ***	6.051 ***
	(6.24)	(8.58)	(5.59)
Inhabitants	0.019 ***	0.041 ***	0.040 ***
	(11.29)	(13.40)	(12.32)
Cumulative precipitation	0.001	−0.004 **	−0.010 ***
	(0.51)	(−2.22)	(−4.63)
Average temperatures	−0.413 ***	−0.571 ***	−0.699 ***
	(−3.16)	(−2.95)	(−3.10)
Hours of sunshine	0.001	0.004 *	−0.000
	(1.02)	(1.93)	(−0.07)
Cities and towns	47.307 ***	52.154 ***	14.618
	(4.95)	(3.54)	(1.19)
Natural environment	6.328 ***	11.737 ***	4.271
	(3.81)	(3.59)	(1.07)
_cons	−43.636 ***	−136.963 ***	−70.169 ***
	(−6.11)	(−8.54)	(−4.02)
N	329.000	329.000	329.000
R^2	0.692	0.781	0.640
F	102.954	163.736	81.480

Table 4 illustrates the exclusion of the indicator non-natural environment from the analysis, due to its high covariance characteristics with the other variables. This table clearly shows the impact of the different factors on carbon emissions at the three points in time. The data from 2001, 2010, and 2020 indicates that GDP has a significant positive effect on CO_2. Furthermore, the indicator is the largest in 2010, and all of them are significant at the 1% level. In terms of the resident population, the larger the resident population, the higher the CO_2 emission. This evidence supports the hypothesis that the resident population exerts a positive and significant effect on CO_2, with the highest level observed in the 2010 data. In terms of other variables, cumulative precipitation has a negative effect on CO_2 emissions in 2010 and 2020; average temperature has a negative effect in all three years; sunshine hours has a positive effect in 2010 but no significant effect in 2020; urbanization and natural environment have unstable effects on CO_2 emissions, with urbanization being significant in 2001 and 2010, and natural environment having a significant positive effect in all three years. Furthermore, the effects of urbanization and the natural environment on CO_2 emissions can be elucidated by examining the scatterplots presented in Appendix B, Figures A1–A3. These plots illustrate the linear relationship between each factor and carbon emissions, thereby providing a more comprehensive understanding of the overall regression results.

3.3.2. Quantitative OLS Regression Analysis of Local Carbon Emission Impact Factors

In the above section, stepwise regression and OLS regression analyses were conducted using two dummy variables, year and region. This allowed for an overall analysis of the data at the national level over time. In this section, the carbon emission impact factor of prefectural cities at the subregion are calculated and analyzed through OLS regression at the whole and individual level over three years. The specific results are presented in Tables 5–8.

Table 5. Localized regression results of carbon emission influencing factors in prefecture-level cities across China. (Note: * $p < 0.1$, ** $p < 0.05$, *** $p < 0.01$).

	CS CO_2	E CO_2	N CO_2	NE CO_2	NW CO_2	SW CO_2
lnGDP	7.739 ***	6.881 ***	14.375 ***	13.926 ***	5.795 ***	5.223 ***
	(13.85)	(7.21)	(3.28)	(11.94)	(8.67)	(6.22)
Inhabitants	0.015 ***	0.048 ***	0.007	0.011 *	0.013 *	0.037 ***
	(6.28)	(15.02)	(0.71)	(1.68)	(1.68)	(15.84)
Cumulative precipitation	0.001	0.002	−0.006	−0.019 ***	−0.016 ***	0.010 **
	(0.81)	(0.85)	(−0.15)	(−4.44)	(−2.85)	(2.23)
Average temperatures	−0.420 *	−0.849 *	0.803	0.693	−0.542	−0.497
	(−1.80)	(−1.78)	(0.50)	(1.50)	(−1.50)	(−1.31)
Hours of sunshine	−0.000	−0.005 **	−0.005	−0.002	−0.000	0.001
	(−0.37)	(−2.16)	(−0.92)	(−1.00)	(−0.12)	(0.43)
Cities and towns	0.897	69.024 ***	48.417	−24.693	82.134 *	27.129 *
	(0.10)	(3.68)	(0.87)	(−0.86)	(1.87)	(1.90)
Natural environment	−4.462	4.686	−5.899	11.623 ***	−2.646	13.568 ***
	(−1.50)	(0.96)	(−0.34)	(2.61)	(−0.39)	(3.34)
_cons	−102.701 ***	−98.020 ***	−192.608 ***	−188.044 ***	−60.016 ***	−91.540 ***
	(−11.63)	(−6.65)	(−3.19)	(−10.17)	(−4.27)	(−6.46)
N	246.000	231.000	72.000	144.000	153.000	141.000
R^2	0.746	0.814	0.652	0.691	0.509	0.818
F	100.027	139.314	17.114	43.511	21.514	85.543

Specifically, Table 5 illustrates the regional impacts. In the Northern Region, economic growth and population size significantly increase CO_2 emissions. Climatic factors such as cumulative precipitation, average temperature, and sunshine hours show no significant effects. Urbanization and the natural environment also exert minimal and statistically insignificant impacts. Similar trends are observed in the Northeast and Northwest Regions, although cumulative precipitation reduces emissions in the Northwest during 2010. In the Southwest, while economic drivers and population continue to increase emissions, climatic effects remain inconsistent, and urbanization along with natural environmental factors display significant impacts under specific conditions.

Table 6. Localized regression results of carbon emission impact factors in prefecture-level cities across the country in 2001. (Note: * $p < 0.1$, ** $p < 0.05$, *** $p < 0.01$).

	CS CO_2	E CO_2	N CO_2	NE CO_2	NW CO_2	SW CO_2
lnGDP	5.707 ***	5.044 ***	6.427	3.935 ***	2.532 ***	1.924
	(7.02)	(3.72)	(1.47)	(3.91)	(5.43)	(1.51)
Inhabitants	0.002	0.019 ***	0.014	0.012 **	0.002	0.020 ***
	(0.87)	(4.38)	(1.49)	(2.58)	(0.48)	(8.32)
Cumulative precipitation	−0.003	0.004	0.020	−0.012 *	0.012 *	0.010 **
	(−1.56)	(0.70)	(0.65)	(−1.88)	(1.99)	(2.40)
Average temperatures	0.594	−1.283	−4.793 **	0.604 **	0.086	−0.899 **
	(1.39)	(−1.61)	(−2.44)	(2.24)	(0.45)	(−2.06)
Hours of sunshine	0.001	−0.026 ***	−0.048 **	−0.010	0.007 **	0.002
	(0.66)	(−3.88)	(−2.86)	(−1.59)	(2.56)	(1.05)
Cities and towns	15.203 *	118.095 ***	58.419	20.199	97.147 ***	352.002 **
	(1.69)	(3.80)	(0.92)	(0.92)	(2.93)	(2.31)
Natural environment	0.223	−0.854	2.810	2.115	−1.232	4.804
	(0.10)	(−0.14)	(0.24)	(0.83)	(−0.50)	(1.08)
_cons	−86.836 ***	−15.311	74.311	−24.002	−52.126 ***	−32.741 *
	(−6.84)	(−0.66)	(1.12)	(−0.98)	(−4.15)	(−1.97)
N	82.000	77.000	24.000	48.000	51.000	47.000
R^2	0.761	0.759	0.865	0.780	0.733	0.851
F	33.617	31.047	14.708	20.256	16.830	31.897

Table 7. Localized regression results of carbon emission impact factors in prefecture-level cities across the country in 2010. (Note: * $p < 0.1$, ** $p < 0.05$, *** $p < 0.01$).

	CS CO_2	E CO_2	N CO_2	NE CO_2	NW CO_2	SW CO_2
lnGDP	12.903 ***	11.230 ***	21.805 *	17.117 ***	7.219 ***	6.500 **
	(6.81)	(4.26)	(2.02)	(5.83)	(5.16)	(2.26)
Inhabitant	0.011 **	0.052 ***	0.003	0.010	0.008	0.039 ***
	(2.01)	(7.64)	(0.16)	(0.86)	(0.73)	(8.13)
Cumulative precipitation	−0.000	0.002	−0.057	0.010	−0.015	0.024 **
	(−0.05)	(0.28)	(−1.15)	(0.88)	(−1.24)	(2.29)
Average temperatures	−0.513	−2.523 **	1.469	−0.914	−0.200	−1.574 *
	(−1.17)	(−2.10)	(0.36)	(−0.95)	(−0.35)	(−2.02)
Hours of sunshine	0.009	−0.027 *	0.000	0.031 **	−0.006	0.010 **
	(1.62)	(−1.79)	(0.00)	(2.12)	(−0.65)	(2.07)
Cities and towns	−5.158	123.778 ***	75.759	−24.589	56.176	475.451 ***
	(−0.36)	(2.90)	(0.70)	(−0.49)	(0.89)	(3.22)
Natural environment	−0.464	6.654	0.352	10.280	2.735	13.056
	(−0.09)	(0.71)	(0.01)	(1.42)	(0.35)	(1.54)
_cons	−195.891 ***	−103.644 *	−307.016	−328.507 ***	−76.690 **	−124.174 ***
	(−6.32)	(−1.89)	(−1.60)	(−5.64)	(−2.32)	(−2.72)
N	82.000	77.000	24.000	48.000	51.000	47.000
R^2	0.811	0.867	0.837	0.745	0.667	0.898
F	45.413	64.401	11.777	16.672	12.280	49.160

Table 8. Localized regression results of carbon emission impact factors in prefecture-level cities across the country in 2020. (Note: * $p < 0.1$, ** $p < 0.05$, *** $p < 0.01$).

	CS CO_2	E CO_2	N CO_2	NE CO_2	NW CO_2	SW CO_2
lnGDP	12.053 ***	7.406 ***	40.444 *	24.397 ***	3.012 *	5.875
	(4.09)	(3.06)	(1.93)	(4.90)	(2.00)	(1.59)
Inhabitant	0.014 *	0.051 ***	−0.033	−0.008	0.032 **	0.045 ***
	(1.99)	(9.49)	(−0.93)	(−0.41)	(2.12)	(7.83)
Cumulative precipitation	−0.003	0.003	0.403	−0.024 **	−0.034 ***	0.008
	(−0.83)	(0.86)	(1.50)	(−2.50)	(−3.24)	(0.99)
Average temperatures	−0.197	−0.642	−5.532	0.436	0.198	−0.071
	(−0.52)	(−1.29)	(−0.91)	(0.46)	(0.25)	(−0.13)
Hours of sunshine	0.001	−0.001	−0.006	−0.001	0.004	0.000
	(0.29)	(−0.44)	(−0.60)	(−0.49)	(1.01)	(0.03)
Cities and towns	−28.271 *	38.433 *	−114.774	−58.302	−1.599	29.424 *
	(−1.83)	(1.92)	(−1.04)	(−1.24)	(−0.02)	(1.87)
Natural environment	−9.252	−2.852	−91.085 *	13.293	−25.227	18.021 **
	(−1.44)	(−0.47)	(−1.95)	(1.46)	(−1.42)	(2.41)
_cons	−171.724 ***	−116.523 ***	−714.300 *	−348.010 ***	−5.144	−113.062 *
	(−3.53)	(−2.79)	(−2.07)	(−4.51)	(−0.16)	(−1.91)
N	82.000	77.000	24.000	48.000	51.000	47.000
R^2	0.707	0.891	0.620	0.729	0.463	0.892
F	25.487	80.350	3.725	15.374	5.306	46.033

Table 6 elaborates on the partial regression results from 2001, where economic development consistently elevates CO_2 emissions, often reaching statistical significance at the 1% or 5% levels. The population's effect is similarly significant, especially in the Eastern and Southwest Regions, which are notable at the 1% significance level. However, the influence of cumulative precipitation and sunshine hours varies, showing both positive and negative impacts that are not always statistically significant. The variability extends to urbanization and natural environment effects, which range from positive to negative, with no consistent pattern.

Further analysis in Table 7 indicates that the coefficients for economic development in 2010 are considerably larger than those in 2001, with GDP showing a significant positive effect across all regions, predominantly at the 1% significance level. The influence of population is also more pronounced in 2010, with significant impacts in multiple regions.

Finally, Table 8 focuses on the 2020 data, revealing that economic development remains a major driver of emissions in the Northern Region, maintaining the trends observed in earlier years. The impact of population and non-natural environmental factors in 2020 is also more substantial compared to previous assessments. Notably, urbanization in the CS region shows a significant negative correlation with emissions, diverging from earlier data, while the natural environment's role in reducing emissions becomes more evident, particularly with its significant negative effect in the Northern Region and a significant positive effect in the Southwest.

4. Discussion

In terms of the changes in carbon emissions and the share of each region between 2001 and 2020, it can be observed that the more developed economy in the Eastern Region of China has led to a faster growth in carbon emissions, which reached its peak and then gradually slowed down in accordance with the requirements of the national situation [60]. On the other hand, in the Northwest, Southwest, and South Central Regions, where more cities with relatively backward economies are struggling to develop, all have achieved 20 years of sustained upward mobility, and their average annual carbon emissions have increased by 8.54%, 5.51%, and 6.05%. This is exemplified by the phenomenon of urbanization, which is occurring at a more rapid pace in these cities, and the pursuit of economic growth in order to catch up with the relatively more developed regions. In addition, over this 20-year period, carbon emissions have made overall progress in an upward trend. The carbon emissions from the country's prefectural cities continue to grow, reflecting the country's rising level of development and the gradual realization of a carbon neutrality policy on that basis. In analyzing the changes in China's carbon emissions, it is also possible to add to and compare the content of studies involving global carbon emissions [61]. The continued growth of carbon emissions in China as a developing country is different from the changes in carbon emissions in the developed countries of Europe, where we can see that Europe is undergoing an energy transition to meet its carbon reduction targets, which the more developed regions of eastern China could emulate [62]. Moreover, the cross-border impacts of carbon emissions in China are more related to economic trade and ecological changes. For example, cross-border co-operation will help EU exporters to mitigate the adverse impacts of the EU Carbon Emission Rights Act and play a key role in global coordination of emission reduction actions. In addition, in the long run, in regions with fragile environments and high per capita incomes, high carbon emission industries are not conducive to the development of the regional economy and industry and the introduction of foreign investment by the regional government, and this will force the regional government to pay attention to environmental regulation and promote the long term sustainable development of the local industry [63,64]. The carbon emissions vary and change at the provincial scale in different regions, which in turn is closely related to factors such as the level of economic development, policy implementation, and industrial structure of the specific region.

Over time, the carbon emissions of prefecture-level city clusters in China have exhibited a clear spatial autocorrelation, with an overall clustering trend. The clustering of carbon emission cold and hotspots in China has been expanding and becoming more tightly clustered simultaneously. Conversely, the addition or reduction of cold and hotspots can also demonstrate the differences and continuity of carbon emissions within the country. The types of carbon emissions clustering in different regions have varied over time, but on the whole they show a certain degree of stability. The absence of H-H type high-value agglomerations in cities in the Pearl River Delta region may be related to the decentralized industrialization and urbanization in the region [65]. Meanwhile, the distribution of carbon emissions in China exhibits an east–west dichotomy, exemplified by the spatial pattern of "hot in the east and cold in the west". The concentration of hot spots is evident in the eastern coastal areas, while cold spots are primarily distributed in the Northwest to Southwest Regions. However, the scope of their agglomeration is expanding and becoming closer simultaneously. At this juncture, the spatial distribution of carbon emissions in China exhibits a growing tendency towards positive autocorrelation. Low-carbon emission cities should prioritize addressing their own challenges, leveraging the insights of neighboring efficient cities, driving the restructuring of their own economy and industry, proactively exploring avenues for carbon emission reduction, and enhancing the efficiency of carbon emission. This is consistent with the findings of Huang et al. [65], which help to elucidate the spatial and temporal CO_2 distribution of carbon emissions in China and provide a scientific basis for the formulation of targeted carbon emission reduction policies.

In terms of influencing factors, economic and population growth can lead to significant increases in carbon emissions. The increase in carbon emissions due to economic development can be derived from changes in affluence or technological development [66]. On the other hand, the impacts behind population growth can be understood from other studies, where population growth, through its impact on demographic processes such as urbanization, population density, age structure, and household size, has an indirect impact on emissions/energy consumption [67]. Taken together, it is found that the economy and population are inextricably linked, implicating the level of consumption as well as the impact of GDP per capita, with an increase in population leading to an increase in consumption demand, which exacerbates the level of carbon dioxide emissions from both production and consumption [68]. The influence of climate factors and urbanization on CO_2 emissions is more complex and unstable. In addition to local policy reasons, there are also a number of unknowns that may affect carbon emissions. Further research is therefore needed to understand the mechanism of its influence. However, with the steady development of the economy and the current slowdown in China's population growth, China's carbon emissions in the future will be more successful in realizing the established national policy requirements.

The analysis of carbon emission impact factors in the national subregional prefecture-level cities revealed that, although there are differences between different regions, the general trend is that the impact of economic development and population growth on CO_2 emissions is generally positive. At the same time, some studies also pointed out [69] that its effect has weakened over time, but economic growth is the most critical factor driving the growth of carbon emissions. From 2005–2010, 2010–2015, and 2015–2020, the carbon emissions driven by economic growth are 5835.51 metric tons, 4735.38 metric tons, and 3137.13 metric tons, respectively. Population growth plays a relatively limited role, contributing 203.48 Mt, 355.45 Mt, and 278.71 Mt in 2005–2010, 2010–2015, and 2015–2020, respectively. The industrial structure of the NWT may favor energy-intensive industries, such as heavy industry or coal mining, which typically generate significant CO_2 emissions [70]. Therefore, with economic expansion and population growth, the expansion of these industries will directly lead to an increase in CO_2 emissions. The significant negative effect of cumulative precipitation may be due to the fact that higher precipitation reduces the operational efficiency of industries such as thermal power plants, which reduces CO_2 emissions. Additionally, higher precipitation may also promote vegetation growth that absorbs carbon dioxide, thereby reducing atmospheric CO_2 concentrations. There are some differences between different regions that need to be further studied and explored.

The regression analyses from 2001 to 2020 reveal a consistent correlation between economic development, population growth, and increased CO_2 emissions, especially in the E and SW Regions. This view is consistent with the findings of previous studies that urbanization leads to the migration of rural populations, which provides human resources for urban development, but also generates large amounts of carbon emissions [71,72]. Moreover, this trend underscores the urgent need for integrated policies that simultaneously address economic expansion and environmental sustainability. The variable impacts of climatic factors on emissions highlight the complex interactions within environmental systems and the necessity for models to better incorporate regional climatic variations for more effective emission management. Urbanization's growing influence on emissions emphasizes the dual challenge of fostering urban growth while minimizing environmental degradation. Promoting sustainable urban practices, such as green consumption and support for eco-friendly industrial transformations, is crucial. The significant economic changes in Region N in 2020, driven by the introduction of new industries, underscore the profound environmental impacts of economic shifts. Additionally, the dynamic interplay between urban planning and demographic changes, particularly in the CS region as discussed by Xu et al. [73], calls for a reevaluation of urbanization strategies to align with sustainable development goals. Moreover, the importance of region-specific approaches, such as those adopted in the Southwest to enhance ecological protection and land use efficiency [74], illustrates the need for adaptive strategies that respect local economic, political, and climatic conditions. This holistic understanding can inform policymakers in designing strategies that not only promote economic growth but also ensure environmental preservation.

These findings underscore the dominant influence of economic activities and demographic growth on CO_2 emissions, while highlighting the variable and often unpredictable effects of climatic and environmental factors across different regions and time periods.

5. Conclusions

In this paper, we analyzed and evaluated CO_2 emissions at the scale of prefecture-level administrative units in China. This study initially examined the changes in carbon emissions and discovered that carbon emissions are rapidly increasing in the economically developed Eastern Region of China, while the less developed Western Region of the country is catching up through accelerated urbanization and economic growth; the spatial and temporal distribution of carbon emissions exhibits a pattern of "hot in the east and cold in the west". Economic expansion and population growth remain the main drivers of carbon emissions growth in each region, while the effects of climatic factors and urbanization are complex and volatile. In the future, the Eastern Region will require a focus on the issues of urbanization progress and population efficiency, while achieving a gradual energy transition. In contrast, the Western Region will need to strengthen ecological protection and improve land use efficiency in order to balance the environmental change caused by carbon emissions. The results and relevant conclusions can serve as a foundation or offer recommendations for China's regional carbon policy formulation and modification.

The limited availability of data poses a significant challenge for the refinement and effectiveness of evaluation index systems, since the carbon emissions data used in this study are the sum of energy-related CO_2 emissions and CO_2 emissions/sequestration from the land use sector, excluding non-CO_2 greenhouse gases, etc. In future studies, the classification and detailed localization of data could greatly enhance the foundational datasets, thereby improving the robustness and relevance of the index system. Such enhancements could involve categorizing sources of carbon emissions, documenting local policy shifts in specific years, and conducting targeted research on climate data. Additionally, expanding the analysis to include more comprehensive comparisons across these variables could significantly enhance our understanding of the factors influencing carbon emissions. This approach would facilitate the identification of critical links that drive regional and temporal differences in emission patterns, providing valuable insights for targeted environmental policy and action.

Author Contributions: Conceptualization, M.S. and B.L.; Methodology, M.S.; Software, M.S.; Validation, M.S.; Formal analysis, M.S.; Investigation, Y.Z.; Resources, X.M. and Y.Z.; Data curation, X.M.; Writing—original draft, M.S.; Writing—review & editing, B.L. and Z.W.; Visualization, M.S.; Supervision, B.L. and Z.W.; Project administration, B.L.; Funding acquisition, J.W. All authors have read and agreed to the published version of the manuscript.

Funding: This paper has been funded by National Natural Science Foundation of China (42330507) and College Student Innovation and Entrepreneurship Training Program (202310022016).

Data Availability Statement: The raw data supporting the conclusions of this article will be made available by the authors on request.

Conflicts of Interest: The authors declare no conflict of interest.

Appendix A

Table A1. VIF certificate.

2001		2010		2020	
VIF	1/VIF	VIF	1/VIF	VIF	1/VIF
4.960	0.202	3.190	0.313	2.480	0.403
4.120	0.242	3.090	0.324	2.170	0.460
3.580	0.279	2.970	0.337	2.090	0.478
2.970	0.336	2.910	0.344	2.060	0.486
2.940	0.340	2.800	0.357	1.990	0.503
2.650	0.378	2.520	0.397	1.930	0.519
2.390	0.419	2.470	0.404	1.050	0.950
3.370		2.850		1.970	

Appendix B

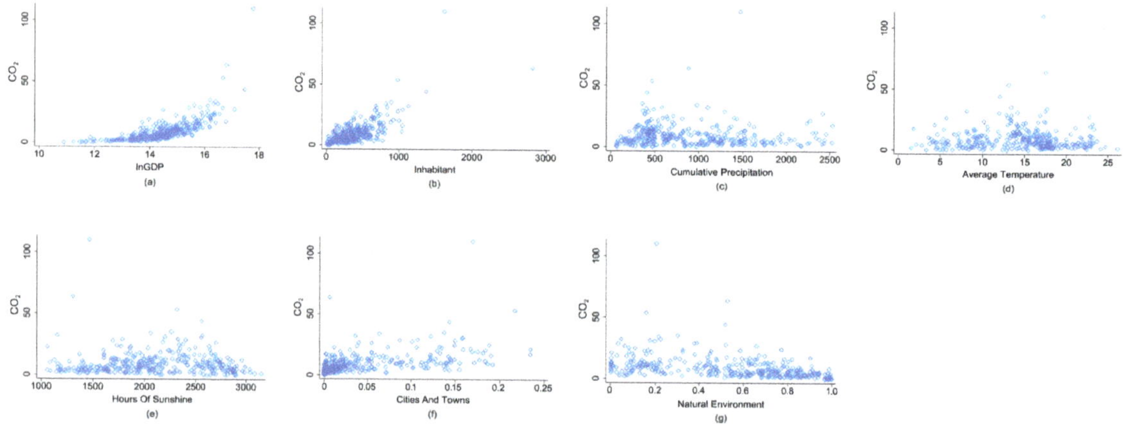

Figure A1. Scatterplot of regression of factors affecting carbon emissions in prefecture-level cities nationwide in 2001. (Note: Scatterplot of linear relationship between factors and carbon emissions, the following data has been collated for the purposes of this study: CO_2 (million tons), lnGDP (million yuan), Inhabitant (ten thousand people), Cumulative precipitation (mm), Average temperature (degrees Celsius °C), Hours of sunshine (hours), Percentage of cities and towns (%), Percentage of natural environment (%)).

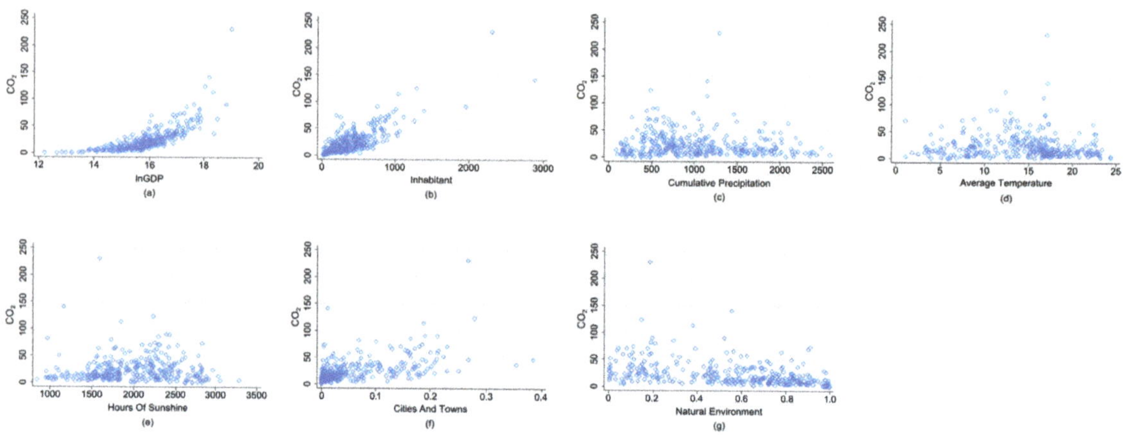

Figure A2. Scatterplot of regression of factors influencing carbon emissions in prefecture-level cities across China in 2010. (Note: Scatterplot of linear relationship between factors and carbon emissions, the following data has been collated for the purposes of this study: CO_2 (million tons), lnGDP (million yuan), Inhabitant (ten thousand people), Cumulative precipitation (mm), Average temperature (degrees Celsius °C), Hours of sunshine (hours), Percentage of cities and towns (%), Percentage of natural environment (%)).

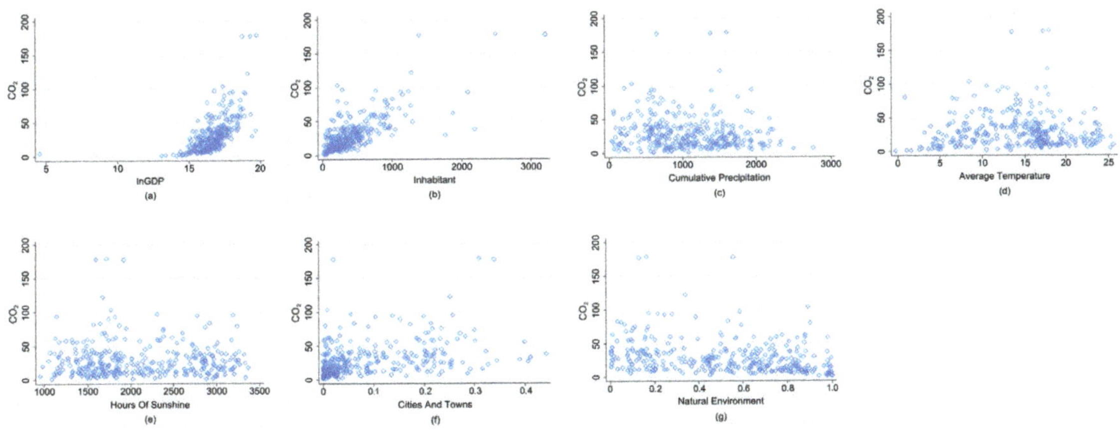

Figure A3. Scatterplot of regression of factors affecting carbon emissions in prefecture-level cities across China in 2020. (Note: Scatterplot of linear relationship between factors and carbon emissions, the following data has been collated for the purposes of this study: CO_2 (million tons), lnGDP (million yuan), Inhabitant (ten thousand people), Cumulative precipitation (mm), Average temperature (degrees Celsius °C), Hours of sunshine (hours), Percentage of cities and towns (%), Percentage of natural environment (%)).

References

1. Ma, R.; Abid, N.; Yang, S. From crisis to resilience: Strengthening climate action in OECD countries through environmental policy and energy transition. *Environ. Sci. Pollut. Res.* **2023**, *30*, 115480–115495. [CrossRef] [PubMed]
2. Newman, R.; Noy, I. The global costs of extreme weather that are attributable to climate change. *J. Nat. Commun.* **2023**, *14*, 6103. [CrossRef] [PubMed]
3. Butsch, C.; Beckers, L.M.; Nilson, E.; Frassl, M.; Brennholt, N.; Kwiatkowski, R.; Söder, M. Health impacts of extreme weather events–Cascading risks in a changing climate. *J. Health Monit.* **2023**, *8* (Suppl. 4), 33. [CrossRef] [PubMed]
4. IPCC. *Climate Change 2013: The Physical Science Basis Working Group I Contribution to the Fifth Assessment Report of the Intergovernmental Panel on Climate Change*; Cambridge University Press: Cambridge, UK, 2014.
5. Huisingh, D.; Zhang, Z.; Moore, J.C.; Qiao, Q.; Li, Q. Recent advances in carbon emissions reduction: Policies, technologies, monitoring, assessment and modeling. *J. Clean. Prod.* **2015**, *103*, 1–12. [CrossRef]
6. Yang, Z.; Zhao, Y. Energy consumption, carbon emissions, and economic growth in India: Evidence from directed acyclic graphs. *Econ. Model.* **2014**, *38*, 533–540. [CrossRef]
7. Liu, H.; Wong, W.K.; Cong, P.T.; Nassani, A.A.; Haffar, M.; Abu-Rumman, A. Linkage among Urbanization, energy Consumption, economic growth and carbon Emissions. Panel data analysis for China using ARDL model. *Fuel* **2023**, *332*, 126122. [CrossRef]
8. Pao, H.T.; Fu, H.C.; Tseng, C.L. Forecasting of CO_2 emissions, energy consumption and economic growth in China using an improved grey model. *Energy* **2012**, *40*, 400–409. [CrossRef]
9. Jian, J.; Fan, X.; He, P.; Xiong, H.; Shen, H. The effects of energy consumption, economic growth and financial development on CO_2 emissions in China: A VECM approach. *Sustainability* **2019**, *11*, 4850. [CrossRef]
10. Zhang, X.P.; Cheng, X.M. Energy consumption, carbon emissions, and economic growth in China. *Ecol. Econ.* **2009**, *68*, 2706–2712. [CrossRef]
11. Tan, X.; Guo, W.; Fan, J.; Guo, J.; Wang, M.; Zen, A.; Su, L.; Sun, Y. Policy framework and technology innovation policy of carbon peak and carbon neutrality. *Bull. Chin. Acad. Sci.* **2022**, *37*, 435–443. [CrossRef]
12. He, J.; Li, Z.; Zhang, X.; Wang, H.; Dong, W.; Chang, S.; Ou, X.; Guo, S.; Tian, Z.; Gu, A.; et al. Comprehensive report on China's long-term low-carbon development strategies and pathways. *Chin. J. Popul. Resour. Environ.* **2020**, *18*, 263–295. [CrossRef]
13. Dong, H.; Liu, Y.; Zhao, Z.; Tan, X.; Managi, S. Carbon neutrality commitment for China: From vision to action. *Sustain. Sci.* **2022**, *17*, 1741–1755. [CrossRef]
14. Walker, W.S.; Gorelik, S.R.; Cook-Patton, S.C.; Baccini, A.; Farina, M.K.; Solvik, K.K.; Ellis, P.W.; Sanderman, J.; Houghton, R.A.; Leavitt, S.M.; et al. The global potential for increased storage of carbon on land. *Proc. Natl. Acad. Sci. USA* **2022**, *119*, e2111312119. [CrossRef] [PubMed]
15. Sun, W.; Liu, X. Review on carbon storage estimation of forest ecosystem and applications in China. *For. Ecosyst.* **2020**, *7*, 4. [CrossRef]

16. Yu, B.; Zhao, G.; An, R.; Chen, J.; Tan, J.; Li, X. Research on China's CO_2 emission pathway under carbon neutral target. *J. Beijing Inst. Technol. (Soc. Sci. Ed.)* **2021**, *23*, 17–24. [CrossRef]
17. Long, Y.; Yoshida, Y. Quantifying city-scale emission responsibility based on input-output analysis–Insight from Tokyo, Japan. *Appl. Energy* **2018**, *218*, 349–360. [CrossRef]
18. Liu, B.; Tian, C.; Li, Y.; Song, H.; Ma, Z. Research on the effects of urbanization on carbon emissions efficiency of urban agglomerations in China. *J. Clean. Prod.* **2018**, *197*, 1374–1381. [CrossRef]
19. Dhakal, S. Urban energy use and carbon emissions from cities in China and policy implications. *Energy Policy* **2009**, *37*, 4208–4219. [CrossRef]
20. Hong, S.; Hui, E.C.M.; Lin, Y. Relationship between urban spatial structure and carbon emissions: A literature review. *Ecol. Indic.* **2022**, *144*, 109456. [CrossRef]
21. Wu, Y.; Zhang, Q. The confrontation and symbiosis of green and development: Coupling coordination analysis between carbon emissions and spatial development in urban agglomerations of China. *Sustain. Cities Soc.* **2024**, *106*, 105391. [CrossRef]
22. Yu, Q.; Li, M.; Li, Q.; Wang, Y.; Chen, W. Economic agglomeration and emissions reduction: Does high agglomeration in China's urban clusters lead to higher carbon intensity? *Urban Clim.* **2022**, *43*, 101174. [CrossRef]
23. Qian, Y.; Wang, H.; Wu, J. Spatiotemporal association of carbon dioxide emissions in China's urban agglomerations. *J. Environ. Manag.* **2022**, *323*, 116109. [CrossRef] [PubMed]
24. Cai, B.; Wang, X.; Huang, G.; Wang, J.; Cao, D.; Baetz, B.W.; Liu, L.; Zhang, H.; Fenech, A.; Liu, Z. Spatiotemporal changes of China's carbon emissions. *Geophys. Res. Lett.* **2018**, *45*, 8536–8546. [CrossRef]
25. Zeng, H.; Shao, B.; Bian, G.; Dai, H.; Zhou, F. Analysis of influencing factors and trend forecast of CO_2 emission in Chengdu-Chongqing urban agglomeration. *Sustainability* **2022**, *14*, 1167. [CrossRef]
26. Luo, H.; Li, Y.; Gao, X.; Meng, X.; Yang, X.; Yan, J. Carbon emission prediction model of prefecture-level administrative region: A land-use-based case study of Xi'an city, China. *Appl. Energy* **2023**, *348*, 121488. [CrossRef]
27. Wang, Q.; Yang, C.H.; Wang, M.L.; Zhao, L.; Zhao, Y.C.; Zhang, Q.P.; Zhang, C.Y. Decoupling analysis to assess the impact of land use patterns on carbon emissions: A case study in the Yellow River Delta efficient eco-economic zone, China. *J. Clean. Prod.* **2023**, *412*, 137415. [CrossRef]
28. Yang, T.; Liu, J.; Mi, H.; Cao, Z.; Wang, Y.; Han, H.; Luan, J.; Wang, Z. An estimating method for carbon emissions of China based on nighttime lights remote sensing satellite images. *Sustainability* **2022**, *14*, 2269. [CrossRef]
29. Yang, D.; Luan, W.; Qiao, L.; Pratama, M. Modeling and spatio-temporal analysis of city-level carbon emissions based on nighttime light satellite imagery. *Appl. Energy* **2020**, *268*, 114696. [CrossRef]
30. Shi, K.; Chen, Y.; Yu, B.; Xu, T.; Chen, Z.; Liu, R.; Li, L.; Wu, J. Modeling spatiotemporal CO_2 (carbon dioxide) emission dynamics in China from DMSP-OLS nighttime stable light data using panel data analysis. *Appl. Energy* **2016**, *168*, 523–533. [CrossRef]
31. Han, F.; Xie, R.; Fang, J.; Liu, Y. The effects of urban agglomeration economies on carbon emissions: Evidence from Chinese cities. *J. Clean. Prod.* **2018**, *172*, 1096–1110. [CrossRef]
32. Yu, X.; Wu, Z.; Zheng, H. How urban agglomeration improve the emission efficiency? A spatial econometric analysis of the Yangtze River Delta urban agglomeration in China. *Environ. Manag.* **2020**, *260*, 110061. [CrossRef] [PubMed]
33. Wang, Y.; Yin, S.; Fang, X.; Chen, W. Interaction of economic agglomeration, energy conservation and emission reduction: Evidence from three major urban agglomerations in China. *J. Energy* **2022**, *241*, 122519. [CrossRef]
34. Chuai, X.; Huang, X.; Wang, W.; Wen, J.; Chen, Q.; Peng, J. Spatial econometric analysis of carbon emissions from energy consumption in China. *J. Geogr. Sci.* **2012**, *22*, 630–642. [CrossRef]
35. Zhang, X.; Liao, Q.; Zhao, H.; Li, P. Vector maps and spatial autocorrelation of carbon emissions at land patch level based on multi-source data. *Front. Public Health* **2022**, *10*, 1006337. [CrossRef] [PubMed]
36. Wang, S.; Huang, Y.; Zhou, Y. Spatial spillover effect and driving forces of carbon emission intensity at the city level in China. *J. Geogr. Sci.* **2019**, *29*, 231–252. [CrossRef]
37. Li, R.; You, K.; Cai, W.; Wang, J.; Liu, Y.; Yu, Y. Will the southward center of gravity migration of population, floor area, and building energy consumption facilitate building carbon emission reduction in China. *Build. Environ.* **2023**, *242*, 110576. [CrossRef]
38. Zou, K.H.; Tuncali, K.; Silverman, S.G. Correlation and simple linear regression. *Radiology* **2003**, *227*, 617–628. [CrossRef]
39. Ma, J.; Cheng, J.C.; Jiang, F.; Chen, W.; Zhang, J. Analyzing driving factors of land values in urban scale based on big data and non-linear machine learning techniques. *Land Use Policy* **2020**, *94*, 104537. [CrossRef]
40. Alizamir, M.; Kim, S.; Kisi, O.; Zounemat-Kermani, M. A comparative study of several machine learning based non-linear regression methods in estimating solar radiation: Case studies of the USA and Turkey regions. *Energy* **2020**, *197*, 117239. [CrossRef]
41. Gunson, N.; Marshall, D.; Morton, H.; Jack, M. User perceptions of security and usability of single-factor and two-factor authentication in automated telephone banking. *Comput. Secur.* **2011**, *30*, 208–220. [CrossRef]
42. Ometov, A.; Bezzateev, S.; Mäkitalo, N.; Andreev, S.; Mikkonen, T.; Koucheryavy, Y. Multi-factor authentication: A survey. *Cryptography* **2018**, *2*, 1. [CrossRef]
43. Saprykin, A.; Chokani, N.; Abhari, R.S. GEMSim: A GPU-accelerated multi-modal mobility simulator for large-scale scenarios. *Simul. Model. Pract. Theory* **2019**, *94*, 199–214. [CrossRef]

44. Elbery, A.; Dvorak, F.; Du, J.; Rakha, H.A.; Klenk, M. Large-scale Agent-based Multi-modal Modeling of Transportation Networks-System Model and Preliminary Results. In Proceedings of the 4th International Conference on Vehicle Technology and Intelligent Transport Systems, Funchal, Portugal, 16–18 March 2018; pp. 103–112.
45. Shan, Y.; Guan, D.; Zheng, H.; Ou, J.; Li, Y.; Meng, J.; Mi, Z.; Liu, Z.; Zhang, Q. China CO_2 emission accounts 1997–2015. *Sci. Data* **2018**, *5*, 170201. [CrossRef] [PubMed]
46. National Bureau of Statistics of China. *Statistical Yearbook of Chinese Cities (2001–2020)*; China Statistics Press: Beijing, China, 2020.
47. National Bureau of Statistics of China. *China Energy Statistics Yearbook (2001–2020)*; China Statistics Press: Beijing, China, 2020.
48. Chen, Y. An analytical process of spatial autocorrelation functions based on Moran's index. *PLoS ONE* **2021**, *16*, e0249589. [CrossRef] [PubMed]
49. Anselin, L. Local indicators of spatial association—LISA. *Geogr. Anal.* **1995**, *27*, 93–115. [CrossRef]
50. He, Y.; Zhou, G.; Tang, C.; Fan, S.; Guo, X. The spatial organization pattern of urban-rural integration in urban agglomerations in China: An agglomeration-diffusion analysis of the population and firms. *Habitat Int.* **2019**, *87*, 54–65. [CrossRef]
51. Jana, M.; Sar, N. Modeling of hotspot detection using cluster outlier analysis and Getis-Ord Gi* statistic of educational development in upper-primary level, India. *Model. Earth Syst. Environ.* **2016**, *2*, 60. [CrossRef]
52. Shacham, M.; Brauner, N. Application of stepwise regression for dynamic parameter estimation. *Comput. Chem. Eng.* **2014**, *69*, 26–38. [CrossRef]
53. Chen, Y.; Shi, R.; Shu, S.; Gao, W. Ensemble and enhanced PM10 concentration forecast model based on stepwise regression and wavelet analysis. *Atmos. Environ.* **2013**, *74*, 346–359. [CrossRef]
54. Yu, T.; Yu, G.; Li, P.Y.; Wang, L. Citation impact prediction for scientific papers using stepwise regression analysis. *Scientometrics* **2014**, *101*, 1233–1252. [CrossRef]
55. Żogała-Siudem, B.; Jaroszewicz, S. Fast stepwise regression based on multidimensional indexes. *Inform. Sci.* **2021**, *549*, 288–309. [CrossRef]
56. Whittingham, M.J.; Stephens, P.A.; Bradbury, R.B.; Freckleton, R.P. Why do we still use stepwise modelling in ecology and behaviour. *J. Anim. Ecol.* **2006**, *75*, 1182–1189. [CrossRef] [PubMed]
57. Fernandes, R.; Leblanc, S.G. Parametric (modified least squares) and non-parametric (Theil–Sen) linear regressions for predicting biophysical parameters in the presence of measurement errors. *Remote Sens. Environ.* **2005**, *95*, 303–316. [CrossRef]
58. Burton, A.L. OLS (Linear) regression. In *The Encyclopedia of Research Methods in Criminology and Criminal Justice*; Barnes, J.C., Forde, D.R., Eds.; John Wiley & Sons: Hoboken, NJ, USA, 2021; pp. 509–514.
59. Steel, R.G.D.; Torrie, J.H. *Principles and Procedures of Statistics*; McGraw-Hill Book Company: New York, NY, USA, 1960.
60. Pan, D.; Li, N.; Li, F.; Feng, K.; Peng, L.; Wang, Z. Mitigation strategy of Eastern China based on energy-source carbon emission estimation. *Acta Sci. Circumst.* **2021**, *41*, 1142–1152. [CrossRef]
61. Nassar, R.; Napier-Linton, L.; Gurney, K.R.; Andres, R.J.; Oda, T.; Vogel, F.R.; Deng, F. Improving the temporal and spatial distribution of CO_2 emissions from global fossil fuel emission data sets. *J. Geophys. Res. Atmos.* **2013**, *118*, 917–933. [CrossRef]
62. Moutinho, V.; Moreira, A.C.; Silva, P.M. The driving forces of change in energy-related CO_2 emissions in Eastern, Western, Northern and Southern Europe: The LMDI approach to decomposition analysis. *J. Renew. Sustain. Energy Rev.* **2015**, *50*, 1485–1499. [CrossRef]
63. Liu, K.; Wu, S.; Guo, N.; Fang, Q. Host Country's carbon emission and cross-border M&A performance: Evidence from listed enterprises in China. *J. Clean. Prod.* **2021**, *314*, 127977. [CrossRef]
64. Shen, H.; Yang, Q.; Luo, L.; Huang, N. Market reactions to a cross-border carbon policy: Evidence from listed Chinese companies. *Br. Account. Rev.* **2023**, *55*, 101116. [CrossRef]
65. Huang, H.; Wei, Z.; Ge, Q.; Guo, Q. Analysis of spatial-temporal evolution and influencing factors of carbon emission efficiency in Chinese cities. *Front. Environ. Sci.* **2023**, *11*, 1119914. [CrossRef]
66. Lee, C.C.; Zhao, Y.N. Heterogeneity analysis of factors influencing CO_2 emissions: The role of human capital, urbanization, and FDI. *Renew. Sustain. Energy Rev.* **2023**, *185*, 113644. [CrossRef]
67. Liddle, B. Impact of population, age structure, and urbanization on carbon emissions/energy consumption: Evidence from macro-level, cross-country analyses. *Popul. Environ.* **2014**, *35*, 286–304. [CrossRef]
68. Zhu, Q.; Peng, X. The impacts of population change on carbon emissions in China during 1978–2008. *Environ. Impact Assess. Rev.* **2012**, *36*, 1–8. [CrossRef]
69. Duman, Z.; Mao, X.; Cai, B.; Zhang, Q.; Chen, Y.; Gao, Y.; Guo, Z. Exploring the spatiotemporal pattern evolution of carbon emissions and air pollution in Chinese cities. *J. Environ. Manag.* **2023**, *345*, 118870. [CrossRef] [PubMed]
70. Jun, L.; Lu, S.; Li, X.; Li, Z.; Cao, C. Spatio-Temporal Evolution Characteristics and Influencing Factors of Industrial Carbon Emission Efficiency in Chinese Prefecture-Level Cities. *Sustainability* **2023**, *15*, 13694. [CrossRef]
71. Wang, Z.; Sun, Y.; Wang, B. How does the new-type urbanisation affect CO_2 emissions in China? An empirical analysis from the perspective of technological progress. *Energy Econ.* **2019**, *80*, 917–927. [CrossRef]
72. Hossain, M.S. Panel estimation for CO_2 emissions, energy consumption, economic growth, trade openness and urbanization of newly industrialized countries. *Energy Policy* **2011**, *39*, 6991–6999. [CrossRef]

73. Xu, L.; Wang, D.; Du, J. Spatial-temporal evolution and influencing factors of urban green and smart development level in China: Evidence from 232 prefecture-level cities. *Int. J. Environ. Res. Public Health* **2022**, *19*, 3939. [CrossRef]
74. Cai, C.; Fan, M.; Yao, J.; Zhou, L.; Wang, Y.; Liang, X.; Liu, Z.; Chen, S. Spatial-temporal characteristics of carbon emissions corrected by socio-economic driving factors under land use changes in Sichuan Province, southwestern China. *Ecol. Inf.* **2023**, *77*, 102164. [CrossRef]

land

MDPI

Article

Research on Improving the Accuracy of SIF Data in Estimating Gross Primary Productivity in Arid Regions

Wei Liu [1,2,3,4,5,6,7], Yu Wang [1,2,3,4,5,6], Ali Mamtimin [1,2,3,4,5,6,*], Yongqiang Liu [7], Jiacheng Gao [1,2,3,4,5,6], Meiqi Song [1,2,3,4,5,6], Ailiyaer Aihaiti [1,2,3,4,5,6], Cong Wen [1,2,3,4,5,6], Fan Yang [1,2,3,4,5,6], Wen Huo [1,2,3,4,5,6], Chenglong Zhou [1,2,3,4,5,6], Jian Peng [8] and Hajigul Sayit [9]

1 Institute of Desert Meteorology, China Meteorological Administration, Urumqi 830002, China; 107552201154@stu.xju.edu.cn (W.L.); wangyu@idm.cn (Y.W.); gaojiach@idm.cn (J.G.); songmg@idm.cn (M.S.); ailiyaer@idm.cn (A.A.); wencong@idm.cn (C.W.); yangfan@idm.cn (F.Y.); huowenpet@idm.cn (W.H.); zhoucl@idm.cn (C.Z.)

2 National Observation and Research Station of Desert Meteorology, Taklimakan Desert of Xinjiang, Urumqi 830002, China

3 Taklimakan Desert Meteorology Field Experiment Station of China Meteorological Administration, Urumqi 830002, China

4 Xinjiang Key Laboratory of Desert Meteorology and Sandstorm, Urumqi 830002, China

5 Wulanwusu National Special Test Field for Comprehensive Meteorological Observation, Urumqi 830002, China

6 Key Laboratory of Tree-Ring Physical and Chemical Research, China Meteorological Administration, Urumqi 830002, China

7 College of Geography and Remote Sensing Sciences, Xinjiang University, Urumqi 830046, China; liuyq@xju.edu.cn

8 Xinjiang Meteorological Technology Equipment Center, Urumqi 830001, China; pengjian@idm.cn

9 Xinjiang Meteorological Society, Urumqi 830002, China; hajigul@idm.cn

* Correspondence: ali@idm.cn

Abstract: Coupling solar-induced chlorophyll fluorescence (SIF) with gross primary productivity (GPP) for ecological function integration research presents numerous uncertainties, especially in ecologically fragile and climate-sensitive arid regions. Therefore, evaluating the suitability of SIF data for estimating GPP and the feasibility of improving its accuracy in the northern region of Xinjiang is of profound significance for revealing the spatial distribution patterns of GPP and the strong coupling relationship between GPP and SIF in arid regions, achieving the goal of "carbon neutrality" in arid regions. This study is based on multisource SIF satellite data and GPP observation data from sites in three typical ecosystems (cultivated and farmlands, pasture grasslands, and desert vegetation). Two precision improvement methods (canopy and linear) are used to couple multiple indicators to determine the suitability of multisource SIF data for GPP estimation and the operability of accuracy improvement methods in arid regions reveal the spatial characteristics of SIF (GPP). The results indicate the following. (1) The interannual variation of GPP shows an inverted "U" shape, with peaks values in June and July. The cultivated and farmland areas have the highest peak value among the sites (0.35 $gC/m^2/month$). (2) The overall suitability ranking of multisource SIF satellite products for GPP estimation in arid regions is RTSIF > CSIF > SIF_OCO2_005 > GOSIF. RTSIF shows better suitability in the pasture grassland and cultivated and farmland areas (R^2 values of 0.85 and 0.84, respectively). (3) The canopy method is suitable for areas with a high leaf area proportion (R^2 improvement range: 0.05–0.06), while the linear method is applicable across different surface types (R^2 improvement range: 0.01–0.13). However, the improvement effect of the linear method is relatively weaker in areas with high vegetation cover. (4) Combining land use data, the overall improvement of SIF (GPP) is approximately 0.11%, and the peak values of its are mainly distributed in the northern and southern slopes of the Tianshan Mountains, while the low values are primarily found in the Gurbantunggut Desert. The annual mean value of SIF (GPP) is about 0.13 $mW/m^2/nm/sr$. This paper elucidates the applicability of SIF for GPP estimation and the feasibility of improving its accuracy, laying the theoretical foundation for the spatiotemporal coupling study of GPP and SIF in an arid region, and providing practical evidence for achieving carbon neutrality goals.

Keywords: solar-induced chlorophyll fluorescence (SIF); gross primary productivity (GPP); applicability; accuracy improvement; spatial features

1. Introduction

Gross primary productivity (GPP) is the comprehensive product of vegetation fixing CO_2 through photosynthesis [1]. As an important factor in terrestrial carbon cycle research, it plays a crucial role in the biosphere [2]. Under the goal of "carbon neutrality", achieving a precise measurement and estimation of GPP is not only of profound significance for understanding the mechanisms of the carbon cycle [3,4], but also plays an important role in determining the comprehensive response of CO_2 to climate change. Currently, the main methods for measuring and estimating GPP include model simulations, ground-based observations, and satellite-based remote sensing. Model simulations primarily refer to methods based on light use efficiency (LUE) models [5,6]. Due to differences in underlying surfaces, variations in vegetation structure, and the comprehensive impact of climate change, the structure and related parameters of LUE models are difficult to accurately construct and quantify, resulting in certain limitations in this method. Ground-based observations primarily refer to the eddy covariance (EC) measurement technique [7]. The EC method measures the net ecosystem exchange (NEE) through flux towers [8], and then decomposes NEE into GPP and ecosystem respiration (Reco) [9]. Due to the limited coverage area of EC measurements, the EC method needs to use related approaches to upgrade the measurement scale when conducting large-scale GPP estimate [10,11]. The upscaled GPP also has certain limitations, such as incomplete driving factors, excessive parameters, and constraints from the original EC flux towers. Therefore, to achieve a precise measurement and estimation of GPP and to promote the realization of the "carbon neutrality" goal, it is necessary to further explore other GPP measurements or alternative methods.

Chlorophyll fluorescence is a light signal produced when chlorophyll molecules move from a low-energy state to a high-energy state and then return to the low-energy state. The wavelength of this fluorescence is approximately in the range of 650–800 nm, which is also referred to as solar-induced chlorophyll fluorescence (SIF). SIF encompasses the spectra contributed by both Photosystem I and Photosystem II. Compared with traditional vegetation indices, SIF can better reveal the dynamic changes and carbon cycling processes of GPP. SIF has become one of the hot research topics in the field of vegetation remote sensing [12]. For example, Gao et al. [13] conducted linear and spatiotemporal analyses using real-time tower-measured GPP from FLUXNET in 2015 and two types of SIF (CSIF and GOSIF), demonstrating that both showed a positive correlation across different spatiotemporal scales. Wang et al. [14] applied SIF and GPP to characterize the spatiotemporal features and regulatory factors of terrestrial ecosystems in China from 2007 to 2018. They compared the spatiotemporal consistency of SIF and GPP between arid and nonarid regions by analyzing different climatic zones. Qiu et al. [15] characterized the response changes of SIF and GPP to drought processes under arid conditions. Wei et al. [16] coupled SIF and GPP satellite products and applied multiple indicators to demonstrate the reduced lag effect of GPP in grasslands in arid regions. Liu et al. [17] demonstrated the advantages of using SIF and GPP data with multiple indicators to represent the water storage sensitivity of desert vegetation in arid and semi-arid regions. Wang et al. [18] integrated remote sensing data with multiple indicators and showed that SIF is more effective in representing the GPP of various systems in arid areas than normalized difference vegetation index and vegetation optical depth. However, given the increasingly important role that SIF satellite products play in GPP estimation, many scholars have begun using models, machine learning, and other methods to generate SIF satellite products with different spatiotemporal characteristics from different platforms. The satellite platforms mainly include Orbiting Carbon Observatory-2 (OCO_2), Sentinel-5P, MODIS, and GOME-2, and the products mainly include CSIF [19,20], RTSIF [21,22], GOSIF [23], SIF_OCO2_005 [24], and others. With the generation of various

SIF satellite products, the comparative analysis of the applicability of different products for GPP estimation in different ecosystems has become a new research topic. However, existing studies mostly focus on single ecosystems. For instance, An et al. [25] assessed the consistency of five SIF satellite products for rubber plantation ecosystems on Hainan Island. Dang et al. [26] verified the feasibility of using SIF instead of GPP to explore the mechanisms affecting humid, arid, and semi-arid regions. Furthermore, because SIF itself only accounts for 2–3% of solar radiation [27], it is necessary to reduce spatial resolution to eliminate certain noise during measurement, resulting in a relatively low spatial resolution of existing SIF satellite products. For instance, MetOp-A/B sensors mounted on the GOME-2 satellite can only obtain monthly scale SIF data with a resolution of 0.5° [28]. Therefore, although SIF can serve as an effective substitute for vegetation photosynthesis and be used for GPP estimation, the overall applicability of multisource SIF satellite products in GPP estimation across different regions and ecosystems requires further comparative validation. The feasibility of improving the accuracy of SIF satellite products (considering various influencing factors, such as different underlying surfaces) and achieving effective GPP estimation requires further comparative research, especially in typical ecosystems in arid regions of China.

The Xinjiang Uygur Autonomous Region is located in the inland northwest of China, far from the ocean, with extremely low precipitation, making it part of China's arid region. It encompasses typical ecosystems such as pasture grasslands, cultivated and farmlands, and coniferous forests. As well as fragile ecosystems like desert vegetation, natural sand dunes, and desert–oasis transition zones, making it exceptionally sensitive to climate change on both a national and global scale. As a typical arid region, which SIF satellite product is most suitable for estimating GPP in this area? Which improvement method can effectively enhance the progress of SIF satellite products in this region? Do the improved SIF satellite products conform to the comprehensive variation characteristics of GPP in arid regions? The resolution of this series of questions is crucial for achieving the "carbon neutrality" goal in arid regions. Unfortunately, these questions have not yet received clear answers.

Given this, this study takes the northern region of Xinjiang as the research area and utilizes indirectly observed GPP data from three typical ecosystems in this region (cultivated and farmlands, pasture grasslands, and desert vegetation). Spatial characteristics, linear regression parameters, GPP sensitivity to influencing factors, and GPP/SIF values under different weather conditions are selected as evaluation criteria. These criteria are used to comprehensively assess the overall applicability of four continuously updated SIF satellite products (CSIF, RTSIF, GOSIF, SIF_OCO2_005) for GPP estimation in typical arid ecosystems. Subsequently, the feasibility of employing canopy and linear accuracy improvement methods to enhance the accuracy of the most suitable SIF satellite product for GPP estimation is verified. Finally, the changes in spatial characteristics of SIF data before and after improvement are revealed, indirectly reflecting the spatial distribution patterns of GPP in arid regions. These studies will provide reliable empirical evidence for the applicability and accuracy improvement of SIF satellite products for GPP estimation in arid regions. Moreover, they fill the gap in the coupling research of SIF satellite products and GPP in this localized arid region of northern Xinjiang, laying a theoretical foundation for further research on GPP influencing mechanisms in Xinjiang and the entire arid region.

2. Data and Methods

2.1. Study Area

The Xinjiang Uygur Autonomous Region (35°~50° N, 73°~96° E) (Figure 1a) is located in northwest China. Its north end is located in the Altay Mountains, its south end is located in the Kunlun Mountains, and its central area has the Tianshan Mountains running through it. It covers a total area of 1.6649 million km^2 [29,30] and is part of China's arid region. Due to the region being far from the ocean and surrounded by mountains, the areas have a large temperature difference and are dry with little rain. The annual average temperature is about

−4.9~14.9 °C [31]. The rivers and lakes in the region (such as the Tarim River and Abi Lake) mainly come from glacial meltwater. Owing to its unique geographical location, the region has diverse land surfaces, including desert vegetation, pasture grasslands, cultivated and farmlands, coniferous forests, and other underlying surfaces, the ecosystem is very typical.

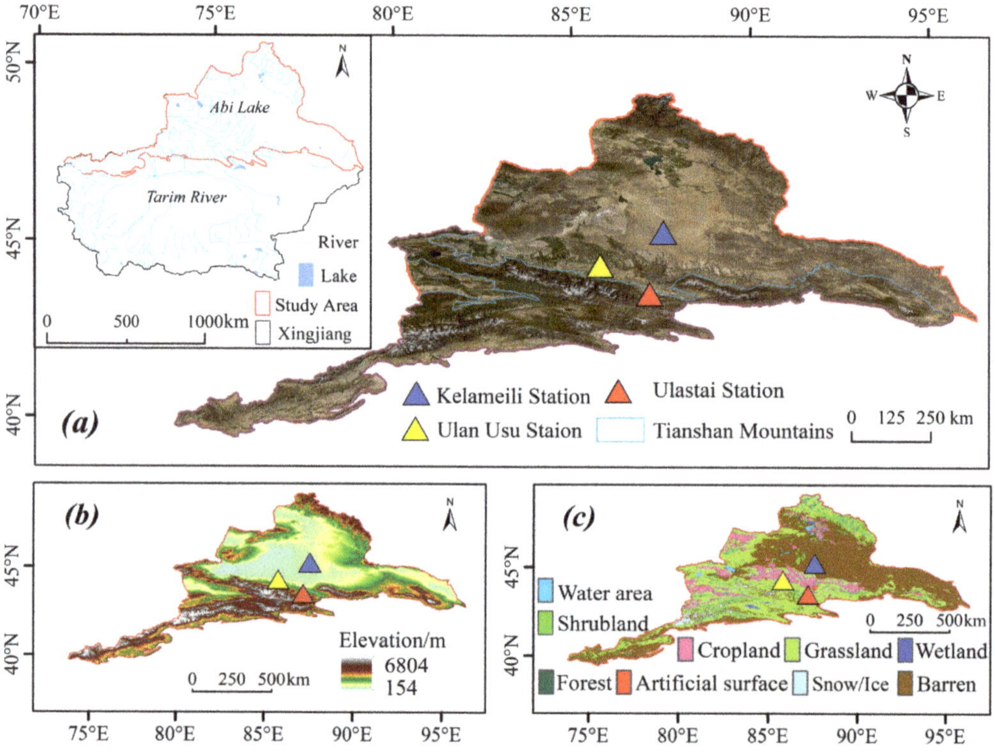

Figure 1. (**a**) Specific locations of the Tianshan Mountains, Ulan Usu Station, Ulastai Station, and Kelameili Station in Xinjiang. (**b**) Schematic representation of the elevations of the study area. (**c**) Schematic representation of the land use types at the study area.

This study focuses on the northern region of Xinjiang as the research area. Three typical ecosystems in this region (cultivated and farmlands, pasture grasslands, and desert vegetation) are selected for the analysis of the coupling between SIF satellite products and GPP in arid regions.

2.2. Data Sources

2.2.1. Site Data

The situ observation data for this study were obtained from the Land-Atmospheric Interaction Observation Stations constructed by the Institute of Desert Meteorology, China Meteorological Administration, Urumqi. There were three stations (Table 1), each equipped with EC systems, radiation observation systems, and gradient tower systems.

Table 1. Basic information of different observation stations.

Station Name	Abbreviation	Coordinates	Elevation	Region	Underlying Surface
Ulan Usu Land-Atmosphere Interaction Observation Station	Ulan Usu Station	44°16′59″ N 85°49′00″ E	406.9 m	The northern slope of Tianshan Mountains	Cultivated and farmland area [32]
Central Tianshan Land-Atmosphere Interaction Observation Station	Ulastai Station	43°28′55″ N 87°12′50″ E	2036 m	The central hinterland of the Tianshan Mountains	Pasture grassland area [33,34]
Kelameili Land-Atmosphere Interaction Observation Station	Kelameili Station	45°14′00″ N 87°35′00″ E	531 m	Gurbantunggut Desert	Desert vegetation area [35,36]

The EC system consisted of an open-path CO_2/H_2O infrared gas analyzer (LI7500, Li-Cor, Lincoln Nebraska, IA, USA) and an ultrasonic three-dimensional anemometer (CSAT3, Campbell Scientific, Logan, UT, USA). The Ultrasonic anemometer can accurately measure pulsating wind speed and acoustic virtual temperature in three different directions, with measurement accuracies of ±4.0 cm/s and ±2.0 cm/s. The data acquisition frequency is 10 Hz/20 Hz and the data output interval is 30 min.

The open-path LI7500 infrared gas analyzer provides accurate measurements of CO_2 concentration and water vapor density in the atmosphere, with measurement accuracies of ±0.01 mmol/mol and ±0.15 mmol/mol. The data output interval is also 30 min.

The data used in this study were continuous observation data in 2020, with observation times synchronized to the local time. All measured data were output through a data collector (CR3000, Campbell Scientific, Logan, UT, USA) at time frequencies of 10 s, 1 min, 30 min, and 1 h. The acquisition frequency of the radiation and gradient observation systems was 1 Hz, whereas that of the EC covariance system was 10 Hz.

2.2.2. Satellite Data

By comprehensively comparing the SIF satellite products that are currently the most widely used and have been proven to have good applicability [13,16–18,25], we have preliminarily selected the SIF satellite product generated based on the two most widely used satellite platforms (OCO_2 and Sentinel-5P). These SIF satellite products have a complete time cycle, time/spatial resolution, and coverage range, which can perfectly match the research area and time period. It mainly includes four satellite products as follows: CSIF [19,20], RTSIF [21,22], GOSIF [23], and SIF_OCO2_005 [24] (Table 2).

The leaf area index (LAI) data for this study were obtained from the HIQ-LAI satellite product data set on the Google Earth Engine platform. This data set was created by Yan et al. [37], using the spatiotemporal informative component analysis (STICA) algorithm, which reanalyzed nearly 22 years of MODIS C6.1 LAI products. The data set has an 8-day temporal scale, and each year includes 46 TIFF format files at a resolution of 500 m.

The land use data for this study were sourced from the Chinese Academy of Sciences Resource and Environment Science Data Registration and Publishing System. This data set was produced by Xu et al. [38], through manual visual interpretation of Landsat 8 remote sensing images to generate the data set. The data set comprises 25 secondary types and offers raster data at resolutions of 1000 m, 100 m, and 30 m, with this study utilizing data at a resolution of 30 m.

This study used the observation data from the aforementioned satellites products in 2020.

Table 2. Basic information on four widely used SIF satellite products.

Product Name	Generation Method	Temporal Resolution	Spatial Resolution	Time Period	Acquisition Platform
CSIF	Based on the combination of SIF from OCO_2 and calibrated MODIS BRDF seven-band surface reflectance, trained artificial neural networks (ANNs), applying ANNs, incorporating weather conditions, and employing machine learning algorithms to generate.	4 days	0.05°	2001–2020	National Tibetan Plateau Data Center (http://data.tpdc.ac.cn, accessed on 18 September 2023)
RTSIF	Generated through machine learning reconstruction of TROPO spheric Monitoring Instrument (TROPOMI) on Copernicus Sentinel-5P mission.	8 days	0.05°	2000–2020	
GOSIF	Generated using data-driven methods based on SIF from OCO_2, MODIS, and meteorological reanalysis data.	8 days	0.05°	2000–2022	Earth System Research Center
SIF_OCO2_005	Utilizes SIF from OCO_2 and calibrated MODIS BRDF seven-band surface reflectance, trained artificial neural networks (ANNs), applying ANNs, incorporating MODIS reflectance, and land cover to predict.	16 days	0.05°	2014–2020	Earth Data Center

2.2.3. Other Auxiliary Data

To improve the accuracy of the SIF data based on the canopy method, various auxiliary parameter data sets were applied concurrently (Table 3).

Table 3. SIF data accuracy improvement auxiliary parameter data set.

Parameter	Name	Value	Reference
L	Canopy continuous radiation intensity	Derived from the RTSIF sensor	-
θs	Solar zenith angle	Atmospheric effects within the SIF satellite spectrum range are negligible and considered as 0	[39]
G(θ)	Geometric mean	0.5	[40]
ω	Absorption value of chlorophyll in the 743–758 nm spectrum range	The change in ω in this spectrum range is minimal and considered as a unit value of 1	[41,42]
E	Solar irradiance of 743–758 nm	1277.3 mW/m^2/nm	[43]
CI	Clumping index	Determine from He L global products based on the type of underlying surface in the research area	[44]

2.3. *Data Processing*

2.3.1. Site Data Processing

The final data collection frequency for EC used in this study is 10 Hz, with a time output frequency of 30 min. The data collection frequency for radiation and gradient observations is 1 Hz, with a time output frequency of 30 min. All original flux observation data are initially in the TOB1 format, which can be converted to the TOB3/5 format for preliminary operations by the LoggerNet software4.0 (Campbell Scientific, Logan, UT, USA). Subsequently, EddyPro7.1 software was employed for data processing, including outlier removal [45], time lag correction [46], coordinate sequence rotation [47], frequency response correction [48], sonic virtual temperature, and density correction [49], thereby obtaining flux data with a time step of half an hour.

Owing to uncontrollable factors such as instrument damage and abrupt weather changes, the preliminary processed flux data may still suffer from issues, such as missing or discontinuous data. Therefore, it is necessary to further comprehensive interpolation using the Max Planck online interpolation tool, which was performed on the half-hourly flux data. Through comprehensive processing, high-quality and continuous half-hourly flux data can be obtained. (The tool was developed by the Max Planck Institute for Biogeochemistry and can be obtained at: https://www.bgc-jena.mpg.de/bgi/index.pHp/Services/REddyProcWeb, accessed on 18 September 2023).

In the data processed by comprehensive interpolation, the CO_2 flux data represents NEE, while GPP and Reco [50] need to be calculated using nighttime and daytime data-splitting methods [7,51]. The calculation formula is as follows:

$$GPP = NEE - Reco \tag{1}$$

Due to SIF being the total product of photosynthesis, it is necessary to finally remove the nighttime value from the GPP half-hourly flux data based on local sunshine time, and convert it into an average energy value GPP data set over 8-day intervals. The ultimate unit for this data set is $gC/m^2/day$. (The sunshine time acquisition can be obtained at: https://richurimo.bmcx.com/xinjiangweiwuerzizhiqu__time__2020_02__richurimo/, accessed on 18 September 2023).

The unit conversion formula is as follows:

$$GPP(gcm^{-2}day^{-1}) = \sum_{i=1}^{48} GPP(co_2/m^2s) \times 1800 \times \frac{12}{44} \tag{2}$$

2.3.2. Satellite Data Processing

Due to the differences in temporal cycles, temporal/spatial resolutions, coverage, and storage formats of each satellite product, comprehensive processing is required. Firstly, the ArcGIS10.8 software iterator tool was used for batch format conversion to create TIFF format satellite product data sets. Secondly, the ArcGIS10.8 software iterator tool was used for mask clipping to obtain coverage consistent with the study area. Subsequently, the ArcGIS10.8 software was used for resampling to obtain a 2020 satellite product data set with consistent spatial resolution (0.05°) in the study area. Finally, the ArcPy10.8 tool and MATLABR2022a software were used for batch extraction of the SIF satellite product raster attribute values corresponding to each site. We set that the SIF satellite products are suitable for arid regions if the spatial attributes of SIF in each ecosystem are reasonable and exhibit annual averages close to the measured GPP.

2.4. Research Methods

2.4.1. Method for Applicability Verification

To verify the applicability of multisource SIF satellite products for GPP estimation in arid regions, linear regression fitting analysis is conducted using measured GPP data from each site and their corresponding SIF data. Our data used for comprehensive calculations have complete and stable time series, with a clear number of variables and the exclusion of outliers and outliers. The adjusted R^2 is used for comprehensive verification and evaluation, which avoids the irrationality, overfitting, bias, and inconsistency of model predictions. We set the suitability of SIF satellite products for arid regions when R^2 is greater than 0.6 in desert vegetation areas and greater than 0.8 in the cultivated and farmland and pasture grassland areas. The parameter calculation formula is as follows:

$$RMSE = \sqrt{\frac{1}{n}\sum_{i=1}^{n} (y_i - \hat{y}_i)^2} \tag{3}$$

$$MB = \frac{1}{n}\sum_{i=1}^{n}(y_i - \hat{y}_i) \tag{4}$$

$$SD = \sqrt{\frac{1}{n}\sum_{i=1}^{n}(x_i - \overline{x}_i)^2} \quad (5)$$

$$R^2 = 1 - \left(\sum_{i=1}^{n}(y_i - \hat{y}_i)^2 / \sum_{i=1}^{n}(y_i - \overline{y}_i)^2\right) \quad (6)$$

where n represents the number of samples, y_i represents the observed values, $\hat{y}_i$ represents the predicted values, and the predicted value was calculated comprehensively based on the slope and intercept.

2.4.2. Method for Response Degrees Verification

To verify the responsiveness of the multisource SIF satellite products to the main influencing factors of GPP in arid regions, Pearson correlation analysis was used to calculate their correlation, and t-tests were performed to determine the confidence level. Our data used for comprehensive calculations have complete and stable time series, with a clear number of variables and the exclusion of outliers and outliers, which avoids bias and inconsistency in comprehensive analysis. We set the SIF satellite product to be applicable in arid regions when it exhibits a high correlation with influencing factors under different ecosystems. The parameter calculation formula is as follows:

$$r = \frac{\sum_{i=1}^{n}(x_i - \overline{x})(y_i - \overline{y})}{\sqrt{\sum_{i=1}^{n}(x_i - \overline{x})^2}\sqrt{\sum_{i=1}^{n}(y_i - \overline{y})^2}} \quad (7)$$

$$t = r/\sqrt{(1 - r^2)/(n - 2)} \quad (8)$$

where n represents the sample size, x_i represents the satellite values, and y_i represents the observed values of influencing factors.

2.4.3. Method for GPP/SIF Verification

To verify the changes in light distribution–sensitive diagnostic indicators (GPP/SIF) under different weather conditions, the Clear Sky Index (CI) method proposed by Gu et al. [52] was used to calculate the weather index of the study area in 2020. Subsequently, the weather conditions were comprehensively classified using the weather condition classification standard proposed by Okogue et al. [53] (Table 4). Finally, the GPP/SIF method derived by Yang et al. [54] was used to calculate the light distribution–sensitive diagnostic values. We set the suitability of SIF satellite products for arid regions when the GPP/SIF ratio in each ecosystem shows fluctuations around the 1:1 line without major abrupt changes. The parameter calculation formula is as follows:

$$CI = R_s/R_0 \quad (9)$$

$$R_0 = R_{sc}(1 + 0.33\cos(360d/365))\sin\beta \quad (10)$$

$$\sin\beta = \sin\phi \cdot \sin\delta + \cos\phi \cdot \cos\delta \cdot \cos\omega \quad (11)$$

where R_{sc} is the solar constant, β is the solar zenith angle, φ is the latitude of the study area, δ is the solar declination angle, and ω is the hourly angle.

Table 4. Study area weather division results.

Weather Conditions	Criteria for Division	Kelameili Station Cultivated and Farmland Area	Ulastai Station Pasture Grassland Area	Ulan Usu Station Desert Vegetation Area
Sunny Day	$0.6 \leq CI < 1$	191 d	177 d	178 d
Cloudy Day	$0 < CI < 0.3$	75 d	69 d	96 d
Overcast Day	$0.3 \leq CI < 0.6$	100 d	120 d	92 d

2.4.4. Method for SIF Precision Improvement

To further explore the optimal accuracy of SIF data in three typical ecosystems in arid regions, we employed two methods to improve the precision of RTSIF satellite products, and then repeated the linear regression fitting analysis process.

Method 1: Accuracy improvement method based on canopy. This method decomposes the incident sunlight into three parts as follows: zero-order canopy transmittance (denoted as t_0), canopy interception rate (denoted as i_0), and escape probability (denoted as *fesc*). Among them, a portion of the sunlight intercepted by the canopy is absorbed (denoted as a) and a portion is scattered (denoted as s). When scattering occurs, collisions occur between the canopies, resulting in a recollision rate (denoted as p). Based on the relationships between these factors and incorporating other parameters (Table 3), the accuracy of SIF data is improved using the following formula:

$$a + s + t_0 = 1 \quad (12)$$

$$P = n - 1/n \times \omega \quad (13)$$

$$R_{obs} = \pi \times L/(E \times \cos\theta_S) \quad (14)$$

$$Le = LAI \times CI \quad (15)$$

$$i_0 = 1 - \mathrm{EXP}(-G(\theta)) \times Le \times \pi \times L/\cos\theta_s \quad (16)$$

$$f_{esc} = R_{obs}/i_0 \times \omega \times \lambda \quad (17)$$

$$\mathrm{SIF}_{improved} = \mathrm{SIF}_{before\ improved} \times \omega \times i_0/R_{obs} \quad (18)$$

where n represents the average number of interactions between the solar radiation and leaf surfaces, R_{obs} is the canopy spectral bidirectional reflectance, and λ is the wavelength. Other parameter information can be found in Table 3.

Method 2: The linear deviation accuracy improvement method [55,56]. This method uses the slope (a) and intercept (b) of the best regression-fitting equation ($y = ax + b$) between the measured GPP and SIF satellite data to comprehensively eliminate bias, thereby improving the accuracy of the SIF data. The calculation formula is as follows:

$$\mathrm{SIF}_{improved} = \mathrm{SIF}_{before\ improved} - [(a-1)GP\ P + b] \quad (19)$$

3. Results

3.1. GPP Various on the Site

From Figure 2, it can be seen that the monthly average GPP of each site exhibited a changing trend of initially increasing and then decreasing during the growing season (March to October), whereas the period outside the growing season showed a relatively flat trend. The overall interannual variation manifested an inverted "U" shape, and the monthly average GPP of each site during the year was sorted according to the underlying surface conditions as follows: cultivated and farmland area > pasture grassland area > desert vegetation area.

Among these, the minimum monthly average GPP value (0.0003 gC/m^2/month) of Ulan Usu Station (cultivated and farmland area) occurred in December, followed by a sharp increase due to the cultivation, after reaching a peak (0.35 gC/m^2/month) in July, then the monthly average GPP value decreased year-on-year. The overall growth rate ranks first among all sites. The monthly average growth rate of GPP at Ulastai Station (pasture grassland area) is second. The overall trend of GPP monthly average change at this station is similar to that of Ulan Usu station, with peak values (0.14 gC/m^2/month) also occurring in July. However, the Ulastai Station is located in the hinterland of Tianshan Mountains, with a relatively high altitude (2036 m) and a relatively low temperature. This results in its annual monthly GPP being slightly lower than the Ulan Usu Station, and its minimum value (0.011 gC/m^2/month) appears in December.

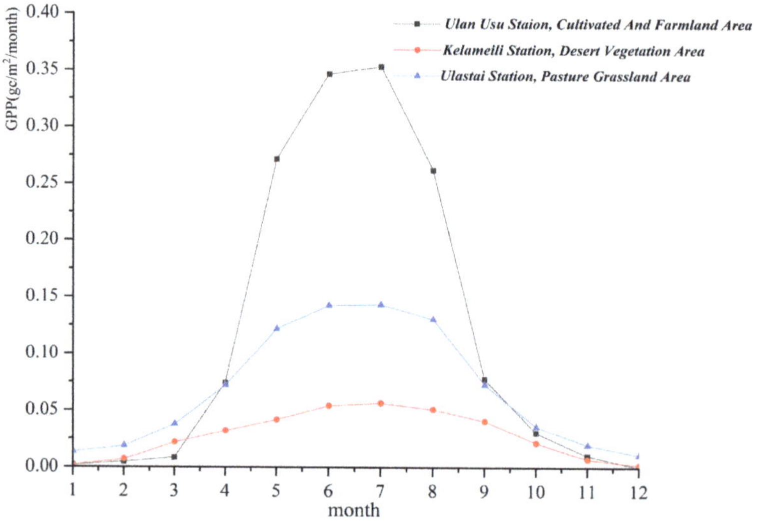

Figure 2. Interannual variation of monthly average GPP at each site in 2020 (excluding nighttime values).

In addition, the annual variation trend of the monthly average GPP at the Kelameili Station (desert vegetation area) with relatively low vegetation coverage is consistent with the above stations. However, due to the short growth cycle of short-lived vegetation, the interannual monthly average GPP at the desert vegetation area is significantly lower. The overall monthly average GPP ranks at the end of each station, with the minimum value (0.0018 gC/m^2/month) occurring in January and a peak (0.057 gC/m^2/month) occurring in July.

3.2. Analysis of GPP Estimation Using Multisource SIF Satellite Products

3.2.1. Analysis of Applicability

From Figure 3, it can be seen that there are differences in the linear regression fit between the RTSIF (Figure 3a), CSIF (Figure 3b), SIF_OCO2_005 (Figure 3c), and GOSIF (Figure 3d) satellite products and their corresponding station GPP data in arid regions. In the pasture grassland area, RTSIF demonstrated the highest R^2 fitting value (0.85), followed by CSIF (0.84), and GOSIF shows the lowest (0.41). The order of R^2 fitting values was the same for the cultivated and farmland area. In the desert vegetation area, the highest R^2 fitting value was for RTSIF (0.62), and the lowest was for SIF_OCO2_005 (0.36). Additionally, there are also differences in the RMSE and SD among the satellites in different underlying surfaces. RTSIF showed the smallest RMSE and SD in the pasture grassland area (0.01 and 0.11, respectively), followed by CSIF (0.01 and 0.13, respectively), and GOSIF showed the largest (0.01 and 24.46, respectively). In the cultivated and farmland area and the desert vegetation area, the order of the RMSE and SD was the same as that in the pasture grassland area.

Furthermore, by comparing the R^2 fitting values of the four satellite products on three typical underlying surfaces with the linear regression optimal value (optimal value is 1), it was found that the difference between the R^2 fitting value and the optimal value was less than 0.38 for RTSIF, less than 0.41 for CSIF, less than 0.49 for SIF_OCO2_005, and less than 0.64 for GOSIF. This indicates that there are differences in the applicability of multisource SIF satellite products for GPP estimation in arid regions. The overall ranking of their applicability is RTSIF > CSIF > SIF_OCO2_005 > GOSIF, with RTSIF having an overall significance greater than 0.5 and a confidence interval of 95%, indicating that RTSIF satellite products have the best suitability.

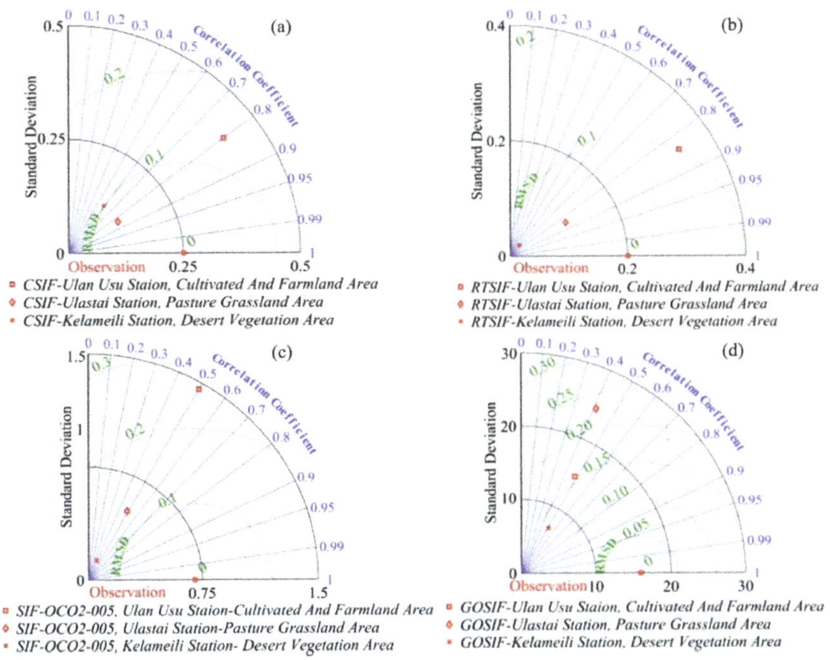

Figure 3. (**a**) The linear regression fitting of 2020 GPP data from three site with corresponding site data of CSIF satellite products. (**b**) The linear regression fitting of 2020 GPP data from three site with corresponding site data of RTSIF satellite products. (**c**) The linear regression fitting of 2020 GPP data from three site with corresponding site data of SIF-OCO-005 satellite products. (**d**) The linear regression fitting of 2020 GPP data from three site with corresponding site data of GOSIF satellite products.

3.2.2. Analysis of Spatial Features

From Figure 4, it can be seen that there are differences in the spatial distribution characteristics of the annual average values of RTSIF, CSIF, SIF_OCO2_005, and GOSIF satellite products in arid regions. The RTSIF, CSIF, and SIF_OCO2_005 satellite products exhibited reasonably distributed spatial patterns with distinct attribute features. The highest values are predominantly distributed in regions with high vegetation cover, such as the Altai Mountains and the northern and southern slopes of the Tianshan Mountains, whereas the lowest values are found in areas with low vegetation cover, such as the Gurbantunggut Desert and the eastern Gobi Desert. Specifically, in the pasture grassland area, the annual mean values are 0.10 $mw/m^2/nm/sr$ (RTSIF), 0.12 $mw/m^2/nm/sr$ (CSIF), and 0.13 $mw/m^2/nm/sr$ (SIF_OCO2_005). In the cultivated and farmland area, the values are 0.26 $mw/m^2/nm/sr$ (RTSIF), 0.18 $mw/m^2/nm/sr$ (CSIF), and 0.24 $mw/m^2/nm/sr$ (SIF_OCO2_005). In the desert vegetation area, the values are 0.03 $mw/m^2/nm/sr$ (RTSIF), 0.01 $mw/m^2/nm/sr$ (CSIF), and 0.04 $mw/m^2/nm/sr$ (SIF_OCO2_005). The annual mean values of RTSIF satellite products are closer to the observed GPP data at the stations.

In contrast, the spatial distribution of the GOSIF satellite products is unreasonable in arid regions, with indistinct attribute features and an overall overestimation tendency. The rationality of the attribute features of these products in arid regions is ranked as follows: RTSIF > CSIF > SIF_OCO2_005 > GOSIF. The RTSIF satellite products most effectively reflects the spatial features of GPP in the study area.

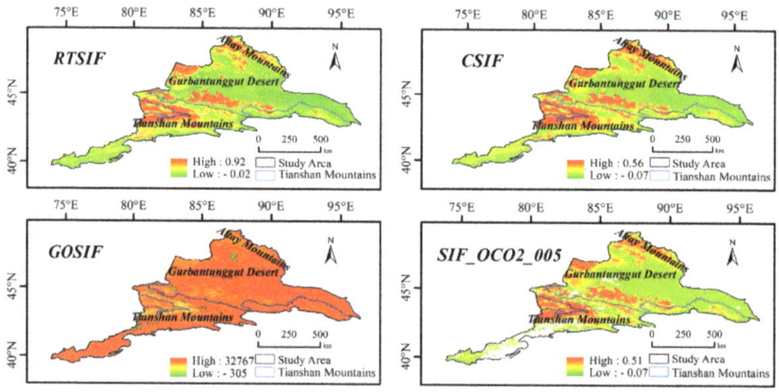

Figure 4. The spatial distribution characteristics of annual mean values of multisource SIF satellite products.

3.2.3. Analysis of the Impact Factor Responsiveness

From Figure 5, it can be seen that the responsiveness of RTSIF, CSIF, SIF_OCO2_005, and GOSIF satellite products to the main influencing factors (photosynthetically active radiation (PAR), soil temperature (Tsoil), air temperature (Tair)) of GPP varies in different ecosystems in arid regions. In the desert vegetation area (Figure 5a), RTSIF exhibited the highest responsiveness to Tsoil, PAR, and Tair, followed by CSIF, and GOSIF showed the weakest response. In the pasture grassland area (Figure 5b), RTSIF showed the highest responsiveness to PAR, followed by SIF_OCO2_005, while GOSIF exhibited the weakest response. For Tsoil and Tair, the highest responsiveness is observed for SIF_OCO2_005, followed by RTSIF, with GOSIF showing the weakest response. In the cultivated and farmland area (Figure 5c), the highest responsiveness to PAR and Tair was observed for RTSIF, followed by SIF_OCO2_005, with GOSIF exhibiting the weakest response. For Tsoil, the highest responsiveness was observed for SIF_OCO2_005, followed by RTSIF, with GOSIF exhibiting the weakest response.

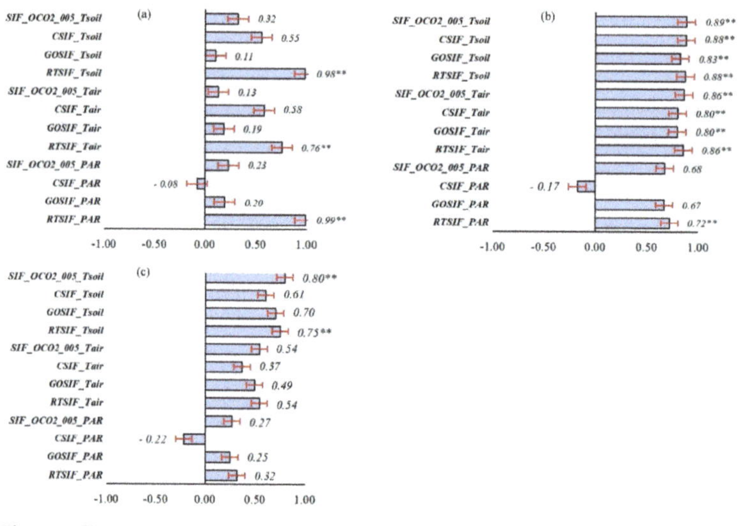

Figure 5. Responsiveness of multisource SIF satellite products to major influencing factors of GPP (** indicates significance at the 0.5 level). (**a**) Kelameili Station, desert vegetation area. (**b**) Ulastai Station, pasture and grassland area. (**c**) Ulan Usu Staion, cultivate land and farmland area.

These results indicate that the responsiveness of the four satellite products in arid regions to the main influencing factors varies greatly among different ecosystems. The overall responsiveness to Tsoil is ranked as follows: SIF_OCO2_005 > RTSIF > CSIF > GOSIF. While the overall responsiveness to Tair and PAR is ranked as follows: RTSIF > SIF_OCO2_005 > CSIF > GOSIF. RTSIF shows an overall significance greater than 0.5 with influencing factors in different ecosystems, indicating that RTSIF satellite products have the highest overall responsiveness.

3.2.4. Analysis of GPP/SIF Values under Different Weather Conditions

From Figure 6, it can be seen that there are differences in GPP/SIF values in the study area under different weather conditions and ecosystems. In the desert vegetation area (Figure 6a), SIF_OCO2_005 and GOSIF exhibited severe overestimation under clear, overcast, and cloudy conditions, while CSIF showed greater overestimation under clear conditions and greater underestimation under overcast and cloudy conditions. RTSIF fluctuated around the 1:1 line, indicating reasonably accurate values. In the pasture grassland area (Figure 6b), SIF_OCO2_005, GOSIF, and CSIF all demonstrated significant overestimation under clear, overcast, and cloudy conditions, whereas RTSIF remained relatively close to the 1:1 line. In the cultivated and farmland area (Figure 6c), SIF_OCO2_005 generally exhibited significant underestimation under clear, overcast, and cloudy conditions, while GOSIF tended to overestimate, and CSIF showed a balance between overestimation and underestimation across different weather conditions. RTSIF maintained fluctuations around the 1:1 line.

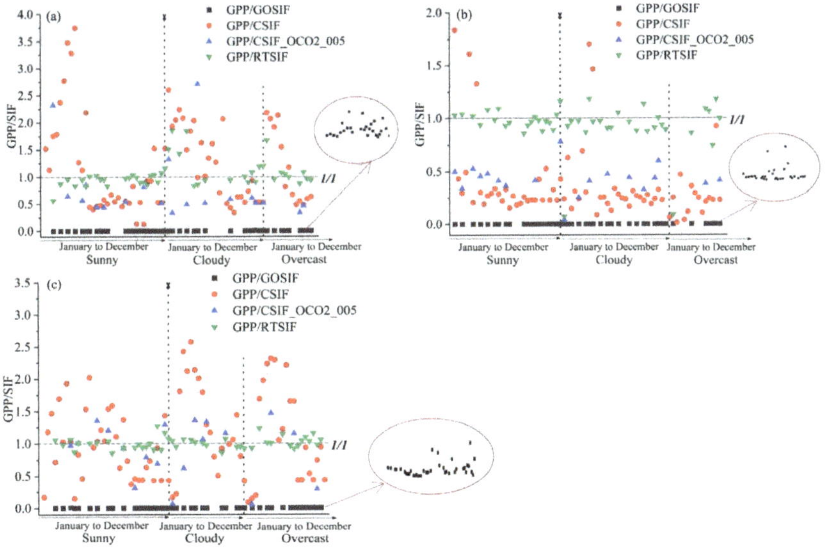

Figure 6. Differences in GPP/SIF values under different weather conditions. (**a**) Kelameili Station, desert vegetation area. (**b**) Ulastai Station, pasture and grassland area. (**c**) Ulan Usu Staion, cultivate land and farmland area.

These results indicate that the GPP/SIF values of the four SIF satellite products were overestimated or underestimated differently under different ecosystem and weather conditions in arid regions. SIF_OCO2_005 and CSIF showed varying degrees of severe overestimation and underestimation, whereas GOSIF demonstrated significant overestimation, and RTSIF maintained relatively reasonable values. The GPP/SIF values ranked overall as follows: RTSIF > CSIF > SIF_OCO2_005 > GOSIF. The RTSIF satellite products exhibited the highest overall rationality in GPP/SIF values.

3.3. SIF Data Accuracy Improvement Analysis

A comparison between multisource SIF satellite products and site data revealed that RTSIF satellite products performed better in estimating GPP than other satellite products in arid regions, but there is still considerable room for improvement in the applicability of RTSIF across different ecosystems. In order to advance research on GPP in arid regions, it is necessary to further refine RTSIF data. Therefore, we applied two correction methods to improve the RTSIF data in arid regions, and evaluated their applicability of two methods across the three ecosystems. Ultimately, the optimal improvement method for different ecosystems is determined, revealing the overall spatial distribution characteristics of SIF in arid regions and indirectly elucidating the spatial characteristics of GPP.

3.3.1. Analysis of Canopy-Based Accuracy Improvement

From Figure 7, it can be seen that the R^2 fitting values after improvement (based on the canopy improvement method) at each site are sorted by underlying surface conditions as follows: cultivated and farmland area (Figure 7a) = pasture grassland area (Figure 7b), with values equal to 0.90, maximizing close to the optimal value. The overall significance is greater than 0.5 with a confidence interval of 95%. Compared to before improvement, the R^2 values after improvement of each underlying surface increased by 0.06 (pasture grassland area) and 0.05 (cultivated and farmland area), respectively. Additionally, the overall improvement ranges of the MB, RMSE, and SD at each site were between 0.0009~0.0013, 0.0012~0.0065, and 0.0348~0.0864, respectively. The overall fitting parameter errors were improved to varying degrees, with the cultivated and farmland area showing the largest overall improvement and the pasture grassland area having the smallest improvement. This indicates that the canopy method has good applicability in areas with high vegetation coverage in arid regions.

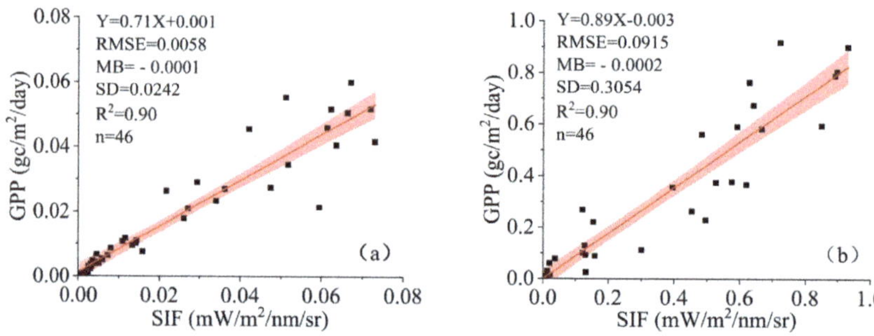

Figure 7. Linear fitting graph of 2020 GPP data and RTSIF corresponding station data for each station after improving based on the canopy method. (**a**) Ulastai Station, pasture and grassland area. (**b**) Ulan Usu Staion, cultivate land and farmland area.

However, owing to the short growth cycle and small leaf area of short-lived vegetation on the underlying surface of the Kelameili Station (desert vegetation area), the LAI in some areas during the corresponding time period cannot be accurately measured, resulting in the inability of the canopy method to accurately achieve the SIF data accuracy improvement for this type of underlying surface. This indicates that the canopy method is highly applicable in areas with dense vegetation in arid regions, whereas its applicability is weaker in areas with sparse vegetation, indicating certain limitations overall.

3.3.2. Analysis of Linear-Based Accuracy Improvement

From Figure 8, it can be seen that the R^2 fitting values after improvement (based on the linear improvement method) at each site are sorted by underlying surface conditions as follows: pasture grassland area (Figure 8b) > cultivated and farmland area (Figure 8c) > desert vegetation area (Figure 8a), with the corresponding values are 0.87, 0.85 and 0.75, respectively. All values were greater than 0.70 and close to the optimal value, indicating that 70% of the dependent variable variability in the underestimated desert vegetation area can be explained by the linear method. In addition, compared with before the improvement, the R^2 fitting values of the cultivated and farmland area, the pasture grassland area, and the desert vegetation area increased by 0.01, 0.02, and 0.13, respectively, showing a significant overall improvement effect.

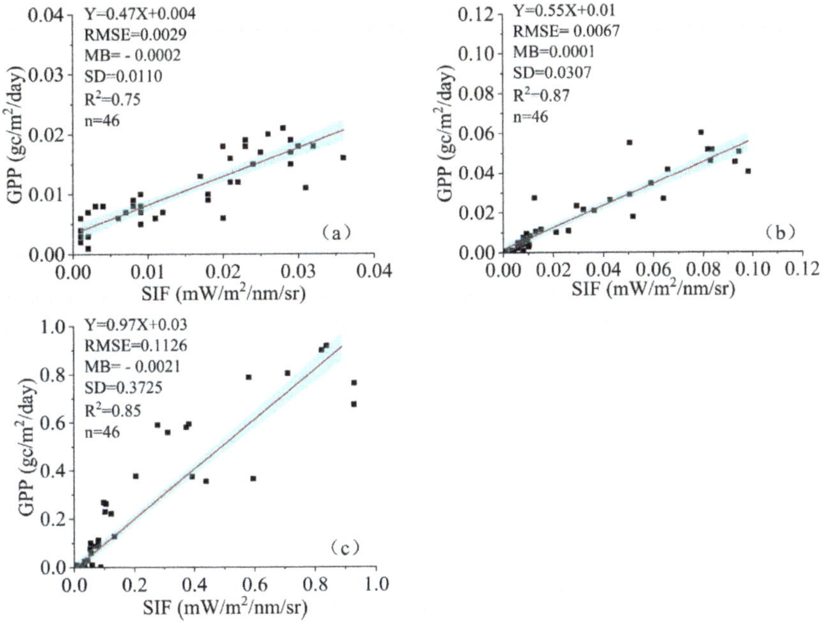

Figure 8. Linear fitting diagram between the 2020 GPP data of each station and the corresponding RTSIF station data after improving based on the linear method. (**a**) Kelameili Station, desert vegetation area. (**b**) Ulastai Station, pasture and grassland area. (**c**) Ulan Usu Staion, cultivate land and farmland area.

In addition, the overall improvement ranges of the MB, RMSE, and SD at each site ranged from 0.0001 to 0.0011, 0.0003 to 0.0146, and 0.0126 to 0.0799, respectively. Various fitting parameters showed different degrees of enhancement, with the greatest improvement observed in the desert vegetation area, and relatively minor improvements in the cultivated and farmland area and the pasture grassland area, the enhanced SIF data can further reflect the characteristics of GPP changes.

Overall, compared to before the improvements, SIF data are now more closely aligned with the measured GPP data. The parameters in the underestimated desert vegetation area showed the greatest overall improvement due to the linear method, whereas the cultivated and farmland area and the pasture grassland area exhibited a relatively smaller overall improvement. This suggests that while the linear method is applicable across different ecosystems, there are limitations when applying this method to improve the accuracy of SIF data in arid regions.

3.3.3. Comparative Analysis of Accuracy improvement Based on Canopy and Linear Methods

From Figure 9, it can be seen that the linear accuracy improvement method can enhance the accuracy of RTSIF data across different ecosystems in arid regions, with overall R^2 values exceeding 0.75 after improvement. However, the canopy accuracy improvement method is only applicable to areas with a higher leaf area index in arid regions, and it cannot achieve improvement in regions with low vegetation cover. Interestingly, the R^2 fitting values decrease overall for the pasture grassland area and cultivated and farmland area after improvement using the linear method, with decreases of 0.03 and 0.05, respectively. This indicates that while the linear method is suitable for various ecosystems in arid regions, its effectiveness in areas with high vegetation cover is not as pronounced as that of the canopy method.

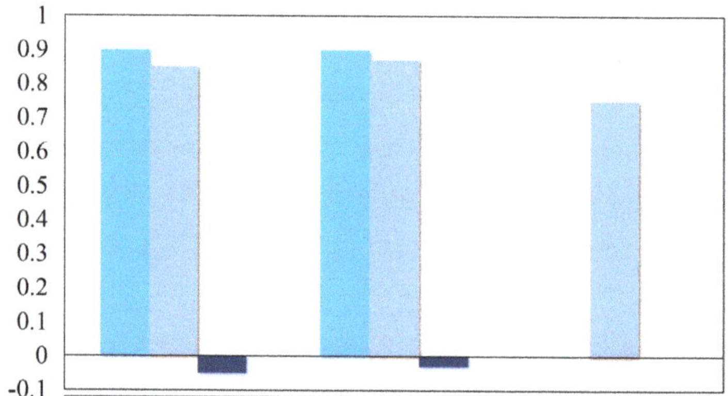

	Ulan Usu Staion	Ulastai Station	Kelameili Station
Canopy correction model	0.9	0.9	0
Linear correction model	0.85	0.87	0.75
Difference Value	-0.05	-0.03	0

Figure 9. The R^2 fitting values for various sites based on two accuracy improvement methods: canopy and linear.

The differences in the error fitting parameters after improvement based on both methods are listed in Table 5. Compared to the canopy method, for the pasture grassland area and the cultivated and farmland area using the linear improvement method results in increased MB, RMSE, and SD errors, with error increases ranging between 0~0.0021, 0.0009~0.0211, and 0.0065~0.0671, respectively. Conversely, for the desert vegetation area, using the linear improvement method results in relatively small MB, RMSE, and SD errors, with error values of 0.0002, 0.0029, and 0.0110, respectively, and the improvement effect is relatively significant. This further emphasizes that while the linear method has broader applicability than the canopy method in arid regions, its effectiveness is relatively weaker in areas with a higher leaf area index.

Table 5. The difference in error fitting parameters for various sites based on two improvement methods.

Station	Parameter	RMSE	|MB|	SD
Ulan Usu Station	Canopy improvement value	0.0915	0.0002	0.3054
	Linear improvement value	0.1126	0.0021	0.3725
	Difference	0.0211	0.0019	0.0671
Ulastai Station	Canopy improvement value	0.0058	0.0001	0.0242
	Linear improvement value	0.0067	0.0001	0.0307
	Difference	0.0009	0	0.0065
Kelameili Station	Canopy improvement value	-	-	-
	Linear improvement value	0.0029	0.0002	0.0110
	Difference	-	-	-

In summary, the linear method can enhance the accuracy of SIF data across different ecosystems in arid regions and further reduce errors between parameters after improvement. However, its effectiveness in areas with high vegetation cover is slightly weaker than that of the canopy method. Therefore, the canopy method can be utilized to improve the accuracy of SIF data in regions with high vegetation cover in arid regions, whereas the linear method can be used, as a supplement, to enhance the accuracy of SIF data for other types of underlying surfaces.After careful verification, we found that the title did not express it clearly, which was due to our negligence. We deeply apologize for this. We have rephrased the title, and the revised title is clear and reasonable. We have marked the modified content in red.

3.4. Spatial Analysis of SIF Data Accuracy Improvement before and after

The spatial characteristics of the average values of RTSIF satellite data before improvement for each quarter (March to May are spring, etc.) are shown in Figure 10 (left column). High values of SIF data are primarily distributed in the cropland area on the northern and southern slopes of the Tianshan Mountains, as well as in the pasture grassland area of the Tianshan and Altai Mountains. Conversely, low values are predominantly found in regions with a smaller leaf area index, such as the Gurbantunggut Desert (desert vegetation area) and the eastern Gobi Desert.

Combining land use data with two precision improvement methods, the final precision improvement results achieved by the RTSIF satellite are shown in Figure 10 (right column). The overall improvement in the mean values of the SIF data for each quarter was 0.11% (0.037% for cultivated and farmland area, 0.028% for pasture grassland area, 0.016% for desert vegetation area, and 0.025% for other area). The most significant improvement was observed in the spring and summer seasons (improvement rate of 0.071%). Specifically, in spring, as snow begins to melt, vegetation (including short-lived vegetation) begins to grow, the contribution of SIF has increased. Which is coherent with the overall enhancement of SIF attributes in the study area after improvement. In particular, significant improvements were observed in regions with a higher leaf area index, such as the northwestern part of the Altai Mountains and both slopes of the Tianshan Mountains. The attributes of the Gurbantunggut Desert also showed a synchronized enhancement, with results falling within a reasonable attribute range for arid regions [57,58].

In summer, most regions have completed snowmelt, and vegetation continues to grow as solar radiation increases. The contribution of the SIF further increases, which is consistent with the overall enhancement of SIF attributes in the study area after improvement. In particular, regions with higher vegetation cover, such as the Altai Mountains and both slopes of the Tianshan Mountains, experience a further increase in the enhancement rate. The attributes in desert areas and similar regions remain within a reasonable range for arid regions [59–61].

In autumn and winter, snow begins to accumulate in most regions, and vegetation growth slows as the overall solar radiation decreases, the contribution of SIF decreases, and the overall change is not significant. This aligns with the overall lower and relatively less pronounced enhancement of the SIF attributes in the study area after improvement. The western part and both slopes of the Tianshan Mountains are areas where the enhancement of SIF attributes is more pronounced during the autumn and winter.

Additionally, after improvement, the annual average SIF (GPP) data for various underlying surfaces in the northern region of Xinjiang was 0.13 mw/m^2/nm/sr (0.26 mw/m^2/nm/sr for cultivated and farmland area, 0.14 mw/m^2/nm/sr for pasture grassland area, 0.034 mw/m^2/nm/sr for desert vegetation area, and 0.11 mw/m^2/nm/sr for other area).

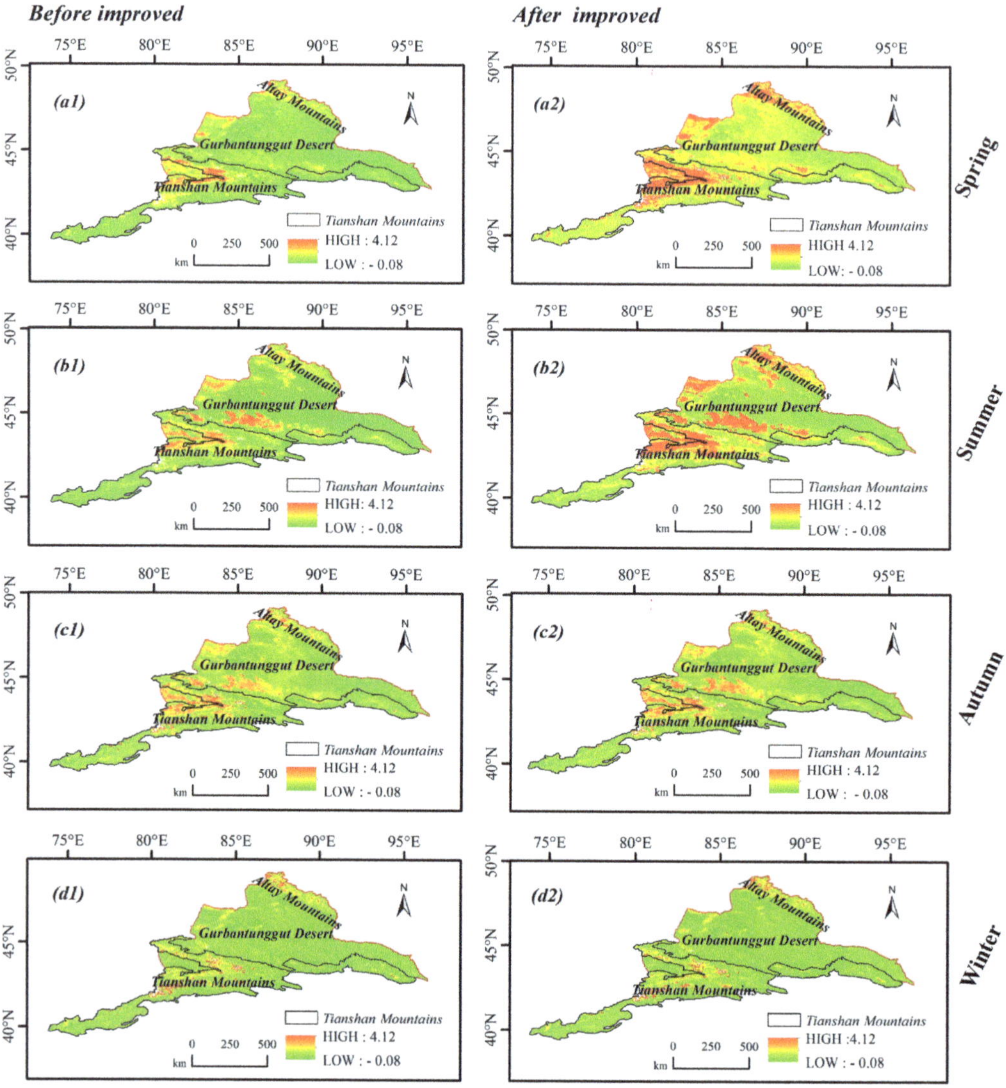

Figure 10. Changes in spatial characteristics of quarterly average values before and after the improvement of SIF satellite product data. (**a1**–**d1**) The spatial variation characteristics of the mean values of each season before improvement, (**a1**) for spring, and so on. (**a2**–**d2**) The spatial variation characteristics of the mean values of each season after improvement, (**a2**) for spring, and so on.

4. Discussion

4.1. Analysis of the Applicability of Multisource SIF Data in Estimating GPP

The overall applicability of using multisource SIF to comprehensively evaluate GPP data in arid regions shows that the overall ranking of the applicability of the four satellites for GPP estimation is as follows: RTSIF > CSIF > SIF_OCO2_005 > GOSIF (based on spatial characteristics, responsiveness to GPP influencing factors, GPP/SIF values under different weather conditions, and other standards). The significance of RTSIF is greater than 0.5, with a confidence interval of 95%. This is because the sensors and generation principles of the four SIF satellite products are different. Among them, CSIF, GOSIF, and SIF_OCO2_005 all come from OCO_2 sensors. This sensor has an early launch time and a narrow spectral band (757–775 nm) [62], leading to a lack of certain physiological and physical significance in the generated satellite data.

However, RTSIF is derived from the TROPOMI sensor on Sentinel-5P, which was launched later and has a more comprehensive spectral band (735–785 nm) [21,22], leading to a more refined understanding of various mechanisms in the obtained SIF data set. Furthermore, CSIF, GOSIF, and SIF_OCO2_005 all utilize the MODIS data set, but MODIS exhibits a noticeable lag effect [63], resulting in a significant deviation from GPP, thus reducing its inversion accuracy to some extent. In contrast, on the basis of the complete sensor band, RTSIF comprehensively considers different weather conditions to generate [21,22], which to some extent improves the inversion accuracy.

Additionally, the applicability of RTSIF in GPP estimation varied across the three ecosystems in the study area. The pasture grassland area exhibits the highest R^2 value of 0.85, with a significance greater than 0.5 and a confidence interval of 95%, indicating the best suitability for GPP estimation. This is attributed to the relatively stable growth cycle of the grassland in Xinjiang, where the green-up period occurs between days 110 and 150 each year, the growing season falls between days 140 and 160, and senescence appears between days 270 and 290. The overall growth cycle of the grassland is stable, with no significant interannual fluctuations in SIF contribution and relatively stable attribute characteristics. Consequently, the suitability of SIF data for GPP estimation in this area was slightly higher than that of the cropland area with a larger leaf area index. This finding is consistent with the conclusions of Dong Tong's [64] research, who utilized machine learning and model construction methods to study the spatiotemporal and phenological characteristics of grasslands in Xinjiang over the last 20 years. In the cultivated and farmland area, such as those cultivating maize and cotton, GPP and SIF are much higher than those of forests and grasslands, with an overall high measurability. However, because of their shorter growth cycle and significant reduction in SIF contribution during the nongrowing season, their suitability for GPP estimation using SIF data is relatively lower than that of the pasture grassland area with an R^2 fitting value of 0.84, but with a significance also greater than 0.5 and a confidence interval of 95%. This is consistent with the conclusion of Chen Xin [65], who used multiple crops for global farmland GPP estimations.

The carbon sink in the desert vegetation area is primarily generated through non-photosynthetic processes, with complex controlling factors and trends. This conclusion is drawn from research by Yang Fan et al. [66], who utilized comparative experiments to demonstrate CO_2 characteristics in the Taklimakan Desert. This indicates that the overall measurability of SIF in the desert vegetation area is relatively low, thus resulting in a weaker applicability of SIF data for GPP estimation on these underlying surfaces. In this study, in the desert vegetation area (Gurbantunggut Desert), the GPP during the growing season of short-lived vegetation shows a trend of increasing initially and then decreasing. Outside the growing season, the interannual variation of GPP shows an inverted "U" shape and indicates a carbon sink. The interannual variations of GPP and SIF are relatively complex, with small overall accumulations of GPP and contributions of SIF. This further validates the aforementioned research conclusion and is consistent with the findings of Gulinur et al. [67] regarding CO_2 fluxes in the Gurbantunggut Desert.

This study validated the applicability of SIF data for GPP estimation in arid regions, the findings of this research are crucial for subsequent comprehensive estimation and feature analysis of GPP in Xinjiang and even the entire arid region based on SIF data.

4.2. Analysis of GPP Estimation Accuracy Based on Improving SIF

The results of precision improvement of SIF data for arid regions indicate that the linear methods can enhance SIF data accuracy for different underlying surfaces (R^2 increased values for the cultivated and farmland area, the pasture grassland area, and the desert vegetation area were 0.01, 0.02, and 0.13, respectively). However, canopy methods are only applicable to regions with a higher leaf area (R^2 increased values for the cultivated and farmland area and pasture grassland area were 0.06 and 0.05, respectively), and cannot be improved for areas with low vegetation cover. The adaptability of canopy methods was validated in a study by Yin Yueqiang [68], who considered the canopy as a factor for global SIF data precision improvement, ultimately reconstructing six sets of SIF data at different resolutions. The adaptability of linear methods was confirmed in a study by Wang Yu et al. [69], where linear methods were used to improve the precision of five solar radiation reanalysis data sets in the eastern Gobi Desert in Xinjiang, eliminating the impact of errors in the reanalysis data on radiation assessments. In addition, in areas of desert vegetation, the error reduction values of MB, RMSE, and SD based on the linear improvement method were 0.0001, 0.0007, and 0.0126, respectively. However, using the linear method to improve the areas of the pasture grassland and cultivated and farmland, except for a decrease in the R^2 fitting value, other fitting parameters, on the contrary, increased overall, with the error increase ranges of the MB, RMSE, and SD between 0.0006~0.0011, 0.0003~0.0146 and 0.0323~0.0799, respectively. This further indicates that the linear method has a wider applicability than the canopy method in arid regions, but the improvement effect is relatively weak in areas with high vegetation coverage. Therefore, the canopy and linear improvement methods can be alternately used to improve the accuracy of SIF satellite products, then the two methods can be integrated to ultimately achieve a precision improvement of SIF data in arid regions.

Furthermore, in conjunction with land use data, SIF data improvement was conducted using the canopy and linear methods, with an overall improvement rate of 0.11% (0.037% for cultivated and farmland area, 0.028% for pasture grassland area, 0.016% for desert vegetation area, 0.025% for other area). After improvement, areas with higher vegetation cover, such as cropland on the northern and southern slopes of the Tianshan Mountains and pasture grassland in the Tianshan and Altai Mountains, exhibited the highest attribute values. Conversely, lower values were mainly distributed in areas with smaller proportions of the leaf area index, such as the Gurbantunggut Desert (desert vegetation area) and the eastern Gobi Desert.

Additionally, the postimprovement annual average SIF (GPP) values for various underlying surfaces in the northern region of Xinjiang are 0.13 $mw/m^2/nm/sr$. Serving as effective proxies for GPP, the spatial distribution of improved SIF better reflects the GPP distribution characteristics in arid regions and further reveals the strong coupling relationship between the GPP and SIF in these areas. The rationality of the attribute values after improvement aligns with the conclusions of numerous scholars studies in arid regions. For example, Li Yue et al. [70] conducted remote sensing monitoring of grassland GPP on the Mongolian Plateau based on SIF data, showing that the annual average values for various grasslands range from 0.11 to 3.48 gC/m^2 (0.14 $mw/m^2/nm/sr$ in this study); Yan Zhirong et al. [71] conducted a study on the spatiotemporal distribution of vegetation GPP in China from 2007 to 2018 based on SIF data, indicating that the annual average values for the desert vegetation area (based on latitude and longitude division) range from 0 to 0.1 gC/m^2 (0.034 $mw/m^2/nm/sr$ in this study); and Song Lian [72] conducted a comparative study on the high-temperature stress mechanisms of crops based on SIF data, showing that the annual values for various crops range from 0.22 to 4.42 gC/m^2 (0.26 $mw/m^2/nm/sr$ in this study).

4.3. Innovation, Limitations, and Prospects

The innovation of this study lies in coupling multiple evaluation indicators (linear regression parameters, satellite spatial characteristics, GPP influencing factor responsiveness, and GPP/SIF values under different weather conditions) to comprehensively compare and analyze the applicability of four commonly used SIF satellite products for GPP estimation in arid regions, and identify which SIF satellite product is most suitable for GPP estimation in arid regions. Additionally, it verifies the feasibility of canopy and linear accuracy improvement methods for SIF accuracy improvement in arid regions based on the most suitable SIF satellite product. By integrating land use data, it also reveals the spatial distribution patterns of SIF (GPP) in arid regions. These results have not been clearly demonstrated in previous studies, thus our research fills the gap in the coupling studies of SIF and GPP in arid regions and lays a theoretical foundation for achieving the "carbon neutrality" goal in these areas.

The limitation of this study is that the climate change in Xinjiang is complex and there are some uncontrollable influencing factors, which may have a certain impact on the comprehensive assessment of applicability. Other methods for improving the accuracy of SIF satellite products in arid regions need to be further developed and validated. There are often biases and uncertainties in data measurement, which result in some abnormal and outlier data, as well as incomplete and nonstationary time series data. These data have a certain impact on the operability of statistical analysis, reasonable estimation of parameters, and effective analysis of dependent variables. Although we have comprehensively processed these data in detail and reasonably, there are still subtle impacts that are inevitable. Additionally, the research results may also be influenced by the limitations of data resolution and remote sensing technology, requiring careful consideration in the application of the results.

Based on the current research results, we will further explore models suitable for GPP inversion in arid regions in the future, coupling the improved SIF with these models to achieve GPP inversion in arid regions, thus analyzing the spatial and temporal patterns of GPP in arid regions. Additionally, we will use measured data of influencing factors such as PAR, Tair, and Tsoil, along with satellite data, to comprehensively analyze the spatial and temporal characteristics of GPP influencing factors/mechanisms in arid regions.

5. Conclusions

This study comprehensively evaluated the applicability of multisource SIF satellite products for GPP estimation in arid regions using various indicators, adopted multiple methods to improve the accuracy of SIF satellite products that are most suitable for GPP estimation in arid areas, and comprehensively analyzed the spatial characteristics of GPP indirectly reflected by SIF data before and after improvement. The final research conclusions are as follows:

(1) The interannual variation of the monthly mean GPP in arid regions shows an inverted "U" shape, with peaks occurring in June and July. During the growing season (March to October), GPP first increases and then decreases, while in the nongrowing season (November to February), GPP fluctuations are not significant.

(2) The overall suitability ranking of multisource SIF satellite products for GPP estimation in arid regions is as follows: RTSIF > CSIF > SIF_OCO2_005 > GOSIF. This has a profound significance for revealing the spatial and temporal patterns of the terrestrial ecosystem carbon cycle in arid regions by coupling multiple factors and provides new approaches for constructing carbon reduction policies in arid regions.

(3) When improving the accuracy of SIF satellite products in arid regions, both the canopy improvement method and the linear improvement method need to be used in combination. This provides practical theory for achieving a more comprehensive and higher accuracy analysis of carbon source/sink spatial and temporal characteristics in arid region terrestrial ecosystems, which is of great significance for achieving "carbon neutrality" in arid regions.

(4) Based on land use data, the spatial characteristics of SIF data in arid regions achieved through the two methods showed a high correlation with vegetation coverage, with the annual mean value of SIF data for each surface after improvement being approximately 0.13 $mw/m^2/nm/sr$.

6. Practical Applications

Based on the research conclusions, the practical applications of this study in arid regions are mainly reflected in the following aspects:

(1) By revealing the interannual variation characteristics of GPP in arid regions, relevant theories can be directly referenced in the subsequent construction of the carbon cycle system in arid regions, thereby avoiding unreasonable interannual variations.
(2) By revealing the most suitable SIF satellite products for GPP estimation in arid regions, the relevant satellites can be directly applied in subsequent analysis of the spatial and temporal patterns of carbon storage in arid regions based on GPP, an important factor of carbon source/sink, thus avoiding repeated comparative validation.
(3) By revealing the methods for improving the accuracy of SIF satellite products in arid regions, these methods can be directly applied in subsequent accuracy improvement of other SIF satellite products in arid regions, thus avoiding repeated exploration and analysis.
(4) By revealing the spatial characteristics of GPP indirectly reflected by SIF in arid regions, accurate carbon reduction policies can be directly constructed based on the spatial patterns to achieve "carbon neutrality" in arid regions, thus avoiding discrepancies between practice and reality.

Author Contributions: W.L.: Conceptualization, data curation, investigation, methodology, resources, project administration, software, visualization, writing—original draft, writing—review and editing. Y.W.: software, supervision. A.M.: conceptualization, funding acquisition, supervision, writing—review and editing, resources. Y.L.: conceptualization, methodology, writing—review and editing, supervision. J.G.: data curation. M.S.: validation. A.A.: data curation. C.W.: validation. F.Y., C.Z., W.H., J.P. and H.S.: formal analysis. All authors have read and agreed to the published version of the manuscript.

Funding: This study received funding support from the following projects. The National Natural Science Foundation of China, grant number: 42305132, funder: Yu Wang; The Youth Innovation Team of the China Meteorological Ad-ministration, grant number: CMA2024QN13, funder: Fan Yang; Research on the Carbon Budget and Influencing Factors of Grassland Ecosystem in the Central Tianshan Mountains, grant number: KJFZ202403, funder: Ali Mamtimin; The Scientific and Technological Innovation Team (Tianshan Innovation Team) project, grant number: 2022TSYCTD0007, funder: Ali Mamtimin; The National Natural Science Foundation of China, grant number: 42375084, funder: Ali Mamtimin; The Special Project for the Construction of Innovation Environment in the Autonomous Region, grant number: PT2203, funder: Ali Mamtimin; And the Special Funds for Basic Scientific Research Business Expenses of Central-level Public Welfare Scientific Research Institutes, grant number: IDM2021005 and IDM2021001, funder: Ali Mamtimin and Yu Wang.

Data Availability Statement: The site data used in this article is confidential and can be obtained from A.M. (ali@idm.cn) upon request.

Acknowledgments: The SIF satellite product data were obtained from the National Tibetan Plateau Scientific Data Center (http://data.tpdc.ac.cn, accessed on 18 September 2023), Earth System Research Center, and the Earth Data Center. The situ observation data were obtained from the Institute of Desert Meteorology, China Meteorological, Urumqi. We extend our sincere appreciation for their substantial contributions to the completion of this paper.

Conflicts of Interest: The author declares that all individuals included in this article have consented to the acknowledgement. There are no known competing financial interests or personal relationships that could have appeared to influence the work reported in this paper.

References

1. Anav, A.; Friedlingstein, P.; Beer, C.; Ciais, P.; Zhao, M. Spatiotemporal patterns of terrestrial gross primary production: A review. *Rev. Geophys.* **2015**, *53*, 785–818. [CrossRef]
2. Jung, M.; Reichstein, M.; Margolis, H.A.; Cescatti, A.; Richardson, A.D.; Arain, M.A.; Arneth, A.; Bernhofer, C.; Bonal, D.; Chen, J.Q.; et al. Global patterns of land-atmosphere fluxes of carbon dioxide, latent heat, and sensible heat derived from eddy covariance, satellite, and meteorological observations. *J. Geophys. Res. Biogeosci.* **2011**, *116*, G3. [CrossRef]
3. Garbulsky, M.F.; Filella, I.; Verger, A.; Pe Uelas, J. Photosynthetic light use efficiency from satellite sensors: From global to Mediterranean vegetation. *Environ. Exp. Bot.* **2014**, *103*, 3–11. [CrossRef]
4. Zhang, X.; Wang, H.; Yan, H.; Ai, J. Remote sensing study on the spatiotemporal variation of China's total primary productivity from 2001 to 2018. *Acta Ecol. Sin.* **2021**, *41*, 6351–6362.

5. Sun, Y.; Frankenberg, C.; Jung, M.; Joiner, J.; Guanter, L.; Köhler, P.; Magney, T. Overview of Solar-Induced chlorophyll Fluorescence (SIF) from the Orbiting Carbon Observatory-2: Retrieval, cross-mission comparison, and global monitoring for GPP. *Remote Sens. Environ.* **2018**, *209*, 808–823. [CrossRef]
6. Xiao, J.; Chevallier, F.; Gomez, C.; Guanter, L.; Hicke, J.A.; Huete, A.R.; Ichii, K.; Ni, W.J.; Pang, Y.; Rahman, A.F.; et al. Remote sensing of the terrestrial carbon cycle: A review of advances over 50 years. *Remote Sens. Environ.* **2019**, *233*, 111383. [CrossRef]
7. Lasslop, G.; Reichstein, M.; Papale, D.; Richardson, A.D.; Arneth, A.; Barr, A.; Stoy, P.; Wohlfahrt, G. Separation of net ecosystem exchange into assimilation and respiration using a light response curve approach: Critical issues and global evaluation. *Glob. Chang. Biol.* **2010**, *16*, 187–208. [CrossRef]
8. Baldocchi, D. TURNER REVIEW No.15. 'Breathing' of the terrestrial biosphere: Lessons learned from a global network of carbon dioxide flux measurement systems. *Aust. J. Bot.* **2008**, *56*, 1–26. [CrossRef]
9. Yu, G.; Fu, Y.; Sun, X.; Wen, X.; Zhang, L. Research progress and development ideas of ChinaFLUX, the Chinese Terrestrial Ecosystem Flux Research Network. *China Sci. D Earth Sci.* **2006**, *S1*, 1–21. [CrossRef]
10. Huang, P. *Estimation of Terrestrial Gross Primary Productivity Based on Solar-Induced Chlorophyll Fluorescence*; Wuhan University of Technology: Wuhan, China, 2019.
11. Tramontana, G.; Jung, M.; Schwalm, C.R.; Ichii, K.; Camps-Valls, G.; Ráduly, B.; Reichstein, M.; Arain, M.A.; Cescatti, A.; Kiely, G.; et al. Predicting carbon dioxide and energy fluxes across global FLUXNET sites with regression algorithms. *Biogeosciences* **2016**, *13*, 4291–4313. [CrossRef]
12. Sun, Z.; Gao, X.; Du, S.; Liu, X. Research progress and prospect of satellite remote sensing products of solar-induced chlorophyll fluorescence. *Remote Sens. Technol. Appl.* **2021**, *36*, 1044–1056.
13. Gao, H.Q.; Liu, S.G.; Lu, W.Z.; Smith, A.R.; Valbuena, R.; Yan, W.D.; Wang, Z.; Li, X.; Peng, X.; Li, Q.Y.; et al. Global Analysis of the Relationship between Reconstructed Solar-Induced Chlorophyll Fluorescence (SIF) and Gross Primary Production (GPP). *Remote Sens.* **2011**, *13*, 2824. [CrossRef]
14. Wang, Y.; Chi, Y.; Zhou, L. 2007-2018 Spatiotemporal Patterns and Climatic Regulation of China's Terrestrial Vegetation Gross Primary Production and Solar-Induced Chlorophyll Fluorescence. *Remote Sens. Technol. Appl.* **2022**, *37*, 692–701.
15. Qiu, R.N.; Han, G.; Li, X.; Xiao, J.F.; Liu, J.G.; Wang, S.H.; Li, S.W.; Gong, W. Contrasting responses of relationship between solar-induced fluorescence and gross primary production to drought across aridity gradients. *Remote Environ.* **2024**, *302*, 113984. [CrossRef]
16. Wei, X.; He, W.; Zhou, Y.; Ju, W.M.; Xiao, J.F.; Li, X.; Liu, Y.B.; Xu, S.H.; Bi, W.J.; Zhang, X.Y.; et al. Global assessment of lagged and cumulative effects of drought on grassland gross primary production. *Ecol. Indic.* **2022**, *136*, 108646. [CrossRef]
17. Liu, X.; Lai, Q.; Yin, S.; Bao, Y.H.; Qing, S.; Mei, L.; Bu, L.X. Exploring sandy vegetation sensitivities to water storage in China's arid and semi-arid regions. *Ecol. Indic.* **2022**, *136*, 108711. [CrossRef]
18. Wang, X. *Understanding Carbon Uptake Using Multi-Scale Novel Remote Sensing Techniques*; The University of Arizona: Tucson, AZ, USA, 2022.
19. Zhang, Y.; Joiner, J.; Alemohammad, S.H.; Zhou, S.; Gentine, P. A global spatially contiguous solar-induced fluorescence (CSIF) dataset using neural networks. *Biogeosciences* **2018**, *15*, 5779–5800. [CrossRef]
20. Zhang, Y. *A Global Spatially Contiguous Solar-Induced Fluorescence (CSIF) Dataset Using Neural Networks (2000–2022)*; National Tibetan Plateau/Third Pole Environment Data Center: Beijing, China, 2021. [CrossRef]
21. Chen, X.; Huang, Y.; Nie, C.; Zhang, S.; Wang, G.Q.; Chen, S.L.; Chen, Z.C. A long-term reconstructed TROPOMI solar-induced fluorescence dataset using machine learning algorithms. *Sci. Data* **2023**, *09*, 1038. [CrossRef]
22. Chen, X.; Huang, Y.; Nie, C.; Zhang, S.; Wang, G.Q.; Chen, S.L.; Chen, Z.C. *Global High-Resolution (8 Days, 0.05°) Solar-Induced Fluorescence Dataset (2001–2020)*; National Tibetan Plateau Data Center: Beijing, China, 2022. [CrossRef]
23. Li, X.; Xiao, J. A global, 0.05-degree product of solar-induced chlorophyll fluorescence derived from OCO_2, MODIS, and reanalysis data. *Remote Sens.* **2019**, *11*, 517. [CrossRef]
24. Yu, L.; Wen, J.; Chang, C.Y.; Frankenberg, C.; Sun, Y. *High Resolution Global Contiguous SIF Estimates from OCO_2 SIF and MODIS*, 2nd ed.; ORNL DAAC: Oak Ridge, TN, USA, 2021. [CrossRef]
25. An, Y.; Zhang, R.; Liu, W.; Wang, P.; Zhang, J.; Wu, L.; Sun, Z. Differential analysis of different SIF products in rubber plantation areas of Hainan Island and their impact on GPP estimation. *J. Trop. Biol.* **2023**, *19*, 074028.
26. Dang, C.Y.; Shao, Z.F.; Huang, X.; Qian, J.X.; Cheng, G.; Ding, Q.; Fan, Y.W. Assessment of the importance of increasing temperature and decreasing soil moisture on global ecosystem productivity using solar-induced chlorophyll fluorescenc. *Glob. Chang. Biol.* **2022**, *28*, 2066–2080. [CrossRef] [PubMed]
27. Frankenberg, C.; Berry, J. Solar Induced Chlorophyll Fluorescence: Origins, Relation to Photosynthesis and Retrieval. *Compr. Remote Sens.* **2018**, *3*, 143–162.
28. Yu, L.; Wen, J.; Chang, C.Y.; Frankenberg, C.; Sun, Y.; Frankenberg, C.; Sun, Y. High-Resolution Global Contiguous SIF of OCO_2. *Geophys. Res. Lett.* **2019**, *46*, 1449–1458. [CrossRef]
29. Li, J.; Pu, Z.; Zhang, S. *Impact of Climate Change on Agriculture in Xinjiang and Regionalization*; China Meteorological Press: Beijing, China, 2018.
30. Hu, R.; Jiang, F.; Wang, Y. Environmental assessment of snow and ice water resources in Xinjiang. *Arid Zone Res.* **2003**, *20*, 187–191.

31. Apar, R.; Yerkejiang, H.; Li, Q.; Huang, J.; Zhang, S.Q.; Changji Prefecture Meteorological Bureau, Xinjiang; Institute of Desert Meteorology, China Meteorological Administration, Urumqi; Xinjiang Agricultural Meteorological Observatory. Analysis of the spatial and temporal characteristics of temperature in Xinjiang. *Hubei Agric. Sci.* **2022**, *61*, 34–41.
32. Zou, C.; Ji, C.; Fan, Z.; Huang, J.; Chen, D.H.; Li, B. Study on water vapor and carbon dioxide fluxes in cotton fields in Ulan Usu oasis. *Chin. Agric. Sci. Bull.* **2014**, *30*, 120–125.
33. Xiao, W.Q.; Mamtimin, M.; Liu, Y.Q.; Wang, Y.; Gao, J.C.; Ajiguli, S.; Gulini, S.M. Evolution of surface radiation balance in the central Tianshan Mountains. *Acta Ecol. Sin.* **2022**, *42*, 4550–4560.
34. Zhang, K.; Wang, Y.; Mamtimin, A.; Liu, Y.Q.; Gao, J.C.; Aihaiti, A.; Wen, C.; Song, M.Q.; Yang, F.; Zhou, C.L.; et al. Temporal and Spatial Variations in Carbon Flux and Their Influencing Mechanisms on the Middle Tien Shan Region Grassland Ecosystem, China. *Remote Sens.* **2023**, *15*, 4091. [CrossRef]
35. Mamtimin, A. *Study on Carbon Balance Characteristics and Influencing Factors in Desert Areas of Xinjiang*; Nanjing University of Information Science and Technology: Nanjing, China, 2015.
36. Gao, J.; Wang, Y.; Aiguli, S.; Mamtimin, A.; Liu, Y.Q.; Zhao, X.S.; Yang, X.H.; Huo, W.; Yang, F.; Zhou, C. Characteristics of surface radiation balance in Gurbantunggut Desert. *Desert Res.* **2021**, *41*, 47–58.
37. Kai, Y.; Wang, J.; Weiss, M.; Myneni, R.B. *A High-Quality Reprocessed MODIS Leaf Area Index Dataset (HiQ-LAI)*, 1st ed.; [Data Set]; Zenodo: Beijing, China, 2023. [CrossRef]
38. Xu, X.; Liu, J.; Zhang, S.; Li, R.; Yan, C.; Wu, S. *China's Multi-Period Land Use Remote Sensing Monitoring Dataset (CNLUCC)*; Resource and Environmental Science Data Registration and Publishing System: Beijing, China, 2018.
39. Long, T.F.; Jiao, W.L.; Zhang, Z.M.; He, G.J. Calculation Method of Satellite Image Pixel Observation Zenith Angle and Azimuth Angle. CN201410352963.8, 6 June 2017.
40. Pan, J. Arithmetic mean, geometric mean and theta function. *J. Harbin Norm. Univ. Nat. Sci. Ed.* **2014**, *30*, 21–22.
41. Feret, J.B.; Francois, C.; Asner, G.P.; Gitelson, A.A.; Martin, R.E.; Bidel, L.P.R.; Ustin, S.L.; Le Maire, G.; Jacquemoud, S. PROSPECT-4 and 5:Advances in the leaf optical properties model separating photosynthetic pigments. *Remote Sens. Environ.* **2011**, *112*, 3030–3043. [CrossRef]
42. Xinjie, L.J.; Luis, G.; Liangyun, L.Y.; Damm, A.; Malenovsky, Z.; Rascher, U.; Peng, D.L.; Du, S.S.; Gastellu-Etchegorry, J.P. Downscaling of solar-induced chlorophyll fluorescence from canopy level to photosystem level using a random forest model. *Remote Sens. Environ.* **2018**, *231*, 110772. [CrossRef]
43. Gueymard, C.A. Parameterized transmittance model for direct beam and circumsolar spectral irradiance. *Sol. Energy* **2001**, *71*, 325–346. [CrossRef]
44. He, L.M.; Chen, J.M.; Pisek, J.; Schaaf, C.B.; Strahler, A.H. Global clumping index map derived from the MODIS BRDF product. *Remote Sens. Environ.* **2012**, *119*, 118–130. [CrossRef]
45. Vickers, D.; Mahrt, L. Quality control and flux sampling problems for tower and aircraft data. *J. Atmos. Ocean. Technol.* **1997**, *14*, 512–526. [CrossRef]
46. Zhuang, J.; Wang, W.; Wang, J. Calculation of eddy correlation flux and comparison analysis of three major software. *Plateau Meteorol.* **2013**, *32*, 78–87.
47. Kaimal, J.C.; Finnigan, J. *Atmospheric Boundary Layer Flows: Their Structure and Measurement*; Oxford University Press: Oxford, UK, 1994.
48. Moore, C. Frequency-response corrections for eddy-correlation systems. *J. Appl. Meteorol.* **1986**, *37*, 17–35. [CrossRef]
49. Schotanus, P.; Nieuwstadt, F.T.M.; Debruin, H.A.R. Temperature-measurement with a sonic anemometer and its application to heat and moisture fluxes. *Bound.-Layer Meteorol.* **1983**, *26*, 81–93. [CrossRef]
50. Zhao, N.; Zhou, L.; Zhuang, J.; Wang, Y.L.; Zhou, W.; Chen, J.J.; Song, J.; Ding, J.X.; Chi, Y.G. Integration analysis of the carbon sources and sinks in terrestrial ecosystems, China. *Acta Ecol. Sin.* **2021**, *41*, 7648–7658.
51. Reichstein, M.; Falge, E.; Baldocchi, D.; Papale, D.; Valentini, R. On the separation of net ecosystem exchange into assimilation andecosystem respiration: Review and improved algorithm. *Glob. Chang. Biol.* **2005**, *11*, 1424–1439. [CrossRef]
52. Gu, L.H.; Fuentes, J.D.; Shugart, H.H.; Staebler, R.M.; Black, T.A. Responses of net ecosystem exchanges of carbon dioxide to changes in cloudiness: Results from two North American deciduous forests. *J. Geophys. Res. Atmos.* **1999**, *104*, 31421–31434.
53. Okogbue, E.C.; Adedokun, J.A.; Holmgren, B. Hourly and daily clearness index and diffuse fraction at a tropical station, Ile-Ife, Nigeria. *Int. J. Climatol.* **2009**, *29*, 1035. [CrossRef]
54. Yang, F. *Predictive Ability of Global Sun-Induced Chlorophyll Fluorescence for Vegetation Gross Primary Production and Its Response to Drought*; Nanjing University of Information Science & Technology: Nanjing, China, 2022. [CrossRef]
55. Glahn, H.R.; Lowry, D.A. The use of model output statistics (MOS) in objective weather forecasting. *J. Appl. Meteorol.* **1972**, *11*, 1203–1211. [CrossRef]
56. Polo, J.; Wilbert, S.; Ruizarias, J.A.; Meyer, R.; Gueymard, C.A.; Suri, M.; Martin, L.; Mieslinger, T.; Blanc, P.; Grant, I. *Integration of Ground Measurements with Model-Derived Data*; Environmental Science, Engineering: Cambridge, MA, USA, 2015.
57. Zhang, Y.L.; Yin, B.F.; Tao, Y.; Li, Y.G.; Zhou, X.B.; Zhang, Y.M. The first rainfall time in early spring and the impact of rainfall on the morphological characteristics and chlorophyll fluorescence of two short-lived plants in the Gurbantunggut Desert. *J. Plant Ecol.* **2022**, *46*, 428–439. [CrossRef]
58. Han, W.; Hamiti, Y.; Li, L.; Li, S.Y.; Zheng, T.T. The response of chlorophyll fluorescence parameters and stem water potential of Populus euphratica and Populus gray to the habitat in the hinterland of the Taklamakan Desert. *Chin. Desert* **2011**, *31*, 1472–1478.

59. Wu, L.Y.; Zhang, J.G.; Chang, W.Q.; Zhang, S.L.; Chang, Q. Diurnal variation characteristics of chlorophyll fluorescence parameters of three desert plants. *Pratacult. Sci.* **2021**, *30*, 203–213.
60. Han, W.; Cao, L.; Haimti, Y.; Xu, X. Response of chlorophyll fluorescence of *Tamarix ramosissima* to dust storm weather. *J. Desert Res.* **2012**, *32*, 86–91.
61. Li, L.; Li, X.Y.; Lin, L.S.; Wang, Y.J.; Xue, W. Application of chlorophyll fluorescence characteristics of 9 Poaceae forages on the photosystem II of the southern edge of the Taklimakan Desert. *Acta Ecol. Sin.* **2011**, *22*, 2599–2603.
62. Frankenberg, C.; O'Dell, C.; Berry, J.; Guanter, L.; Joiner, J.; Köhler, P.; Pollock, R.; Taylor, T.E. Prospects for chlorophyll fluorescence remote sensing from the Orbiting Carbon Observatory-2. *Remote Sens. Environ.* **2014**, *147*, 1–12. [CrossRef]
63. Liu, X.T.; Zhou, L.; Shi, H.; Wang, S.Q.; Chi, Y.G. Comparative analysis of phenological characteristics of temperate coniferous and broad-leaved mixed forests based on multiple remote sensing vegetation indices, chlorophyll fluorescence, and CO_2 flux data. *Acta Ecol. Sin.* **2018**, *38*, 13. [CrossRef]
64. Dong, T. Study on the Spatiotemporal Evolution Characteristics of Arid Climate and Its Impact on Grassland Phenology in Xinjiang. Ph.D. Thesis, Xinjiang Agricultural University, Urumqi, China, 2023.
65. Chen, X. Research on Global Total Primary Productivity Estimation of Multiple Crop Types. Master's Thesis, Nanjing University of Information Science & Technology, Nanjing, China, 2023.
66. Yang, F.; Huang, J.; Zheng, X.; Zheng, X.; Huo, W.; Zhou, C.; Wang, Y.; Han, D.L.; Gao, J.C.; Mamtimin, M.; et al. Evaluation of carbon sink in the Taklimakan Desert basedon correction of abnormal negative CO_2 flux of IRGASON. *Sci. Total Environ.* **2022**, *838*, 155988. [CrossRef] [PubMed]
67. Amar, G.; Mamtimin, A.; Wang, Y.H.; Wang, Y.; Gao, J.C.; Yang, F.; Song, M.Q.; Aihaiti, A.; Wen, C.; Liu, J.J. Factors controlling and variations of CO_2 fluxes during the growing season in Gurbantunggut Desert. *Ecol. Indic.* **2023**, *154*, 110708. [CrossRef]
68. Yin, Y. Reconstruction of Global Solar-Induced Chlorophyll Fluorescence Remote Sensing Data Considering Canopy Correction. Master's Thesis, Nanjing Normal University, Nanjing, China, 2021.
69. Wang, Y.; Zhao, X.S.; Mamtimin, A.; Sayit, H.; Abulizi, S.; Maturdi, A.; Yang, F.; Huo, W.; Zhou, C.L.; Yang, X.H.; et al. Evaluation of Reanalysis Datasets for Solar Radiation with In Situ Observations at a Location over the Gobi Region of Xinjiang, China. *Remote Sens.* **2021**, *13*, 4191. [CrossRef]
70. Li, Y.; Sun, Z.G. Spatiotemporal dynamics analysis of grassland productivity in the Mongolian Plateau based on remote sensing of chlorophyll fluorescence. *Jiangsu Agric. Sci.* **2021**, *49*, 219–226.
71. Yan, Z.; Liu, L.Y.; Jing, X. A Study on the Temporal and Spatial Changes of Vegetation Fluorescence and Climate Response Models in China from 2007 to 2018. *Remote Sens. Technol. Appl.* **2022**, *37*, 702–712.
72. Song, L. Research on Remote Sensing Monitoring of Chlorophyll Fluorescence under High-Temperature Stress of Crops. Ph.D. Thesis, Nanjing University, Nanjing, China, 2023.

land

MDPI

Article

Spatio-Temporal Diversification of per Capita Carbon Emissions in China: 2000–2020

Xuewei Zhang [1,2], Yi Zeng [3], Wanxu Chen [2,3,*], Sipei Pan [4], Fenglian Du [1,*] and Gang Zong [5,*]

1 School of Economics and Management, Inner Mongolia University, Hohhot 010020, China; 111989047@imu.edu.cn
2 Institute of Geographic Sciences and Natural Resources Research, Chinese Academy of Sciences, Beijing 100101, China
3 Department of Geography, School of Geography and Information Engineering, China University of Geosciences, Wuhan 430078, China; zy415@cug.edu.cn
4 College of Public Administration, Nanjing Agricultural University, Nanjing 210095, China; pamsp@cug.edu.cn
5 School of Economics and Management, Beijing University of Technology, Beijing 100124, China
* Correspondence: cugcwx@cug.edu.cn (W.C.); fldu@imu.edu.cn (F.D.); 15847206038@163.com (G.Z.)

Abstract: Exploring the low-carbon transition in China can offer profound guidance for governments to develop relevant environmental policies and regulations within the context of the 2060 carbon neutrality target. Previous studies have extensively explored the promotion of low-carbon development in China, yet no studies have completely explained the mechanisms of the low-carbon transition in China from the perspective of per capita carbon emissions (PCEs). Based on the statistics and carbon emissions data of 367 prefecture level cities in China from 2000 to 2020, this study employed markov chain, kernel density analysis, hotspots analysis, and spatial regression models to reveal the spatiotemporal distribution patterns, future trends, and driving factors of PCEs in China. The results showed that China's PCEs in 2000, 2010, and 2020 were 0.72 ton/persons, 1.72 ton/persons, and 1.91 ton/persons, respectively, exhibiting a continuous upward trend, with evident regional heterogeneity. PCEs in northern China and the eastern coastal region were higher than those of southern China and the central and southwestern regions. The PCEs in China showed obvious spatial clustering, with hot spots mainly concentrated in Inner Mongolia and Xinjiang, while cold spots were mainly in some provinces in southern China. The transition of PCEs in China exhibited a strong stability and a 'club convergence' phenomenon. A regression analysis revealed that the urbanization level and latitude had negative effects on PCEs, while the regional economic development level, average elevation, average slope, and longitude showed positive effects on PCEs. These findings have important implications for the promotion of the low-carbon transition and the effective achievement of the "dual carbon" goal.

Keywords: low-carbon transition; driving mechanisms; spatial autocorrelation; spatial regression; per capita carbon emissions; China

1. Introduction

The phenomenon of global warming, coupled with the melting of glaciers and a series of extreme weather events, has a significant impact on human survival and social activities, thereby posing ongoing challenges to social development [1,2]. The acceleration of CO_2 emissions has been unequivocally demonstrated as the dominant contributing factor to global warming [3,4]. Climate change has been induced by the explosive growth of greenhouse gases, primarily carbon emissions (CEs). How to tackle climate issues is the biggest environmental challenge in the world [5]. According to Climate Watch (2024) [6], China was the leading contributor to greenhouse gas emissions in 2020, accounting for 27%. In the ranking, the United States accounted for 11%, India accounted for 7%, while the European Union accounted for 6% of the emissions. The achievement of the Paris Agreement in 2015 has accelerated the global shift towards a low-carbon economy, thereby prompting significant transformations across various aspects of the world economy and society [7]. The Intergovernmental Panel on Climate Change reports underscore the importance of integrating renewable energy, enhancing energy efficiency, and fostering sustainable urban development. Initiatives such as the Global Carbon Project and the International Energy Agency's sustainable development scenarios provide pathways and policy recommendations for achieving low-carbon transformation. As the most rapidly developing country in the world, China has become the largest emitter of CEs globally. In 2022, China's CO_2 emissions reached 12.1 billion tons, accounting for 30% of the world, placing the country under severe pressure to reduce CEs. Although China ranks first in total CEs, the United States has the highest per capita carbon emissions (PCEs). Reducing CO_2 emissions has become a common goal for all countries [8]. In response, China is making efforts to carry forward low-carbon transition to reduce CO_2 emissions.

The evolution of CEs has been the subject of the existing research. Examining CEs from a per capita perspective is particularly important for understanding the unequal distribution. China exhibits significant differences across regions in terms of natural conditions, economic development, and population density. By approaching CEs from a per capita perspective, we can achieve a more precise measurement of regional contributions and responsibilities in CEs, reflecting the intensity and efficiency of CEs in each region. Hickel (2020) highlighted that China, despite its significant contribution to cumulative emissions, has four times the population of the United States [9]. If this difference is taken into account, the United States should be more responsible for CEs than China. Matthews (2016) quantified climate debts by the principle of atmospheric commons from a per capita perspective between 1960 (or 1990) and 2013 and determined whether countries' PCEs are above or below the global average [10]. It was observed that if a country's PCEs exceed the global average, it can be considered to be in debt. Conversely, if a country's PCEs are below the global average, it can be considered to be in credit. Therefore, exploring CEs from a per capita perspective is based on fairness and equity. Thus, it is imperative to elaborate the spatio-temporal patterns and driving mechanism of PCEs in China, which would provide significantly insights for understanding the low-carbon transition and formulating effective action plans to achieve "dual carbon" targets.

The reduction in CEs has emerged as a key area of interest and investigation. The primary focuses of current research concerning CEs are on three aspects: CEs measurement [11–13], spatial distribution patterns and discrepancy [14,15], and driving mechanisms [16–18]. The degree of CEs output has been widely measured in many fields, such as the energy [19,20], manufacturing [21], service [22], tourism [23], and transportation industries [24]. The study of CEs has been progressively integrated with that of land use [25], thereby significantly enhancing the depth and breadth of research into CEs. In terms of scale, this research has included countries [26,27], urban agglomerations [28], river basins [29], provinces [30], and counties [31]. Moreover, there is a growing trend in researching spatial patterns and heterogeneity related to CEs. Scholars employed quantitative methods to quantify spatial autocorrelation and discrepancies, such as Moran's *I*, the Theil index, and Dagum's Gini index [32,33]. Moran's *I* is extensively applied to quantitatively describe spatial autocorrelation [8,34]. For example, Chen et al. (2022) confirmed that building CEs had a positive spatial autocorrelation [35]. Wang et al. (2022) argued that the distribution of PCEs demonstrated self-reinforcing agglomeration in spatial distribution [36]. Furthermore, the spatio-temporal patterns and disparities were assessed using the Theil index and Dagum's Gini coefficient. For example, Ma et al. (2022) employed Dagum Gini coefficient to assess the PCEs of commercial buildings in different regions, revealing that intra-regional disparities were smaller than inter-regional disparities [37].

The investigation of the driving factors contributing to CEs has been a topic of considerable interest and debate, drawing upon both theoretical deduction and empirical evidence. Existing academic studies have discussed the impacts of influencing factors, such as financial development [38], economic development [39,40], foreign direct investment [41], urbanization level [42], high-speed railways [43], and environmental policies [44]. The findings of previous studies indicated that factors such as economic development, population growth, and traffic congestion were positively associated with CEs. In addition, economic growth may also induce the unintended consequence of environmental pollution, thus resulting in higher CEs [45]. In the early stage, the conventional economic model resulted in a considerable amount of CEs. The process of urbanization has been identified as a significant driver of CEs [46]. As a consequence of urbanization, the rising demand for clothing, food, shelter, and transportation leads to different lifestyles, consequently increasing CEs. From this perspective, the process of urbanization inevitably increases CEs [47].

Meanwhile, when analyzing which factor exerts the greatest influence on CEs, a variety of analytical techniques are available for this purpose, including structural decomposition analysis (SDA) [48], Stochastic Impacts by Regression on Population, Affluence and Technology (STIRPAT), the logarithmic mean Divisia index (LMDI), and spatial econometric methods. For instance, Xu et al. (2021) took Guangdong province in China as the research object, and revealed that the factors of consumption structure, per capita consumption, and population caused the largest increase in CEs [48]. Yu et al. (2023) used the STIRPAT method to analyze the factors of household CO_2 emissions and identified that the total population, household size, unemployment rate, and urbanization level were the most significant factors [49]. Using a spatial econometric model, Liu et al. (2023) revealed that rapid economic growth and traditional industrial structure transform had promotion effects on CEs, while government expenditure, population clustering, and scientific innovation had inhibitory effects on CEs [50].

By the way of conclusion, previous research has provided a comprehensive examination of CEs, establishing a robust theoretical and empirical foundation. However, the previous research is not without limitations. Few studies have completely explained the mechanisms of the low-carbon transition in China from the perspective of PCEs. Furthermore, the study of how geographical factors influence PCEs has received only sporadic attention to date. To bridge the research gaps, this study aimed to analyze the spatiotemporal diversification of PCEs across different regions in China from 2000 to 2020, identify the trends and spatial autocorrelation in PCEs, assess the impact of economic and geographical factors on PCE levels in various regions, and provide insights into effective strategies for regional low-carbon transitions. So, we aimed to construct a framework to characterize the spatio-temporal patterns of PCEs, thereby expanding the depth and scope of research on PCEs. Secondly, we aimed to analyze the dynamic evolution of PCEs, recognizing that the CEs varied significantly across regions due to different influencing factors. Therefore, it is essential to explore this spatial heterogeneity, which can be visualized through spatial patterns. Finally, we attempted to explore the driving factors from both socio-economic and natural perspectives. While previous studies have identified various driving factors affecting CEs, there has been a tendency to focus on socio-economic factors, with relatively little attention paid to the role of natural and geographical factors. However, there is evidence that these factors exert a significant influence on CEs. Therefore, we integrated the effects of the urbanization rate, gross domestic product (GDP) density, average slope, average digital elevation model (DEM), latitude, and longitude factors to explore the effects on CEs. Revealing their driving mechanisms will help us to uncover the primary factor affecting China's low-carbon transition, thereby contributing to achieving the goals of carbon peaking and neutrality.

2. Materials and Methods

2.1. Study Area

This study analysed 367 units at the prefecture level and above in mainland China (including a few county-level administrative districts; data for Hong Kong, Macao, and Taiwan were not yet available), including 4 municipalities directly under the central government, 293 prefectural-level cities, 7 districts, 30 autonomous prefectures, 3 leagues, and 30 provincial-level direct-administration units, for a total of 367 computational units (Figure 1). To scientifically reflect the socio-economic development of different regions in China, this study area was divided into four major economic subregions: Northeast, Central, Western, and Eastern China. The northeastern region mainly included the municipal units of Liaoning, Jilin, and Heilongjiang provinces; the eastern region includes the municipal units of Beijing, Tianjin, Hebei, Shanghai, Jiangsu, Zhejiang, Fujian, Shandong, Guangdong, and Hainan provinces; the central region includes the municipal units of Shanxi, Anhui, Jiangxi, Henan, Hubei, and Hunan provinces; and the rest is the western region.

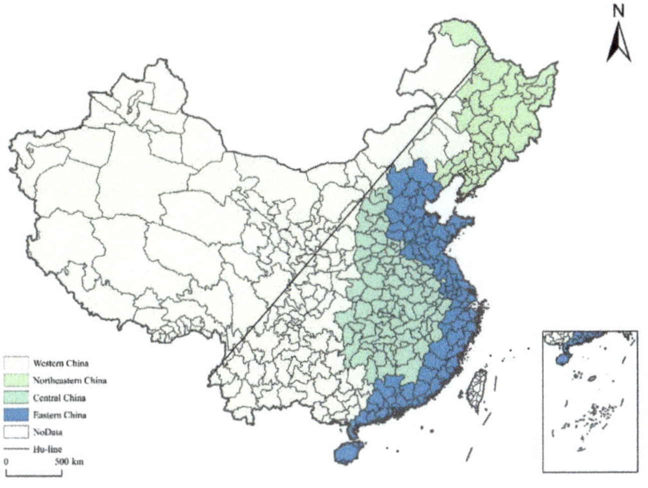

Figure 1. Study area.

2.2. Data Sources

This study selected 367 cities in China as the research sample, covering 30 provinces, municipalities, and autonomous regions. The CE data were derived from the Center for Global Environmental Research (https://db.cger.nies.go.jp/dataset/ODIAC/, accessed on 2 August 2024). The population data were mainly derived from the main data bulletins of the fifth, sixth, and seventh national censuses in 2000, 2010, and 2020, respectively. The land use data, DEM, and slope were derived from EPS data platform (https://www.epsnet.com.cn/index.html#/Index, accessed on 2 August 2024) and Data Center for Resources and Environmental Sciences (https://www.resdc.cn/, accessed on 2 August 2024). The GDP data were derived from the Statistical Yearbooks. In addition, for areas within the Xinjiang Uygur Autonomous Region where the seventh census data was unavailable, this study supplemented the missing values using linear regression.

2.3. Per Capita Carbon Emissions

In this context, PCEs represent the ratio of CEs to population [51–53]. The calculation equations were as follows:

$$PCE = CEs/Population \tag{1}$$

$$PCEC = {}^{CEs_{t2}}\!\Big/_{Population_{t2}} - {}^{CEs_{t1}}\!\Big/_{Population_{t1}} \tag{2}$$

where $PCEC$ represents the change in PCEs; CEs_{t1} and CEs_{t2} represent CEs of a unit at time $t1$ and $t2$, and $Population_{t1}$ and $Population_{t2}$ represent the population of a unit at time $t1$ and $t2$.

2.4. Markov Chain

The markov chain provides a mechanism for explaining how the probability distribution of transitions from one state to another eventuates [54]. Due to its steady-state analysis, the markov chain can be utilized to predict the long-term patterns and trends of geographic phenomena with a high degree of accuracy. In this study, we used markov chain to predict the trend of PCEs in each region, and the probabilistic transfer model constructed by the markov chain explaining the PCE conversion state in each region, reflecting the future interconversion of the values of different magnitudes. In general, the state type of PCEs at moment t is represented by a 1 × k state probability vector as Et = [E1,t, E2,t, . . ., Ek,t], and the transition process of PCE state types can be represented by a k × k Markov transition matrix M. Based on the similar counts for each type of county, PCEs at the county level were categorized into four types using quartiles (0.25/0.5/0.75), labeled as k = 1, 2, 3, and 4, respectively.

2.5. Kernel Density Analysis

As one kind of nonparametric estimation, kernel density estimation was characterized by relative smoothness and unbiasedness, which can accurately describe the distribution of random variables [55,56]. The calculation equations were as follows:

$$f(x) = \frac{1}{nh}\sum_{i=1}^{n} K\left(\frac{D_i - D}{h}\right) \tag{3}$$

$$K(t) = \frac{1}{\sqrt{2\pi}}e^{-\frac{t^2}{2}} \tag{4}$$

where $K(x)$ is the kernel function; D_i is the independent and equally distributed observed values; D is mean value; n is the number of observed values; and h is the bandwidth. This study used Gaussian kernel function to effectively explore the changing law of PCEs in various regions so as to reflect the heterogeneity and the dynamic trend of PCEs in each region.

2.6. Hotspots Analysis

This study intended to use the Moran's I index to reveal the spatial autocorrelation characteristics of PCEs. To further reflect the spatial agglomeration of PCEs and changes in PCEs, the Getis-Ord Gi* index was used to measure the statistically significant hot and cold spots of PCEs [35]. The Getis-Ord Gi* index can be used for measuring the spatial clustering, which can test whether variables exhibit the feature of high-value cluster or low-value cluster, thereby identifying the locations of hot and cold spots, and spatial outliers [57,58].

2.7. Spatial Regression

Spatial autocorrelation is a common phenomenon among geographical elements, and non-spatial models may overlook the spatial dependence between factors, which may lead to potential biases in identifying underlying influencing factors. To illustrate, an element is influenced not only by the level of the element in the study unit, but also by the level of the element in adjacent or distant units. Based on this, this study intended to use the least squares method (OLS), the spatial lag model (SLM), and spatial error model (SEM) to reveal the mechanism of China's low-carbon transition [59]. The cross-sectional benchmark model was expressed as Equation (5):

$$PCE_i = \alpha_0 + \alpha_1 LUI_i + \alpha_2 GDPD_i + \alpha_3 ASLO_i + \alpha_4 ADEM_i + \alpha_5 LAT_i + \alpha_6 LON_i + \varepsilon_i \tag{5}$$

where i represents a city, and LUI is the development and utilization intensity of land [60,61], which represents the land urbanization level. The function, intensity, and efficiency of CEs are influenced by important driving forces such as land urbanization and associated human activities [62]. $GDPD$ is GDP density, representing economic growth, which is an important factor. GDP promotes PCEs, while this effect weakens with the growth of GDP [63,64]. *ASLO* represents average slope, and *ADEM* represents average DEM. These two factors represent the topographic elements [20,51]. The *ASLO* and *ADEM* factors exert notable effects on CEs by influencing urban expansion [65,66]. In this study, we also added longitude (LON) and latitude (LAT) factors to reveal the rules of CEs in terms of latitude and longitude.

SLM was expressed as shown in Equation (6):

$$PCE_i = \rho\sum_{i=1}^{n} W_{ij} \times PCE_i + \beta X_i + \varepsilon_i \tag{6}$$

where ρ is the spatial autoregressive coefficient, W_{ij} is the spatial weight matrix, β is a vector of coefficients X_i, and X_i represents the set of independent variables.

SEM was expressed as shown in Equation (7):

$$PCE_i = \beta X_i + \varepsilon_i, \varepsilon_i = \lambda\sum_{j=1}^{n} W_{ij}\varepsilon_i + \varphi_i \tag{7}$$

where λ is the spatial autocorrelation coefficient of the error term.

By combining multiple methods, this study aimed to evaluate the influencing factors to provide a reference for integrated regional green development (see Figure 2).

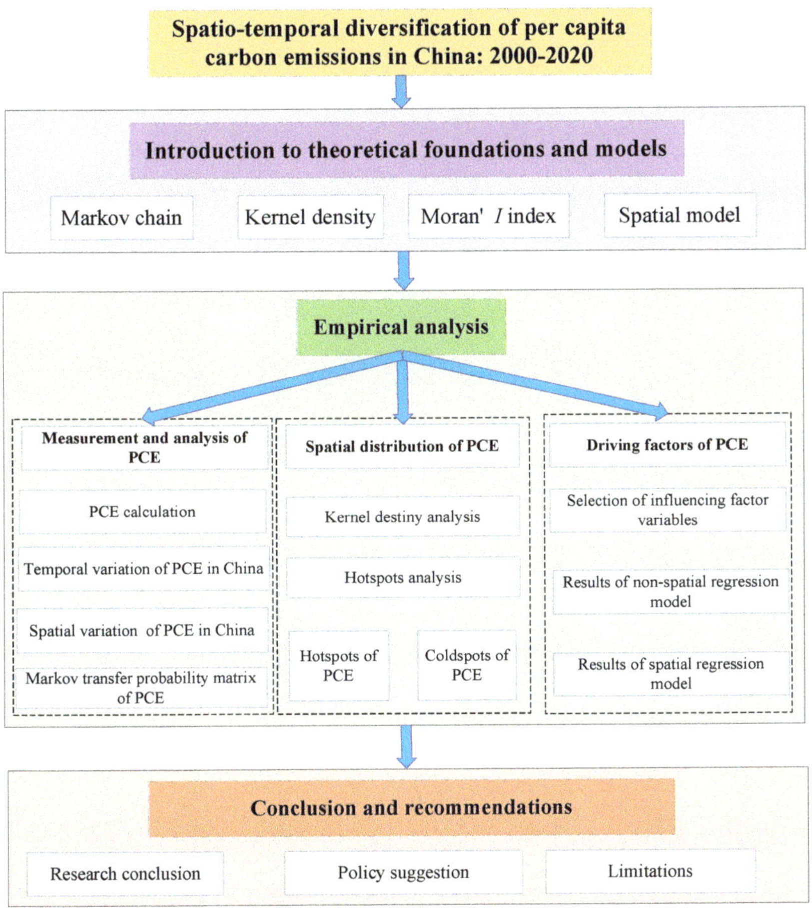

Figure 2. Research design.

3. Results

3.1. PCEs in China

China's PCEs in 2000, 2010, and 2020 were 0.72 ton/persons, 1.72 ton/persons, and 1.91 ton/persons, respectively. These figures demonstrated a continuous increasing trend, with notable regional variations in PCEs. The PCEs were the highest in the northeast and the lowest was in the central region with values of 0.53 ton/persons in 2000, 1.23 ton/persons in 2010, and 1.43 ton/persons in 2020. Additionally, the PCEs southeast of Hu-line were was lower than those in the northwest. Specifically, in 2000, 2010, and 2020 the PCEs southeast of Hu-line were 0.72 ton/persons, 1.63 ton/persons, and 1.81 ton/persons, respectively, while that northwest of Hu-line were 1.09 ton/person, 2.81 ton/persons, and 3.06 ton/persons, respectively (Figure 3). From the perspective of spatial distribution, the PCEs in southern China were obviously lower than those in northern China. Also, the eastern coastal areas had higher PCEs than central and southwest China. Seeing the change from 2010 to 2020, the PCEs in different regions exhibited an increasing tendency (Figure 4).

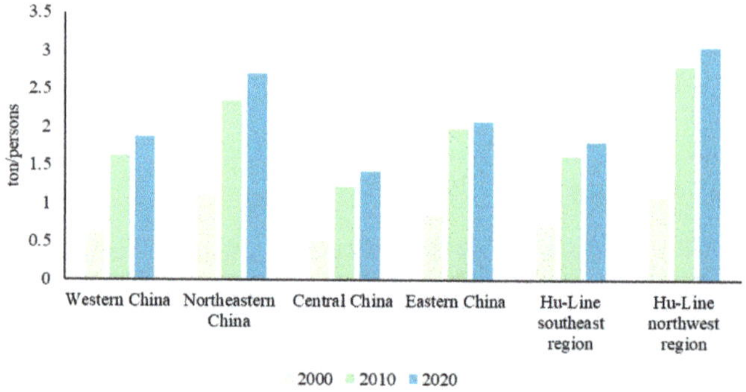

Figure 3. PCEs in different regions in 2000, 2010, and 2020.

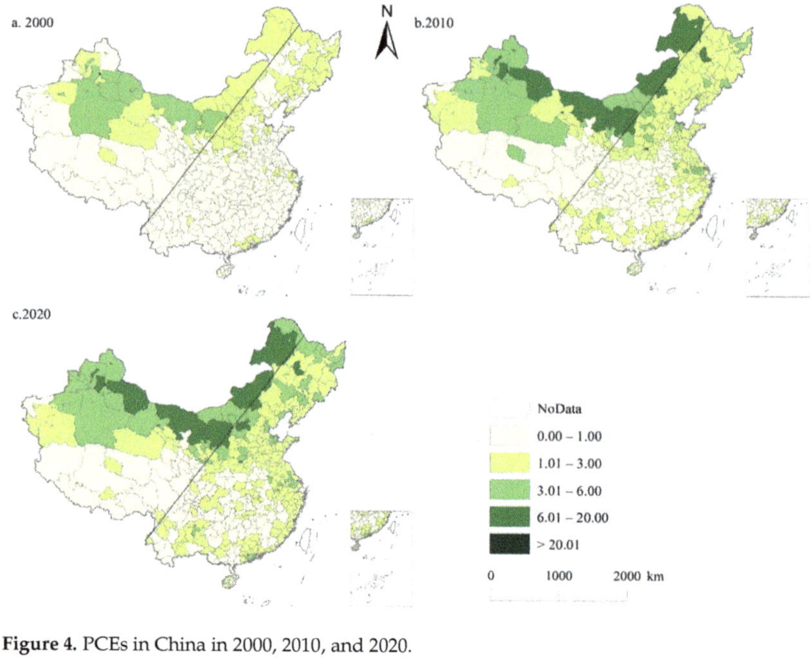

Figure 4. PCEs in China in 2000, 2010, and 2020.

3.2. Spatiotemporal Dynamics of PCEs

The markov transfer probability matrix was employed to predict future PCEs with a time span of 10 years (Table 1). In overall spatial terms, PCEs were expected to exhibit stable growth over the next 10 years, with PCEs showing a tendency to converge to higher values. Specifically, this tendency was particularly pronounced in the northeast, followed by the central region. It was also observed that, with the exception of the northeast and the east, the shift probabilities on the main diagonal were higher than those off the diagonal, indicating relatively stable PCE levels. The probability of shifting from high to higher values in the northeast and eastern regions suggested that higher growth rates in PCEs exist in these regions.

Table 1. Markov transfer probability matrix of PCEs.

Variables	Spatial Extent	Time Span	Type	L	ML	MH	H	Number
PCE	China	10	L	0.720	0.274	0.004	0	211
			ML	0.011	0.655	0.327	0.005	180
			MH	0	0.016	0.661	0.322	183
			H	0	0	0.025	0.975	160
	Eastern China	10	L	0.724	0.241	0.034	0	58
			ML	0.035	0.482	0.428	0.053	56
			MH	0	0.060	0.460	0.480	50
			H	0	0	0.100	0.900	40
	Central China	10	L	0.714	0.285	0	0	49
			ML	0	0.553	0.446	0	47
			MH	0	0.026	0.736	0.236	38
			H	0	0	0	1	40
	Western China	10	L	0.581	0.383	0.034	0	86
			ML	0	0.686	0.313	0	67
			MH	0	0.014	0.720	0.264	68
			H	0	0	0.046	0.953	65
	Northeastern China	10	L	0.545	0.409	0.045	0	22
			ML	0	0.473	0.526	0	19
			MH	0	0	0.350	0.650	20
			H	0	0	0	1	11

Notes: L stands for low-level type; ML stands for medium–low-level type; MH stands for medium–high-level type; and H stands for high-level type.

3.3. Kernel Density

The changes in the study elements over the course of the study can be encapsulated in the contour plots of the kernel density of the normalized PCEs. The contour lines should remain close to the 45° diagonal if no drastic changes have occurred. Figure 5a,b,d showed the kernel density of PCEs over a 10-year time span for northeast, eastern, and central China, respectively, and it was found that there were shifts to higher or lower values of PCEs in these regions, with a more pronounced shift towards higher values in eastern and central China. Figure 5c shows the trend of the shift in western China over a 10-year time span, revealing that the PCEs shifted to higher values, while in some regions with higher values, the PCEs shifted to lower values. The changes in PCEs in China, as shown in Figure 5e, indicated that some peaks of the kernel density contours were basically located near 45°, showing that there was some degree of stability in PCEs, as well as shifts from low to high and from high to low.

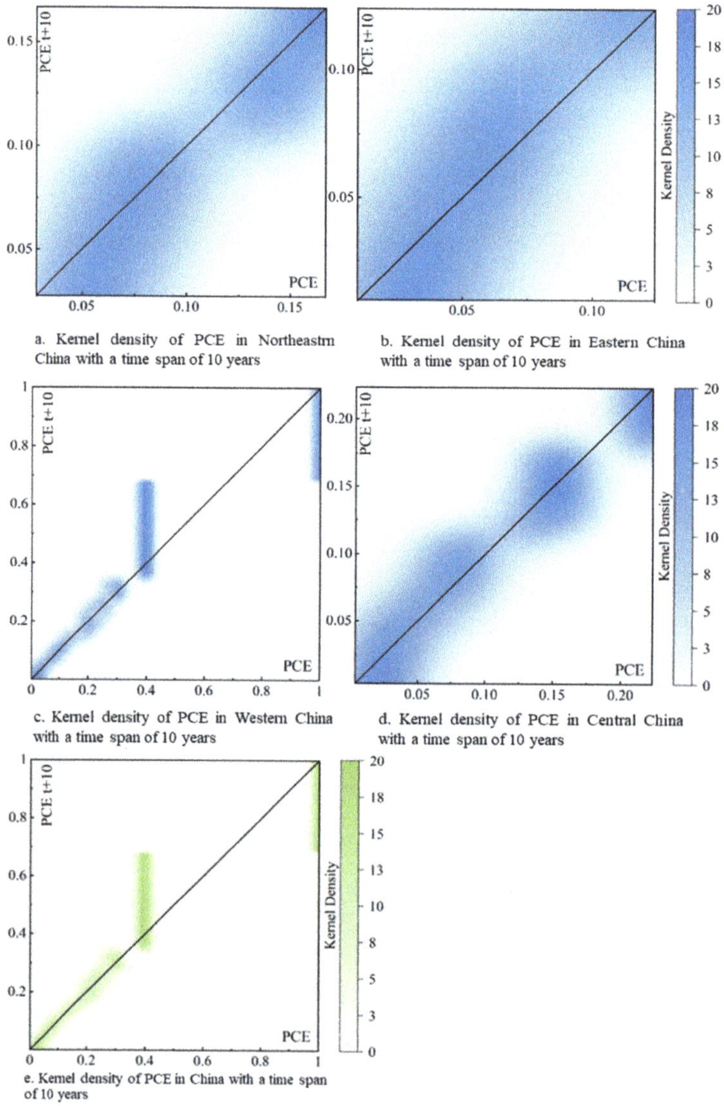

Figure 5. Kernel density contour map of PCEs.

3.4. Change in PCEs in China

From 2000 to 2010, the number of cities with a reduction in PCEs accounted for 0.544%, and only a small proportion of prefecture-level cities achieved a decreases in PCEs during that period. Correspondingly, a large number of prefecture-level cities had an evident increase in PCEs. For example, the proportion of prefecture-level cities with PCEs increasing by more than 1 ton/persons was 40.871%. However, the number of cities with a decrease in PCEs from 2010 to 2020 was 10.354%, which was higher than that of the previous decade. It is noteworthy that the proportion of cities with an increase in PCEs greater than 1 ton/persons was 1.907%, which was evidently lower than that in the previous decade. Most cities (29.700%) had PCEs increases of less than 1 ton/persons, indicating that China has undergone a drastic transition in terms of PCEs, and there has been remarkable progress towards the low-carbon transition (Figure 6).

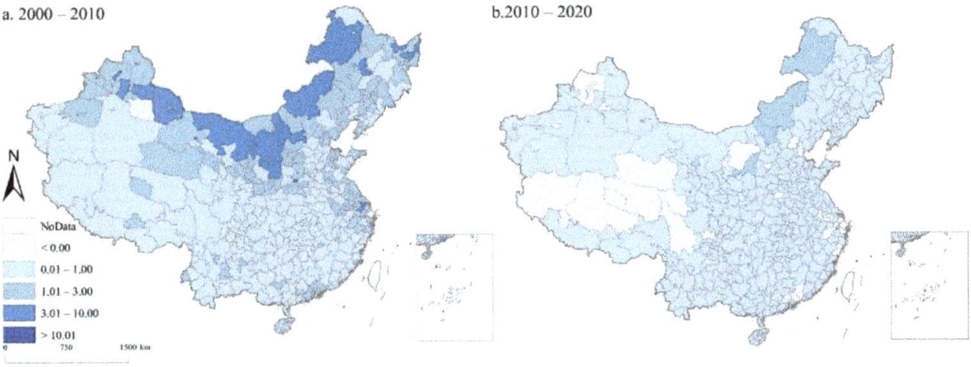

Figure 6. Change in PCEs in China during 2000–2020.

To gain further insight into the distribution of PCEs, we calculated the Moran's *I* indexes of PCEs, which were 0.258, 0.316, and 0.316 in 2000, 2010, and 2020, respectively, and the *p*-value was significant at the level of 0.001, obviously showing a significant spatial agglomeration pattern in PCEs. The specific spatial distribution characteristics can be obtained from the hot spot distribution map. We found that the hot spots were mainly distributed in Inner Mongolia and Xinjiang, while the cold spots were primarily distributed in some provinces of southern China (e.g., Sichuan, Chongqing, Guizhou, Hubei, Hunan, Jiangxi, and Hainan). From 2000 to 2020, it is evident that the PCE hot spots gradually expanded eastward into Inner Mongolia and northeastern China (Figure 7).

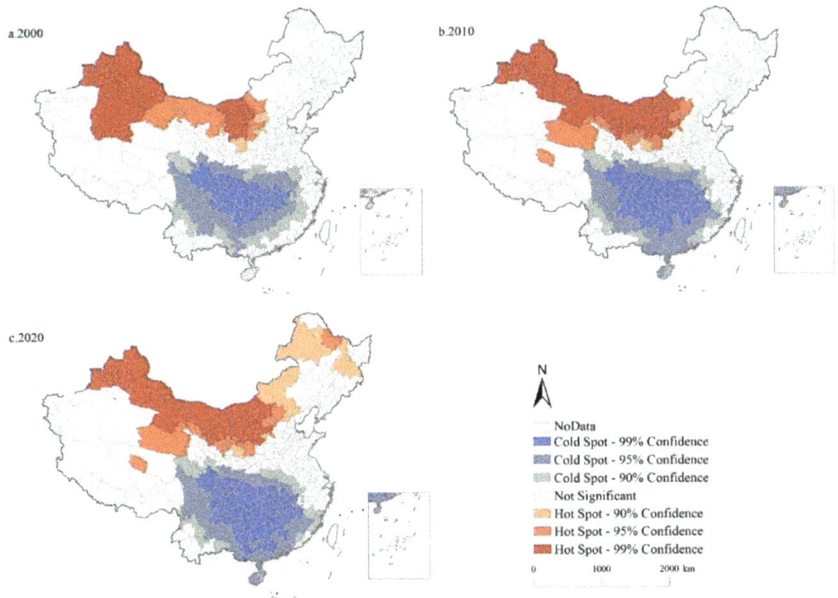

Figure 7. Hotspots of PCEs in China during 2000–2020.

To further reveal the hotspots of changes in PCEs, we calculated the spatial autocorrelation index of PCE changes during 2000–2010 and 2010–2020. The spatial autocorrelation indexes of PCE changes during 2000–2010 and 2010–2020 were 0.354 and 0.179, respectively, and the p-value was significant at the level of 0.001. The hot spot analysis results demonstrated that the hot spots of PCE changes during 2000–2010 were distributed in Inner Mongolia, Northern Xinjiang, and Northern Qinghai Province. The cold spots of PCE changes were distributed in southern China, showing a similar pattern to those of the PCEs. The cold spots of PCE changes in China during 2010–2020 were distributed in the west of Xinjiang and the Pearl River Delta, while the hot spots were distributed in northeast China (Figure 8).

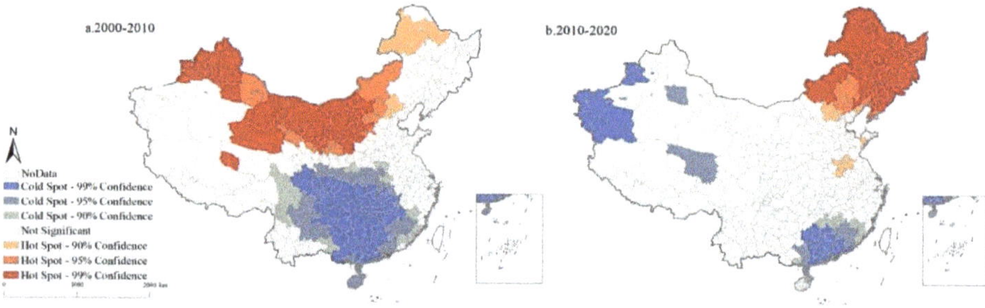

Figure 8. Hotspots of PCE changes in China during 2000–2020.

3.5. *Driving Mechanism of Low-Carbon Transition*

To reveal the driving mechanism of the low-carbon transition in China, the OLS, SEM, and SLM were employed in this study. This study first carried out OLS model (Table 2) to diagnose. The coefficients of LUI in 2000, 2010, and 2020 were −3.239, −8.440, and −7.027, respectively, which indicated that the LUI had a significant negative impact on PCEs. The coefficients of GDPD in 2000, 2010, and 2020 were 26.842, 44.587, and 4.659, which indicated that GDPD had a significant positive impact on PCEs. Higher GDPD typically indicates more economic activity and a greater population density, thus leading to higher energy consumption and CEs. The coefficients of ASLO in 2000, 2010, and 2020 were −1.209, −2.707, and −2.993, respectively, indicating that the ASLO had a significant negative impact on PCEs. Steeper slopes may limit large-scale agricultural or construction activities, potentially leading to lower levels of energy consumption and CEs. The coefficients of ADEM in 2000, 2010, and 2020 were −2.076, −3.205, and −3.131, illustrating that the ADEM had a negative impact on PCEs. Regions with higher DEM, characterized by mountainous and hilly landscapes, were less suitable for extensive agricultural and industrial operations, leading to a reduction in energy consumption and CEs. The coefficients of LAT in 2000, 2010, and 2020 were 0.068, 0.152, and 0.156, respectively, indicating that the LAT had a significant positive impact on PCEs. Higher latitudes often have colder climates, which may lead to increased energy consumption for heating, thus increasing PCEs. The coefficients of LON in 2000, 2010, and 2020 were −0.043, −0.066, and −0.056, respectively, indicating that the LON had a negative impact on PCEs. This is largely because that the eastern regions were the most economically developed areas in China, boasting advanced technological levels that were conducive to reducing CEs. The results showed that the Moran's I indexes of PCEs in 2000, 2010, and 2020 were 3.225, 4.291, and 4.341, respectively, and the p-value was significant at the level of 0.05. Therefore, the results obtained using the OLS regression model cannot fully and scientifically explain the relationship between them. When considering spatial lag and spatial error terms, the SLM and SEM performed better.

The coefficient of LUI was negative, indicating that the urbanization promotes low-carbon transition (Table 3). The coefficient of the GDPD was positive, but it was significant only in the SEM and SLM models of 2010. The coefficient of the ASLO was negative, but it was significant only in the SEM model of 2010 and the SLM and SEM models of 2020. The coefficient of the ADEM was negative in all models, indicating that PCE decreases with increasing altitude. PCEs in China increased significantly with rising latitude, while PCEs decreased significantly with rising longitude. In addition, PCEs were influenced not only by their own unit factors but also by those of adjacent units, as indicated by the spatial lag terms being significant at the 0.001 level. An average 1% increase in PCEs in surrounding areas in 2000, 2010, and 2020 would result in increases of 0.238%, 0.290%, and 0.304% increases in PCEs in their own units. In addition, the spatial error terms were also statistically significant at the 0.001 level, indicating that PCE changes were not only influenced by the aforementioned factors, but also by other factors.

Table 2. Regression results of OLS.

Variables	2000	2010	2020
LUI	−3.239 *** (0.950)	−8.440 *** (1.818)	−7.027 *** (1.816)
GDPD	26.842 * (13.565)	44.587 *** (11.510)	4.659 * (2.115)
ASLO	−1.209 * (0.578)	−2.707 * (1.109)	−2.993 ** (1.127)
ADEM	−2.076 *** (0.574)	−3.205 ** (1.120)	−3.131 ** (1.143)
LAT	0.068 *** (0.010)	0.152 *** (0.020)	0.156 *** (0.020)
LON	−0.043 *** (0.008)	−0.066 *** (0.016)	−0.056 *** (0.016)
Constant	5.883 *** (1.029)	10.269 *** (2.033)	8.721 *** (2.088)
Moran's *I* (error)	3.225 **	4.291 ***	4.341 ***
LM (lag)	8.744 **	16.081 ***	16.967 ***
Robust LM (lag)	1.946	2.522	3.339
LM (error)	7.324 **	13.865 ***	14.408 ***
Robust LM (error)	0.526	0.305	0.779
LM (SARMA)	9.271 **	16.387 ***	17.747 ***
Measures of fit			
Log likelihood	−607.080	−846.958	−853.786
AIC	1228.160	1707.920	1721.570
SC	1255.500	1735.250	1748.910
R-Squared	0.248	0.272	0.262
N	367	367	367

Notes: *** $p \leq 0.001$, ** $p \leq 0.01$, * $p \leq 0.05$. Standard deviation is shown in parentheses.

Table 3. Regression results of SLM and SEM.

Variables	SLM	SEM	SLM	SEM	SLM	SEM
	2000		2010		2020	
LUI	−2.470 ** (0.937)	−2.468 * (1.109)	−6.279 *** (1.795)	−6.749 ** (2.163)	−5.061 ** (1.771)	−5.034 ** (2.153)
GDPD	21.547 (13.206)	19.399 (13.780)	34.852 ** (11.219)	38.262 ** (12.964)	3.428 (2.032)	3.013 (2.162)
ASLO	−0.988 (0.565)	−1.200 (0.656)	−2.105 (1.075)	−2.639 * (1.288)	−2.299 * (1.090)	−2.818 * (1.319)

Table 3. *Cont.*

Variables	SLM	SEM	SLM	SEM	SLM	SEM
	2000		2010		2020	
ADEM	−1.858 ** (0.574)	−2.316 *** (0.685)	−2.814 ** (1.098)	−3.76 ** (1.387)	−2.716 * (1.116)	−3.641 * (1.434)
LAT	0.055 *** (0.011)	0.072 *** (0.013)	0.114 *** (0.022)	0.162 *** (0.027)	0.115 *** (0.022)	0.167 *** (0.027)
LON	−0.038 *** (0.009)	−0.051 *** (0.010)	−0.058 *** (0.016)	−0.085 *** (0.021)	−0.050 ** (0.016)	−0.075 *** (0.022)
Constant	5.085 *** (1.067)	6.259 *** (1.297)	8.599 *** (2.062)	11.161 *** (2.708)	7.322 *** (2.075)	9.366 *** (2.826)
Spatial lag term	0.238 *** (0.070)		0.290 *** (0.068)		0.304 *** (0.067)	
Spatial error term		0.257 *** (0.071)		0.310 *** (0.069)		0.326 *** (0.068)
Measures of fit						
Log likelihood	−602.400	−602.588	−839.089	−839.433	−845.333	−845.680
AIC	1220.800	1219.18	1694.180	1692.870	1706.670	1705.360
SC	1252.040	1246.510	1725.420	1720.210	1737.910	1732.700
R-Squared	0.275	0.276	0.315	0.315	0.309	0.309
N	367	367	367	367	367	367

Notes: *** $p \leq 0.001$, ** $p \leq 0.01$, * $p \leq 0.05$. Standard deviation is shown in parentheses.

The OLS regression analysis results (Table 4) indicated that the Moran's *I* indexes of the 2000, 2010, and 2020 settlements in the eastern region were 4.070, 2.709, and 3.959, respectively, with a significant *p*-value at the 0.01 level, which indicated that the model's settlements in the eastern region had a strong spatial autocorrelation, so the spatial lag and spatial error terms were considered, and the SLM and SEM were used to improve the goodness of fit (Tables 5 and 6). The SLM and SEM regression results analysis in Tables 5 and 6 indicated that the main factors affecting PCEs varied across the four regions of China. In the western region, LUI and GDPD were the main factors significantly impact on PCEs. LUI exhibited a significant negative effect, while GDPD had a significant positive effect. The reason was that urbanization often led to a shift from high-carbon-emission industries and agriculture to low-energy-consuming service and high-tech industries, thereby resulting in reduced PCEs. However, economic growth often lead to increased energy consumption and production, which can result in higher CEs. In the northeastern region, LUI and ASLO were the main factors, which both had significant negative effects on PCEs. This can be attributed to the abundant forest resources in northeastern China. The steep areas with dense vegetation act as carbon sinks, absorbing CO_2 from the atmosphere and thereby reducing the overall carbon footprint. In the central region, LUI and GDPD were the main factors which had significant effect on PCEs. LUI had a significant negative effect, while GDPD had a significant positive effect. In the eastern region, GDPD and LON were the main factors, which both had significant positive effects on PCEs.

Table 4. Regression results of OLS for different regions of China.

Variables	Western China			Northeastern China			Central China			Eastern China		
	2000	2010	2020	2000	2010	2020	2000	2010	2020	2000	2010	2020
LUI	−1.592	−10.968 *	−10.676 *	−0.667	−6.731	−5.800	−3.502 *	−10.198 **	−9.015 *	−2.423 ***	−5.624	−1.775
	(2.550)	(4.397)	(4.456)	(1.738)	(4.465)	(5.667)	(1.398)	(3.276)	(3.681)	(1.284)	(2.963)	(3.003)
GDPD	50.300	361.097 **	90.605 **	90.515	66.990	2.065	53.920	91.123 ***	15.105 *	17.800	24.656 ***	0.988
	(228.985)	(115.654)	(30.547)	(49.639)	(52.383)	(29.286)	(37.676)	(24.616)	(7.083)	(4.536)	(5.562)	(0.945)
ASLO	−1.234	−1.811	−2.133	−1.606	−6.625	−5.410	−1.699 *	−4.247 *	−3.269	−0.275 *	−2.146	−2.965
	(1.099)	(2.028)	(2.008)	(1.724)	(4.490)	(5.302)	(0.820)	(1.864)	(2.140)	(0.720)	(1.593)	(1.797)
ADEM	−1.954 *	−3.098	−3.235	2.107	0.443	−2.997	2.021	5.196	3.328	−3.360	−3.237	0.309
	(0.991)	(1.871)	(1.879)	(3.246)	(8.574)	(9.434)	(1.667)	(3.762)	(4.264)	(1.699)	(3.875)	(4.229)
LAT	−0.038	−0.017	0.002	0.036	0.135 *	0.140	0.113	0.123 *	0.134 *	−0.025	0.006	0.061
	(0.020)	(0.035)	(0.035)	(0.024)	(0.062)	(0.072)	(0.026)	(0.058)	(0.067)	(0.014)	(0.033)	(0.036)
LON	0.095 ***	0.213 ***	0.228 ***	0.037	0.002	−0.014	0.113 ***	0.221 ***	0.279 ***	0.037 ***	0.072 *	0.025
	(0.029)	(0.052)	(0.052)	(0.038)	(0.100)	(0.115)	(0.029)	(0.069)	(0.078)	(0.011)	(0.028)	(0.027)
Constant	3.568	3.787	1.644	−4.647	−10.126	−9.627	0.028	−13.254	−16.710*	4.350 **	2.614	−4.425
	(2.702)	(5.116)	(5.141)	(3.249)	(8.346)	(9.460)	(3.079)	(6.893)	(7.909)	(1.632)	(3.801)	(3.825)
Moran's *I* (error)	1.915	1.676	1.711	−1.055	−1.000	−1.404	1.706	2.182 *		1.688	2.709 **	3.959 ***
LM (lag)	1.331	1.567	1.504	3.379	3.790 *	3.353	1.333	0.960	1.191	12.149 ***	5.503	8.524 ***
Robust LM (lag)	0.109	1.324	0.750	0.318	0.791	0.529	2.591	0.028	2.044	1.911	2.654	0.263
LM (error)	1.500	1.002	1.097	3.065	3.099	4.139 *	0.545	1.253	0.529	10.240 ***	3.525	9.558 ***
Robust LM (error)	0.278	0.759	0.343	0.004	0.100	1.315	1.803	0.321	1.381	0.001	0.675	1.298
LM (SARMA)	1.609	2.326	1.847	3.384	3.890	4.669	3.136	1.282	2.573	12.151 **	6.179 *	9.822 **
Measures of fit												
Log likelihood	−296.557	−383.134	−382.904	−12.723	−46.081	−52.108	−46.992	−117.917	−129.776	−32.805	−115.618	−126.148
AIC	607.113	780.268	779.808	39.447	106.162	118.218	107.985	249.833	273.552	79.611	245.235	266.296
SC	627.853	801.008	800.548	50.531	117.247	129.302	125.246	267.095	290.814	97.986	263.61	284.671
R-Squared	0.257	0.319	0.326	0.294	0.315	0.264	0.508	0.504	0.439	0.235	0.349	0.250
N	143	143	143	36	36	36	87	87	87	102	102	102

Notes: *** $p \leq 0.001$, ** $p \leq 0.01$, * $p \leq 0.05$. Standard deviation is shown in parentheses.

Table 5. Regression results of SLM for different regions of China.

Variables	Western China			Northeastern China			Central China			Eastern China		
	2000	2010	2020	2000	2010	2020	2000	2010	2020	2000	2010	2020
LUI	−0.874	−9.350 *	−9.289 *	−0.583	−6.453	−4.865	−3.156 *	−9.456 **	−8.175 *	−2.007	−4.197	−1.361
	(2.467)	(4.279)	(4.325)	(1.341)	(3.406)	(4.404)	(1.366)	(3.259)	(3.606)	(1.136)	(2.756)	(2.721)
GDPD	23.714	335.386 **	87.693 **	49.497	32.640	−12.276	52.691	87.954 ***	14.178 *	13.715 ***	18.992 ***	0.501
	(221.392)	(111.95)	(29.534)	(38.662)	(39.884)	(22.700)	(35.743)	(24.093)	(6.747)	(4.015)	(5.279)	(0.855)
ASLO	−1.171	−1.675	−1.966	−2.343	−7.222 *	−6.313	−1.424	−3.755 *	−2.729	−0.201	−1.484	−2.227
	(1.066)	(1.971)	(1.951)	(1.336)	(3.443)	(4.119)	(0.796)	(1.815)	(2.065)	(0.639)	(1.484)	(1.643)
ADEM	−1.790	−2.802	−2.885	1.702	−2.903	−4.449	1.561	4.349	2.405	−2.803	−2.815	0.440
	(0.977)	(1.833)	(1.840)	(2.518)	(6.543)	(7.315)	(1.596)	(3.620)	(4.071)	(1.499)	(3.585)	(3.827)
LAT	−0.037	−0.024	−0.005	0.061 ***	0.200 ***	0.218 ***	0.024	0.108	0.115	−0.019	−0.002	0.039
	(0.019)	(0.034)	(0.034)	(0.019)	(0.048)	(0.058)	(0.024)	(0.057)	(0.065)	(0.013)	(0.031)	(0.033)
LON	0.082 **	0.179 ***	0.191 ***	0.038	0.023	0.0204	0.096 **	0.192 **	0.241 **	0.027 **	0.054 *	0.015
	(0.030)	(0.056)	(0.057)	(0.031)	(0.082)	(0.094)	(0.032)	(0.073)	(0.083)	(0.010)	(0.027)	(0.025)
Constant	3.408	4.265	2.362	−6.997 **	−17.122 *	−19.265 *	−3.247	−11.318	−14.162	3.301 *	2.720	−2.592
	(2.625)	(4.949)	(4.970)	(2.572)	(6.741)	(7.809)	(2.976)	(6.820)	(7.832)	(1.452)	(3.514)	(3.490)
Spatial lag term	0.163	0.160	0.160	−0.743 ***	−0.763 ***	−0.733 ***	0.189	0.160	0.182	0.435 ***	0.316 **	0.387 ***
	(0.111)	(0.109)	(0.109)	(0.211)	(0.203)	(0.212)	(0.146)	(0.143)	(0.147)	(0.099)	(0.111)	(0.109)
Measures of fit												
Log likelihood	−295.739	−382.252	−382.037	−9.469	−42.470	−48.971	−46.287	−117.397	−129.135	−26.245	−112.564	−121.633
AIC	607.478	780.503	780.075	34.938	100.942	113.942	108.575	250.794	274.27	68.491	241.128	259.266
SC	631.181	804.206	803.777	47.606	113.61	126.61	128.302	270.521	293.997	89.490	262.128	280.265
R-Squared	0.270	0.331	0.338	0.478	0.507	0.451	0.519	0.513	0.451	0.361	0.403	0.340
N	143	143	143	36	36	36	87	87	87	102	102	102

Notes: *** $p \leq 0.001$, ** $p \leq 0.01$, * $p \leq 0.05$. Standard deviation is shown in parentheses.

Table 6. Regression results of SEM for different regions of China.

Variables	Western China			Northeastern China			Central China			Eastern China		
	2000	2010	2020	2000	2010	2020	2000	2010	2020	2000	2010	2020
LUI	−0.275	−9.211 *	−8.918 *	−0.142	−5.562	−3.579	−3.511 *	−10.718 ***	−9.135 *	−2.045	−4.122	−2.234
	(2.653)	(4.527)	(4.544)	(1.191)	(3.115)	(3.765)	(1.403)	(3.360)	(3.692)	(1.258)	(2.999)	(2.962)
GDPD	−15.509	27.583 **	85.35 **	34.575	−3.413	−21.060	53.435	97.235 ***	14.347 *	11.673	18.752 ***	0.305
	(222.797)	(113.220)	(29.620)	(35.537)	(32.089)	(20.643)	(36.688)	(24.908)	(6.892)	(3.804)	(5.803)	(0.842)
ASLO	−1.410	−2.156	−2.372	−2.748 *	−7.250 *	−6.033	−1.587	−3.885	−3.069	−0.805	−2.083	−3.670
	(1.177)	(2.128)	(2.117)	(1.119)	(3.008)	(3.620)	(0.829)	(1.934)	(2.164)	(0.769)	(1.715)	(1.964)
ADEM	−2.078	−3.158	−3.221	2.559	−2.851	−3.830	1.745	4.237	2.683	−2.527	−2.621	0.790
	(1.110)	(2.033)	(2.045)	(1.960)	(5.208)	(5.436)	(1.672)	(3.863)	(4.286)	(1.535)	(3.686)	(3.882)
LAT	−0.046 *	−0.031 ***	−0.010	0.052 ***	0.150 ***	0.166 ***	0.027	0.122	0.133	−0.012	0.015	0.071
	(0.023)	(0.039)	(0.040)	(0.011)	(0.031)	(0.035)	(0.027)	(0.066)	(0.071)	(0.022)	(0.043)	(0.051)
LON	0.099 **	0.213	0.230 ***	0.001	−0.075	−0.067	0.117 ***	0.238 ***	0.289 ***	0.027	0.059	0.023
	(0.033)	(0.058)	(0.058)	(0.024)	(0.064)	(0.072)	(0.030)	(0.071)	(0.079)	(0.014)	(0.032)	(0.033)
Constant	3.776	4.445	2.082	−5.238 **	−8.728	−11.849 *	−3.917	−13.374	−16.717 *	2.941	1.047	−5.023
	(3.166)	(5.760)	(5.800)	(1.801)	(4.683)	(4.956)	(3.246)	(7.724)	(8.344)	(2.384)	(4.728)	(5.387)
Spatial error term	0.207	0.164	0.168	−0.923 ***	−0.810 ***	−0.957 ***	0.134	0.213	0.133	0.513 ***	0.336 **	0.449 ***
	(0.110)	(0.112)	(0.112)	(0.199)	(0.214)	(0.193)	(0.156)	(0.149)	(0.156)	(0.096)	(0.115)	(0.104)
Measures of fit												
Log likelihood	−295.458	−382.430	−382.149	−7.776	−40.236	−47.226	−46.686	−117.175	−129.477	−25.921	−113.195	−120.578
AIC	604.917	778.861	778.299	29.553	94.472	108.452	107.374	248.351	272.954	65.843	240.391	255.156
SC	625.657	799.601	799.039	40.637	105.557	119.537	124.635	265.612	290.216	84.218	258.766	273.531
R-Squared	0.275	0.330	0.337	0.555	0.555	0.545	0.513	0.517	0.445	0.380	0.397	0.364
N	143	143	143	36	36	36	87	87	87	102	102	102

Notes: *** $p \leq 0.001$, ** $p \leq 0.01$, * $p \leq 0.05$. Standard deviation is shown in parentheses.

4. Discussion

4.1. Interpretation of Findings

Theoretical and empirical research on PCEs made a significant contribution to the promotion of China's low-carbon transition. In this research, we analyzed the various spatiotemporal patterns and projected future state of PCEs. Our findings revealed an upward trend in China's PCEs, with northern China's PCEs being significantly higher than those of southern China, and the east coastal region's PCEs exceeding that of the central and southwestern regions. With economic growth, an increasing number of individuals have migrated to urban areas, which has subsequently resulted in a rise in CEs [67]. Northern China is currently the largest energy base, and energy consumption and greenhouse gas emissions have been rising sharply due to the rapid development of energy-intensive and polluting industries. For instance, Xue et al. (2011) confirmed that fossil energy production regions produce significantly more CEs than other regions [68]. Wang et al. (2022) demonstrated that the PCE level was positively correlated with the geographical and scale distribution of cities [69]. Compared to other regions, the eastern region has higher PCEs and CEs, demonstrating a connection between the region's economic level and PCE level.

In the markov transition probability matrix, the highest probability was found in the H-type, reaching 97.5%, with an average stability probability of 75.27%, indicating that China's PCEs were stable and exhibited a "club convergence" phenomenon. At the same time, the low-level (0.720) and high-level (0.975) convergence probabilities were greater than the medium–low-level (0.655) and medium–high-level (0.661) convergence probabilities, indicating that low- and high-level regions tended to stabilize at their own levels, while medium–low and medium–high levels tended to transfer to low and high levels. From the perspective of probability values, the probabilities of downward transfer were 32.7% and 32.2%, while those of upward transfer were 1.10% and 1.60%, indicating that the types with lower and higher intensities showed a good trend of downward transfer. The result obtained was consistent with the findings of Cui et al. (2022), Liu et al. (2023), and Wang et al. (2019) [16,50,52]. A dynamical procedure underlying the transition matrix showed that a discernible "Matthew effect" was observed in PCEs, whereby cities with both low and high emissions exhibited tendencies to maintain their initial state throughout the transfer process. This illustrated that the region may become trapped in a development path locked into specific trajectories, making it challenging to swiftly reduce PCEs through technological means.

By conducting a comprehensive review on China's CEs, we can clearly identify the differences between this study and previous research. These differences primarily focus on the following three main aspects. (1) Regional level. Previous studies may focus on the disparities in CEs across different regions. This study analyzed 367 municipal units at the prefecture level. There were presently many studies in the literature on different regions. Jin et al. (2024) [70] took the urban agglomeration of the Yangtze River Economic Belt's as the study areas of CEs. Bei et al. (2024) [71] conducted the study of CEs at Wuhan. Yan et al. (2023) [72] considered 30 provincial units (provinces, cities, and autonomous regions) in mainland China as their basic research units. (2) Research methods. This study used markov chain, kernel density, and spatial economic regression analyses. Feng et al. (2024) [73] and Chen and Bi (2022) [74] used Geodetector models to empirically analyze the driving mechanisms of CEs. Li et al. (2024) [75] used the LMDI model to investigate each province's CEs drivers. Chen et al. (2023) [76] used kernel density estimation, a spatial autocorrelation analysis, and the spatial-temporal LMDI model to explore the spatiotemporal patterns and driving mechanism of CEs. (3) Driving factors. This study utilized a spatial economic model to explore the driving factors of CEs. Wei et al. (2021) [77] verified that GDP was the dominant factor affecting CEs efficiency. Wang et al. (2023) [78] revealed that technology and population size played important roles in CEs reductions. Jiang et al. (2024) [79] deduced that the scale of urban construction presented different promoting or inhibiting effects in different regions. Zhang et al. (2024) [80] demonstrated that technological progress was the main factor influencing CEs.

4.2. Driving Mechanisms of Low-Carbon Transition

Based on the results of the OLS, SLM, and SEM, we revealed the driving mechanism of China's low-carbon transition and explored how socioeconomic and geographical factors that have significantly influenced low-carbon transition. The LUI was proven to have a significantly negative impact on PCEs, which indicated that the increase in urbanization promoted the low-carbon transition, which was inconsistent with previous studies showing increasing CEs due to urbanization [81]. It is a widely accepted hypothesis that the LUI can increase PCEs. On the contrary, some studies have pointed out that urbanization can indeed lead to a reduction in CEs [82,83]. The reason for this was that urbanization promoted the transformation of industrial structures and thus contributes to CE efficiency. The coefficients of the LUI affecting PCEs were −3.239, −8.440, and −7.027 in 2000, 2010, and 2020, which showed that urbanization was more crucial to the reduction in PCEs in 2010 and 2020 than in 2000. In the early stage, due to the problem of blindly pursuing quantity rather than quality under the traditional urbanization and economic model, the reduction in CEs has been inhibited [84]. However, as urbanization levels increased and shifted towards equity, a green economy, and efficiency as core objectives have contributed to the reduction in CEs. Moreover, several studies have corroborated the assertion that urbanization can result in a reduction in CEs, albeit with notable spatial heterogeneity. For instance, Li et al. (2023) [85] revealed that a higher level of urbanization reduced the CEs intensity and explored the significant implications of urbanization on the low-carbon transition by promoting green technological innovation. GDPD was also proven to be positively correlated with CEs. This is because economic growth was usually accompanied by the development of industrialization and manufacturing, which often required more energy, especially fossil fuels, thereby releasing large amounts of CEs [86]. Many scholars have found that economic growth inevitably promoted CEs, leading to global warming and glacier melting, which posed serious challenges to sustainable economic development. For example, Guo and Fang (2023) [87] revealed that CEs was positively correlated with economic growth but present a fluctuating trend. Similarly, Li et al. (2022) [88] concluded that there was a positive effect on GDP per capita and total CO_2. The difference of this study was investigating the impact of GDPD on PCEs, with GDP being categorized by land area.

The regression results also demonstrated that the low-carbon transition was driven by the joint action of socio-economic systems and natural systems, rather than being solely influenced by either socio-economic factors or geographical factors. This finding underscores the comprehensiveness of the driving mechanism and emphasized the importance of formulating region-specific policies. The European Union emissions trading system (EU ETS) was launched in 2005, which has demonstrated significant success in reducing greenhouse gas emissions across 31 countries, accounting for over 40% of the EU's total greenhouse gas emissions. The EU ETS has led to substantial emission reductions while increasing the regulated firms' revenues and fixed assets, as verified by Dechezleprêtre et al. (2023) [89]. The Regional Greenhouse Gas Initiative (RGGI) in the United States was officially implemented in 2009 and established a cap-and-trade program for CEs in the power sector. The RGGI not only achieved significant emission reductions but also generated economic benefits by reinvesting auction proceeds into energy efficiency and renewable energy projects. However, the RGGI has caused significant emissions reductions within regulated states and emissions leakages in nearby unregulated states (Yan, 2021) [90]. In future research, we would like to explore this driving mechanism by taking China's environmental policy into consideration.

4.3. Policy Implications of Low-Carbon Transition

4.3.1. Promoting the Low-Carbon Transition

A significant proportion of prefecture-level cities exhibited a noticeable increase in PCEs. China's CEs have remained high level due to the decision that was made for its economic development mode. As rising CEs contributed to global warming, China, as the biggest developing country, is enacting a low-carbon transition and has made remarkable achievements in curbing the growth of CEs. New-type urbanization is the strategic fulcrum of future development, which is crucial for promoting the low-carbon transition. New urbanization has the potential to establish a favorable policy environment and offer significant scientific and technological support, which can facilitate the low-carbon transition. Throughout the progress of new urbanization [91,92], it is of paramount importance to underscore the significance of enhancing the quality of urbanization [93].

4.3.2. Focusing on the Spatial Agglomeration Effects of PCE

PCEs showed a relatively obvious distribution pattern of spatial agglomeration, and the degree of this continued to strengthen. The PCEs of each city were susceptible to influences from neighboring cities. Therefore, it is crucial for each city to not only prioritize its own PCEs but also to collaborate with neighboring cities. By establishing regional cooperation mechanisms, cities can share information and technology related energy and environmental protection, jointly implement emission reduction measures, as well as collectively assume the responsibility of promoting coordinated economic development and environmental protection. Considering the differences across regions, for provinces such as Chongqing, Sichuan, Guizhou, Anhui, and Jiangxi, which are experiencing rapid economic growth, it is recommended to optimize their industrial structure, support low-energy consuming and clean industries, and promote the shift from high-speed to high-quality economic development. For resource-rich provinces such as Inner Mongolia and Xinjiang, which have resource-endowed advantages, it is suggested that these governments appropriately increase their carbon allowances and set higher thresholds for their high-energy-consuming enterprises, thereby promoting the introduction of high-energy-efficiency technologies from the east to the west.

4.3.3. Coordinating Economic Growth and CEs

In terms of driving factors, economic growth acts as a "double-edged" sword, not only increasing CEs but also providing momentum for the low-carbon transition. The contradiction between the increase in GDP and CEs still exists. However, with the improvement of quality, economic progress can stimulate technological innovation and accelerate emission reductions. Enhancing the quality of economic growth quality would inevitably result in a reduction in CEs and the promotion of the low-carbon transition. Economic growth should leverage the advantages of agglomeration effects and scale effects.

4.3.4. Formulating Differential Strategy

Based on the different regions of CEs and their spatial-temporal characteristics, we proposed a targeted strategy to reduce PCEs. At the national level, the focus should be on improving the efficiency of fossil energy use and replacing fossil energy with non-fossil energy sources, such as developing hydropower, wind power, and solar power. Additionally, carbon sinks can be increased by using trees or other plants to absorb the carbon dioxide emitted into the atmosphere. At the regional level, for the eastern and central regions, the priority should be to strengthen green technology upgrades, leverage their advantages in low-carbon technology research and development, and actively explore innovative systems and mechanisms for low-carbon economic development. For the northeastern regions, it is essential to promote the transformation of resource-depleted cities and the adjustment of old industrial bases in conjunction with the development of a low-carbon economy. Considering the reality that the northeastern region has more per capita arable land and forestry resources, agriculture and forestry should play a significant role in the low-carbon economic development in this region. For the western region, the focus should be on reducing CEs in resource development, distinctive industry development, and infrastructure construction. In terms of increasing carbon sinks, efforts should be made to strengthen carbon sink construction in ecological functional zones, integrating the construction of carbon sinks in the western region with the establishment of a national ecological security barrier.

5. Conclusions

This study aimed to investigate the distribution patterns, future trends, and driving factors of PCEs in 367 cities in China. The results revealed that the PCEs presented a continuous upward trend, with obvious differences across regions. Therefore, CEs reduction policies should be formulated based on a per capita perspective and regional differences. The PCEs also showed obvious spatial clustering, with hot spots mainly concentrated in Inner Mongolia and Xinjiang, while cold spots were mainly distributed in some provinces in southern China. Meanwhile, China's low-carbon transition has achieved noteworthy outcomes, with the majority of prefecture-level cities (29.700%) having PCEs of less than 1 ton/persons. The overall PCE level in China remained stable despite the dynamic changes observed over the ten-year period. In particular, the shift of the PCEs of eastern and central China to higher values was obvious. The regression analysis indicated that China's PCEs undergone changes due to a confluence of factors. The LUI and latitude had negative effects on PCEs in prefecture-level cities, and regional economic development, elevation, slope, and longitude had positive effects. The results can serve as a basis for developing CEs policies, which are of strategic importance for promoting sustainable development and formulating rational CE reduction policies.

However, this study still has several limitations. First, we acknowledge that it is insufficient to use data from 2000, 2010, and 2020 to explore the distribution patterns, future trends, and driving factors of China's PCEs. This is because the PCE calculations were based on data extracted from the bulletins of the fifth, sixth, and seventh national censuses, which covered a ten-year period. In future studies, we will use annual data to supplement or verify the findings of the studied ten-year period. Second, although this study verified the importance of natural geographical factors and socioeconomic factors in influencing the low-carbon transition, due to data availability limitations, other driving forces, such as the digital economy and trading policies, were not taking into account. In the future, we will consider more driving factors to analyze the mechanism.

Author Contributions: Conceptualization, X.Z. and Y.Z.; Methodology, X.Z. and Y.Z.; Software, X.Z., Y.Z. and W.C.; Resources, Y.Z. and W.C.; Data curation, X.Z. and G.Z.; Writing—original draft, X.Z. and Y.Z.; Writing—review and editing, S.P. and F.D.; Supervision, X.Z. and F.D. All authors have read and agreed to the published version of the manuscript.

Funding: This study was supported by the Innovative Research Group Project of the National Natural Science Foundation of China (Grant number: 42121001). This study was also supported by the Natural Science Foundation of China (Grant numbers: 72364027; 42371258; 72164030). This study was supported in part by the China Postdoctoral Science Foundation (Grant number: 2023M733466), the Social Science Foundation of China (Grant number: 22VMZ013), the Natural Science Foundation of Inner Mongolia (Grant No. 2023QN07008), and the Social Science Foundation of Inner Mongolia (Grant No. 2024EY51).

Data Availability Statement: The original contributions presented in the study are included in the article, further inquiries can be directed to the corresponding authors.

Conflicts of Interest: The authors declare no conflicts of interest.

References

1. Bakhtyar, B.; Ibrahim, Y.; Alghoul, M.A.; Aziz, N.; Fudholi, A.; Sopian, K. Estimating the CO_2 abatement cost: Substitute price of avoiding CO_2 emission (SPAE) by renewable energy's feed in tariff in selected countries. *Renew. Sustain. Energy Rev.* **2014**, *35*, 205–210. [CrossRef]

2. Gökmenoğlu, K.; Taspinar, N. The relationship between CO_2 emissions, energy consumption, economic growth and FDI: The case of Turkey. *J. Int. Trade. Econ. Dev.* **2016**, *25*, 706–723. [CrossRef]
3. Allen, M.R.; Frame, D.J.; Huntingford, C.; Jones, C.D.; Meinshausen, N. Warming caused by cumulative carbon emissions towards the trillionth tonne. *Nature* **2009**, *458*, 1163–1166. [CrossRef] [PubMed]
4. Höhne, N.; den Elzen, M.; Rogelj, J.; Metz, B.; Fransen, T.; Kuramochi, T.; Olhoff, A.; Alcamo, J.; Winkler, H.; Fu, S.; et al. Emissions: World has four times the work or one-third of the time. *Nature* **2020**, *579*, 25–28. [CrossRef] [PubMed]
5. Meinshausen, M.; Meinshausen, N.; Hare, W.; Raper, S.C.; Frieler, K.; Knutti, R.; Frame, D.J.; Allen, M.R. Greenhouse-gas emission targets for limiting global warming to 2 C′. *Nature* **2009**, *458*, 1158–1162. [CrossRef]
6. Climate Watch. Historical GHG Emissions. 2024. Available online: https://www.climatewatchdata.org/ghg-emissions (accessed on 26 August 2024).
7. Söderholm, P.; Hildingsson, R.; Johansson, B.; Khan, J.; Wilhelmsson, F. Governing the transition to low-carbon futures: A critical survey of energy scenarios for 2050. *Futures* **2011**, *43*, 1105–1116. [CrossRef]
8. Wang, S.B.; Liu, H.M.; Pu, H.X.; Yang, H. Spatial disparity and hierarchical cluster analysis of final energy consumption in China. *Energy* **2020**, *197*, 117195. [CrossRef]
9. Hickel, J. Quantifying national responsibility for climate break down:an equality-based attribution approach for carbon dioxide emissions in excess of the planetary boundary. *Lancet Planet. Health* **2020**, *4*, e399–e404. [CrossRef]
10. Matthews, H.D. Quantifying historical carbon and climate debts among nations. *Nat. Clim. Chang.* **2016**, *6*, 60–64. [CrossRef]
11. Adriana, D.B.; Michela, G.; Nicolò, S.; Baccelli, O.; Croci, E.; Molteni, T. Impact of circular measures to reduce urban CO_2 emissions: An analysis of four case studies through a production- and consumption-based emission accounting method. *J. Clean. Prod.* **2022**, *380*, 134932.
12. Ali, U.; Guo, Q.B.; Zhanar, N.; Nurgazina, Z.; Sharif, A.; Kartal, M.T.; Depren, S.K.; Khan, A. Heterogeneous impact of industrialization, foreign direct investments, and technological innovation on carbon emissions intensity: Evidence from Kingdom of Saudi Arabia. *Appl. Energy* **2023**, *336*, 120804. [CrossRef]
13. Wang, Z.; Zeng, Y.; Wang, X.; Gu, T.; Chen, W. Impact of urban expansion on carbon emissions in the urban agglomerations of Yellow River Basin, China. *Land* **2024**, *13*, 651. [CrossRef]
14. Shi, Y.; Han, B.; Han, L.; Wei, Z. Uncovering the national and regional household carbon emissions in China using temporal and spatial decomposition analysis models. *J. Clean. Prod.* **2019**, *232*, 966–979. [CrossRef]
15. Huo, T.; Xu, L.; Feng, W.; Cai, W.; Liu, B. Dynamic scenario simulations of carbon emission peak in China's city-scaleurban residential building sector through 2050. *Sci. Total Environ.* **2021**, *159*, 112612.
16. Wang, S.J.; Huang, Y.Y. Spatial spillover effect and driving forces of carbon emission intensity at city level in China. *Acta Geogr. Sin.* **2019**, *74*, 1131–1148. (In Chinese) [CrossRef]
17. Li, S.L.; Wang, Z.Z. The effects of agricultural technology progress on agricultural carbon emission and carbon sink in China. *Agriculture* **2023**, *13*, 793. [CrossRef]
18. Chen, C.H.; Luo, Y.Q.; Zhou, H.; Huang, J.B. Understanding the driving factors and finding the pathway to mitigating carbon emissions in China's Yangtze River Delta region. *Energy* **2023**, *278*, 127897. [CrossRef]
19. Wang, S.; Xie, Z.; Wu, R.; Feng, K. How does urbanization affect the carbon intensity of human well-being? A global assessment. *Appl. Energy* **2022**, *312*, 118798. [CrossRef]
20. Zhou, Y.Q.; Zou, S.; Duan, W.L.; Chen, Y.N.; Takara, K.R.; Di, Y.F. Analysis of energy carbon emissions from agroecosystems in Tarim River Basin, China: A pathway to achieve carbon neutrality. *Appl. Energy* **2022**, *325*, 119842. [CrossRef]
21. Huang, Y.M.; Zhang, Y.N. Digitalization, positioning in global value chain and carbon emissions embodied in exports: Evidence from global manufacturing production-based emissions. *Ecol. Econ.* **2023**, *205*, 107674. [CrossRef]
22. Wang, S.J.; Fang, C.L.; Wang, Y. Spatiotemporal variations of energy-related CO_2 emissions in China and its influencing factors: An empirical analysis based on provincial panel data. *Renew. Sustain. Energy Rev.* **2016**, *55*, 505–515. [CrossRef]
23. Ehigiamusoe, K.U.; Shahbaz, M.; Vo, X.V. How Does Globalization Influence the Impact of Tourism on Carbon Emissions and Ecological Footprint? Evidence from African Countries. *J. Travel. Res.* **2023**, *62*, 1010–1032. [CrossRef]
24. Jan, S.; Liu, Y.; Stefan, S. Carbon emissions from European land transportation: A comprehensive analysis. *Transp. Res. Part D Transp. Environ.* **2023**, *121*, 103851.
25. Zhang, C.Y.; Zhang, H.T.; Chen, M.N.; Fang, R.Y.; Yao, Y.; Zhang, Q.P.; Wang, Q. Spatial-temporal characteristics of carbon emissions from land use change in Yellow River Delta region. *China Ecol. Indic.* **2022**, *136*, 108623. [CrossRef]
26. Dogan, E.; Turkekul, B. CO_2 emissions, real output, energy consumption, trade, urbanization and financial development: Testing the EKC hypothesis for the USA. *Environ. Sci. Pollut. Res.* **2016**, *23*, 1203–1213. [CrossRef]
27. Erdogan, S. Dynamic nexus between technological innovation and building sector carbon emissions in the BRICS Countries. *J. Environ. Manag.* **2021**, *293*, 112780. [CrossRef] [PubMed]
28. Chen, X.L.; Di, Q.B.; Jia, W.H.; Hou, Z.W. Spatial correlation network of pollution and carbon emission reductions coupled with high-quality economic development in three Chinese urban agglomerations. Sustain. *Cities Soc.* **2023**, *94*, 104552. [CrossRef]
29. Wu, H.; Yang, Y.; Li, W. Analysis of spatiotemporal evolution characteristics and peak forecast of provincial carbon emissions under the dual carbon goal: Considering nine provinces in the Yellow River basin of China as an example. *Atmos. Pollut. Res.* **2023**, *14*, 101828. [CrossRef]

30. Fan, J.J.; Wang, J.L.; Qiu, J.X.; Li, N. Stage effects of energy consumption and carbon emissions in the process of urbanization: Evidence from 30 provinces in China. *Energy* **2023**, *276*, 127655. [CrossRef]
31. Liu, Q.F.; Song, J.P.; Dai, T.Q.; Shi, A.; Xu, J.H.; Wang, E.R. Spatio-temporal dynamic evolution of carbon emission intensity and the effectiveness of carbon emission reduction at county level based on nighttime light data. *J. Clean. Prod.* **2022**, *362*, 132301. [CrossRef]
32. Liu, H.; Lei, J. The impacts of urbanization on Chinese households' energy consumption: An energy input-output analysis. *J. Renew. Sustain. Energy* **2018**, *10*, 015903. [CrossRef]
33. Wang, S.J.; Liu, X.P. China's city-level energy-related CO_2 emissions: Spatiotemporal patterns and driving forces. *Appl. Energy* **2017**, *200*, 204–214. [CrossRef]
34. Zhang, Y.J.; Hao, J.F.; Song, J. The CO_2 emission efficiency, reduction potential and spatial clustering in China's industry: Evidence from the regional level. *Appl. Energy* **2016**, *174*, 213–223. [CrossRef]
35. Chen, W.X.; Chi, G.Q. Ecosystem services trade-offs and synergies in China, 2000–2015. *Int. J. Environ. Sci. Technol.* **2022**, *20*, 3221–3236. [CrossRef]
36. Chancel, L. Global carbon inequality over 1990–2019. *Nat. Sustain.* **2022**, *5*, 931–938. [CrossRef]
37. Ma, T.; Liu, Y.S.; Yang, M. Spatial-temporal heterogeneity for commercial building carbon emissions in China: Based the dagum gini coefficient. *Sustainability* **2022**, *14*, 5243. [CrossRef]
38. Tao, M.M.; Sheng, M.S.; Wen, L. How does financial development influence carbon emission intensity in the OECD countries: Some insights from the information and communication technology perspective. *J. Environ. Manag.* **2023**, *335*, 117553. [CrossRef]
39. Jebli, M.B.; Kahia, M. The interdependence between CO_2 emissions, economic growth, renewable and non-renewable energies, and service development: Evidence from 65 countries. *Clim. Chang.* **2020**, *162*, 193–212. [CrossRef]
40. Akram, R.; Umar, M.; Gan, X.L.; Chen, F. Dynamic linkages between energy efficiency, renewable energy along with economic growth and carbon emission. A case of MINT countries an asymmetric analysis. *Energy Rep.* **2022**, *8*, 2119–2130. [CrossRef]
41. Nasir, M.A.; Huynh, T.L.D.; Tram, H.T.X. Role of financial development, economic growth & foreign direct investment in driving climate change: A case of emerging ASEAN. *J. Environ. Manag.* **2019**, *242*, 131–141.
42. Chen, W.X.; Wang, G.Z.; Xu, N.; Ji, M.; Zeng, J. Promoting or inhibiting? New-type urbanization and urban carbon emissions efficiency in China. *Cities* **2023**, *140*, 104429. [CrossRef]
43. Tang, Z.P.; Wang, L.H.; Wu, W. The impact of high-speed rail on urban carbon emissions: Evidence from the Yangtze River Delta. *J. Transp. Geogr.* **2023**, *110*, 103641. [CrossRef]
44. Wang, K.K.; Su, X.W.; Wang, S.H. How does the energy-consuming rights trading policy affect China's carbon emission intensity? *Energy* **2023**, *276*, 127579. [CrossRef]
45. Li, R.; Wang, Q.; Liu, Y.; Jiang, R. Per-capita carbon emissions in 147 countries: The effect of economic, energy, social, and trade structural changes. *Sustain. Prod. Consum.* **2021**, *27*, 1149–1164. [CrossRef]
46. Yu, M.; Meng, B.; Li, R. Analysis of China's urban household indirect carbon emissions drivers under the background of population aging. *Struct. Chang. Econ. Dynam.* **2022**, *60*, 114–125. [CrossRef]
47. Ahmed, M.; Shuai, C.; Ahmed, M. Influencing factors of carbon emissions andtheir trends in China and India: A machine learning method. *Environ. Sci. Pollut. Res.* **2022**, *29*, 48424–48437. [CrossRef]
48. Xu, W.H.; Xie, Y.L.; Xia, D.H.; Ji, L.; Huang, G.H. A multi-sectoral decomposition and decoupling analysis of carbon emissions in Guangdong province, China. *J. Environ. Manag.* **2021**, *298*, 113485. [CrossRef] [PubMed]
49. Yu, Y.; Dai, Y.Q.; Xu, L.Y.; Zheng, H.Z.; Wu, W.H.; Chen, L. A multi-level characteristic analysis of urban agglomeration energy-related carbon emission: A case study of the Pearl River Delta. *Energy* **2023**, *263*, 125651. [CrossRef]
50. Liu, X.J.; Jin, X.B.; Luo, X.L.; Zhou, Y.K. Quantifying the spatiotemporal dynamics and impact factors of China's county-level carbon emissions using ESTDA and spatial econometric models. *J. Clean. Prod.* **2023**, *410*, 137203. [CrossRef]
51. Chen, W.X.; Zeng, J.; Li, N. Change in land-use structure due to urbanisation in China. *J. Clean. Prod.* **2021**, *321*, 128986. [CrossRef]
52. Cui, Y.; Khan, S.U.; Deng, Y.; Zhao, M. Spatiotemporal heterogeneity, convergence and its impact factors: Perspective of carbon emission intensity and carbon emission per capita considering carbon sink effect. *Environ. Impact Assess. Rev.* **2022**, *92*, 106699. [CrossRef]
53. Wang, M.; Feng, C. Tracking the inequalities of global per capita carbon emissions from perspectives of technological and economic gaps. *J. Environ. Manag.* **2022**, *315*, 115144. [CrossRef]
54. Wu, S.; Zhang, K. Influence of urbanization and foreign direct investment on carbon emission efficiency: Evidence from urban clusters in the Yangtze River Economic Belt. *Sustainability* **2021**, *13*, 2722. [CrossRef]
55. Wang, X.; Wang, C.X.; Shi, J.L. Evaluation of urban resilience based on Service-Connectivity-Environment (SCE) model: A case study of Jinan city, China. *Int. J. Disaster Risk Reduct.* **2023**, *95*, 103828. [CrossRef]
56. Yang, L.Y.; Fang, C.L.; Chen, W.X.; Zeng, J. Urban-rural land structural conflicts in China: A land use transition perspective. *Habitat. Int.* **2023**, *138*, 102877. [CrossRef]
57. Getis, A.; Aldstadt, J. Constructing the Spatial Weights Matrix Using a Local Statistic. *Geogr. Anal.* **2004**, *36*, 90–104. [CrossRef]
58. Getis, A.; Griffith, D.A. Comparative spatial filtering in regression analysis. *Geogr. Anal.* **2002**, *34*, 130–140. [CrossRef]
59. Chen, W.X.; Wang, G.Z.; Zeng, J. Impact of urbanization on ecosystem health in Chinese urban agglomerations. *Environ. Impact Assess. Rev.* **2023**, *98*, 106964. [CrossRef]

60. Felipe-Lucia, M.R.; Soliveres, S.; Penone, C.; Fischer, M.; Ammer, C.; Boch, S.; Boeddinghaus, R.S.; Bonkowski, M.; Buscot, F.; Fiore-Donno, A.M.; et al. Land-use intensity alters networks between biodiversity, ecosystem functions, and services. *Proc. Natl. Acad. Sci. USA* **2020**, *117*, 28140–28149. [CrossRef]
61. Xu, F.; Chi, G.Q. Spatiotemporal variations of land use intensity and its driving forces in China, 2000–2010. *Reg. Environ. Chang.* **2019**, *19*, 2583–2596. [CrossRef]
62. Zhou, Y.; Chen, M.X.; Tang, Z.P.; Mei, Z. Urbanization, land use change, and carbon emissions: Quantitative assessments for city-level carbon emissions in Beijing-Tianjin-Hebei region, Sustain. *Cities Soc.* **2021**, *66*, 102701. [CrossRef]
63. Wang, W.Z.; Liu, L.C.; Liao, H.; Wei, Y.M. Impacts of urbanization on carbon emissions: An empirical analysis from OECD countries. *Energy Policy* **2021**, *151*, 112171. [CrossRef]
64. Zhang, H.M.; Xu, L.; Zhou, P.; Zhu, X.D. Coordination between economic growth and carbon emissions: Evidence from 178 cities in China. *Econ. Anal. Policy* **2024**, *81*, 164–180. [CrossRef]
65. Braimoh, A.K.; Onishi, T. Spatial determinants of urban land use change in Lagos, Nigeria. *Land Use Policy* **2007**, *24*, 502–515. [CrossRef]
66. Liu, X.; Xu, H.Z.; Zhang, M. The effects of urban expansion on carbon emissions: Based on the spatial interaction and transmission mechanism. *J. Clean. Prod.* **2024**, *434*, 140019. [CrossRef]
67. Li, K.; Lin, B.Q. Impacts of urbanization and industrialization on energy consumption/CO_2 emissions: Does the level of development matter? *Renew. Sustain. Energy Rev.* **2015**, *52*, 1107–1122. [CrossRef]
68. Xue, B.; Li, C.; Liu, Z.; Geng, Y.; Xi, F. Analysis on CO_2 emission and urbanization at global level during 1970–2007. *Clim. Chang. Res.* **2011**, *7*, 423–427.
69. Wang, S.J.; Wang, Z.H.; Fang, C.L. Evolutionary characteristics and drivers of carbon emission performance of Chinese cities. *Sci. Sin. Terrae* **2022**, *52*, 1613–1626.
70. Jin, Y.Z.; Zhang, K.R.; Li, D.Y.; Wang, S.Y.; Liu, W.Y. Analysis of the spatial–temporal evolution and driving factors of carbon emission efficiency in the Yangtze River economic Belt. *Ecol. Indic.* **2024**, *165*, 112092. [CrossRef]
71. Bei, L.; Yang, W.; Wang, B.; Gao, Y.W.; Wang, A.N.; Lu, T.F.; Liu, H.T.; Sun, L.S. Characteristics of residents' carbon emission and driving factors for carbon peaking: A case study in Wuhan, China. *Energy Sustain. Dev.* **2024**, *81*, 101471. [CrossRef]
72. Yan, D.S.; Liu, C.G.; Li, P.X. Effect of carbon emissions and the driving mechanism of economic growth target setting: An empirical study of provincial data in China. *J. Clean. Prod.* **2023**, *415*, 137721. [CrossRef]
73. Feng, X.H.; Wang, S.S.; Li, Y.; Yang, J.Y.; Lei, K.G.; Yuan, W.K. Spatial heterogeneity and driving mechanisms of carbon emissions in urban expansion areas: A research framework coupled with patterns and functions. *Land Use Policy* **2024**, *143*, 107209. [CrossRef]
74. Chen, C.Y.; Bi, L.L. Study on spatio-temporal changes and driving factors of carbon emissions at the building operation stage-A case study of China. *Build. Environ.* **2022**, *219*, 109147. [CrossRef]
75. Li, S.Y.; Yao, L.L.; Zhang, Y.C.; Zhao, Y.X.; Sun, L. How do driving factors of carbon emissions and scenario forecasting differ across provinces in China? Investigation and analysis. *Environ. Sustain. Indic.* **2024**, *22*, 100390.
76. Chen, H.D.; Du, Q.X.; Huo, T.F.; Liu, P.R.; Cai, W.G.; Liu, B.S. Spatiotemporal patterns and driving mechanism of carbon emissions in China's urban residential building sector. *Energy* **2023**, *263*, 126102. [CrossRef]
77. Wei, W.; Zhang, X.; Zhou, L.; Xie, B.; Zhou, J.; Li, C. How does spatiotemporal variations and impact factors in CO_2 emissions differ across cities in China? Investigation on grid scale and geographic detection method. *J. Clean. Prod.* **2021**, *321*, 128933. [CrossRef]
78. Wang, Z.; Shao, H.-A.-O. Spatiotemporal differences in and influencing factors of urban carbon emission efficiency in China's Yangtze River Economic Belt. *Environ. Sci. Pollut. Res.* **2023**, *30*, 121713–121733. [CrossRef] [PubMed]
79. Jiang, H.; Yin, J.; Wei, D.; Luo, X.; Ding, Y.; Xia, R. Industrial carbon emission efficiency prediction and carbon emission reduction strategies based on multi-objective particle swarm optimization-backpropagation: A perspective from regional clustering. *Sci. Total Environ.* **2024**, *906*, 167692. [CrossRef] [PubMed]
80. Zhang, N.; Sun, F.; Hu, Y. Carbon emission efficiency of land use in urban agglomerations of Yangtze River Economic Belt, China: Based on three-stage SBM-DEA model. *Ecol. Indic.* **2024**, *160*, 111922. [CrossRef]
81. Xu, J.H.; Li, Y.Y.; Hu, F.; Wang, L.; Wang, K.; Ma, W.H.; Ruan, N.; Jiang, W.Z. Spatio-temporal variation of carbon emission intensity and spatial heterogeneity of influencing factors in the Yangtze River Delta. *Atmosphere* **2023**, *14*, 163. [CrossRef]
82. Adams, S.; Nsiah, C. Reducing carbon dioxide emissions: Does renewable energy matter? *Sci. Total Environ.* **2019**, *693*, 133288. [CrossRef] [PubMed]
83. Kwakwa, P.A.; Adzawla, W.; Alhassan, H.; Oteng-Abayie, E.F. The effects of urbanization, ICT, fertilizer usage, and foreign direct investment on carbon dioxide emissions in Ghana. *Environ. Sci. Pollut. Res. Int.* **2023**, *30*, 23982–23996. [CrossRef] [PubMed]
84. Pan, J.N.; Huang, J.T.; Chiang, T.F. Empirical study of the local government deficit, land finance and real estate markets in China. *China Econ. Rev.* **2015**, *32*, 57–67. [CrossRef]
85. Li, X.Q.; Zheng, Z.J.; Shi, D.Q. New urbanization and carbon emissions intensity reduction: Mechanisms and spatial spillover effects. *Sci. Total Environ.* **2023**, *905*, 167172. [CrossRef] [PubMed]
86. Abraham, D.; Huseyin, O.; Mehdi, S. The effect of GDP, renewable energy and total energy supply on carbon emissions in the EU 27: New evidence from panel GMM. *Environ. Sci. Pollut. Res.* **2023**, *30*, 28206–28216.

87. Guo, X.M.; Fang, C.L. How does urbanization affect energy carbon emissions under the background of carbon neutrality? *J. Environ. Manag.* **2023**, *327*, 103851.
88. Li, W.Y.; Ji, Z.S.; Dong, F.G. Spatio-temporal evolution relationships between provincial CO_2 emissions and driving factors using geographically and temporally weighted regression model. *Sustain. Cities Soc.* **2022**, *81*, 103836. [CrossRef]
89. Dechezleprêtre, A.; Nachtigall, D.; Venmans, F. The joint impact of the European Union emissions trading system on carbon emissions and economic performance. *J. Environ. Econ. Manag.* **2023**, *118*, 102758. [CrossRef]
90. Yan, J.C. The impact of climate policy on fossil fuel consumption: Evidence from the Regional Greenhouse Gas Initiative (RGGI). *Energy Econ.* **2021**, *100*, 105333. [CrossRef]
91. Zhang, W.S.; Xu, Y.; Lei, J.; Streets, D.G.; Wang, C. Direct and spillover effects of new-type urbanization on CO_2 emissions from central heating sector and EKC analyses: Evidence from 144 cities in China. *Resour. Conserv. Recy.* **2023**, *192*, 106913. [CrossRef]
92. Zhang, W.S.; Xu, Y.; Streets, D.G.; Wang, C. Can new-type urbanization realize low-carbon development? A spatiotemporal heterogeneous analysis in 288 cities and 18 urban agglomerations in China. *J. Clean. Prod.* **2023**, *420*, 138426. [CrossRef]
93. Guo, X.M.; Fang, C.L.; Mu, X.F.; Chen, D. Coupling and coordination analysis of urbanization and ecosystem service value in Beijing-Tianjin-Hebei urban agglomeration. *Ecol. Indic.* **2022**, *137*, 108782.

land

MDPI

Article

Response of Hydrothermal Conditions to the Saturation Values of Forest Aboveground Biomass Estimation by Remote Sensing in Yunnan Province, China

Yong Wu [1,2], Binbing Guo [1,2], Xiaoli Zhang [1,2], Hongbin Luo [1,2], Zhibo Yu [1,2], Huipeng Li [1,2], Kaize Shi [3], Leiguang Wang [1,4], Weiheng Xu [1,4] and Guanglong Ou [1,2,*]

1 Key Laboratory of State Administration of Forestry and Grassland on Biodiversity Conservation in Southwest China, Southwest Forestry University, Kunming 650233, China; yongwu@swfu.edu.cn (Y.W.); guobinbing@swfu.edu.cn (B.G.); karojan@swfu.edu.cn (X.Z.); luohongbin@swfu.edu.cn (H.L.); yuzhibo@swfu.edu.cn (Z.Y.); lhp@swfu.edu.cn (H.L.); leiguangwang@swfu.edu.cn (L.W.); xwh@swfu.edu.cn (W.X.)

2 Key Laboratory for Forest Resources Conservation and Utilization in the Southwest Mountains of China, Ministry of Education, Southwest Forestry University, Kunming 650233, China

3 Yunnan Institute of Forest Inventory and Planning, Kunming 650051, China; niejing@swfu.edu.cn

4 Institute of Big Data and Artificial Intelligence, Southwest Forestry University, Kunming 650224, China

* Correspondence: olg2007621@swfu.edu.cn; Tel.: +86-139-0885-7108

Abstract: Identifying the key climate variables affecting optical saturation values (OSVs) in forest aboveground biomass (AGB) estimation using optical remote sensing is crucial for analyzing OSV changes. This can improve AGB estimation accuracy by addressing the uncertainties associated with optical saturation. In this study, *Pinus yunnanensis* forests and Landsat 8 OLI imagery from Yunnan were used as case studies to explain this issue. The spherical model was applied to determine the OSVs using specific spectral bands (Blue, Green, Red, Near-Infrared (NIR), and Short-Wave Infrared Band 2 (SWIR2)) derived from Landsat 8 OLI imagery. Canonical correlation analysis (CCA) uncovered the intricate relationships between climatic variables and OSV variations. The results reveal the following: (1) All Landsat 8 OLI spectral bands showed a negative correlation with the *Pinus yunnanensis* forest AGB, with OSVs ranging from 104.42 t/ha to 209.11 t/ha, peaking in the southwestern region and declining to the lowest levels in the southeastern region. (2) CCA effectively explained 93.2% of the OSV variations, identifying annual mean temperature (AMT) as the most influential climatic factor. Additionally, the mean temperature of the wettest quarter (MTQ) and annual precipitation (ANP) were significant secondary determinants, with higher OSV values observed in warmer, more humid areas. These findings offer important insights into climate-driven OSV variations, reducing uncertainty in forest AGB estimation and enhancing the precision of AGB estimations in future research.

Keywords: optical saturation variations; *Pinus yunnanensis*; climate; aboveground biomass estimation; Landsat 8 optical imagery

1. Introduction

Forests are among the most significant terrestrial ecosystems, playing a fundamental role in climate regulation, managing the carbon cycle, and controlling water resources [1,2]. The carbon cycle is at the forefront of global climate discussions, and the accurate estimation of forest aboveground biomass (AGB) is essential for evaluating carbon storage and emissions [3,4]. Traditional methods for estimating forest AGB are focused on localized studies and specific ecological assessments [5]. By contrast, remote sensing technologies enable broader application across large and topographically challenging areas, offering comprehensive coverage, high spatial and temporal resolution, and the ability to monitor ecosystem changes dynamically [6,7]. These advanced technologies provide crucial insights into forest AGB at a macro scale, deepening our understanding of ecosystem functions and their influence on global carbon cycles.

Despite the numerous benefits of using remote sensing techniques in forest AGB estimation, uncertainty continues to be a critical challenge in improving AGB estimation accuracy, arising from factors such as remote sensing data, estimation models, forest heterogeneity, and the data saturation problem [6,8,9]. The issue of saturation is especially prominent when using optical imagery for forest AGB estimation in regions with high forest heterogeneity [10]. This occurs when spectral reflectance values become less responsive to further increases in AGB after surpassing a certain threshold, a challenge first noted in the literature in 1968 [11]. Research has shown that this saturation is largely caused by the decreased sensitivity of the red band, which is absorbed by chlorophyll as the canopy density increases [12,13]. Furthermore, studies demonstrate that the near-infrared (NIR) band shows a linear reduction in reflectance as stand age and AGB increase [14]. This issue is compounded by mature vegetation, where spectral indices like the NDVI struggle to estimate AGB with precision, particularly at the peak of the growing season [15]. However, utilizing seasonal NDVI time series instead of single NDVI measurements has been shown to enhance AGB estimation accuracy and mitigate saturation issues [16].

Notably, the application of the short-wave infrared band (SWIR2) of Landsat TM in Zhejiang province, using a spherical model, allowed for the precise estimation of AGB across various vegetation types. This approach not only significantly reduced residual errors, but also improved the accuracy of optical saturation value (OSV) estimations, which were found to vary considerably across different forest types [17]. These findings have spurred increased research efforts aimed at reducing the uncertainty associated with the saturation problem and identifying OSVs across diverse ecosystems and geographical regions [6,10]. However, despite these advances, the underlying variation patterns of OSVs and the environmental factors influencing OSVs remain insufficiently explored. Recent studies have demonstrated that climate factors provide greater explanatory power than soil and topography in accounting for the variations in OSVs, as evidenced by research on oak forests in Yunnan Province [18]. Nevertheless, the extent to which climate factors drive OSV variability across broader forest ecosystems has yet to be fully elucidated. This underscores the need for further investigation into the relationship between climate dynamics and OSV variations, which could provide critical insights into the climatic controls of saturation thresholds in forest AGB estimation.

Climate is a fundamental driver of vegetation distribution and biomass allocation, particularly within forest ecosystems [19]. These climatic forces exert profound influence over the growth, survival, and spatial arrangement of plant species, with temperature, precipitation, and solar radiation acting as the primary regulators of photosynthetic processes and biomass accumulation [19,20]. Optimal climatic conditions enhance photosynthetic efficiency, thereby promoting the accumulation of forest AGB, while fluctuations in temperature or moisture availability can lead to significant variations in vegetation structure and biomass distribution [21,22]. For instance, shifts in seasonal precipitation and temperature patterns, driven by climatic variability, can substantially alter forest composition, species distribution, and biomass storage [23]. Thus, a comprehensive understanding of climatic role is essential for boosting the precision of AGB estimations, particularly in regions with high climatic variability. Beyond influencing vegetation dynamics, climate notably impacts the accuracy of AGB estimates obtained through remote sensing. Changes in temperature and precipitation modify the spectral reflectance characteristics of forests, which consequently affect OSVs estimation, a crucial element in minimizing uncertainty in forest AGB assessments [24]. As forest ecosystems adapt to fluctuating climatic conditions, the resulting OSV variability underscores the necessity for models tailored to specific regions, addressing climate-related differences in biomass distribution [18]. Therefore, elucidating the relationship between climate and OSV variability is imperative for refining forest AGB estimation models across diverse forested landscapes.

Yunnan province, located in southwestern China, offers an ideal natural setting for investigating the influence of climate on AGB estimation. It is renowned as the Kingdom of Plants due to its exceptional biodiversity, with a vast range of vegetation spanning from tropical to temperate ecosystems [25]. *Pinus yunnanensis* trees are widely distributed and the indigenous evergreen coniferous trees in Yunnan [26,27]. The region's complex topography and diverse climatic conditions make it a crucial focal point for forest AGB research. However, these characteristics, particularly the high variability in terrain and forest heterogeneity, present significant challenges to accurately estimating AGB through remote sensing [28]. The optical saturation problem emerges as a key source of uncertainty in such heterogeneous forests, complicating the relationship between spectral data and biomass estimates [18]. Thus, gaining insights into how Yunnan's unique climate interacts with OSV variations is crucial for enhancing the precision of remote sensing models in this area. By analyzing the climatic factors driving OSV variability, this study seeks to improve AGB estimation methods in Yunnan, offering findings that may apply to other regions with complex ecological and climatic conditions [29–31].

Overall, it is essential to elucidate the relationship between climatic factors, particularly the various temperature and precipitation variables, and OSVs in order to identify the key climatic drivers influencing OSV variations. In this study, the *Pinus yunnanensis* forests, located in different vegetation sub-regions of Yunnan, were selected as the research focus. Landsat 8 OLI imagery was employed to estimate the OSVs of these forests using a spherical model. Furthermore, the spatial distribution patterns of OSVs were analyzed, and the key climatic variables influencing OSV variations were identified. The aims of this research are as follows:

To explain the OSV variation patterns for *Pinus yunnanensis* forest AGB estimation.

To determine the key climatic factors driving OSV variations.

2. Materials and Methods

This study adhered to the methodological flowchart shown in Figure 1, with the following steps: (1) obtaining data on the distribution of *Pinus yunnanensis* forests from the Forest Management Inventory (FMI); (2) calculating the *Pinus yunnanensis* forests' AGB across eight sub-regions classified by the Yunnan flora system; (3) collecting and processing Landsat 8 OLI imagery along with relevant climate data; (4) extracting the original spectral bands; (5) analyzing the correlation between forest AGB and the original bands; (6) applying the spherical model to calculate OSVs; (7) examining OSV variations in response to climatic factors; and (8) identifying the key climatic variables that influence OSV variations.

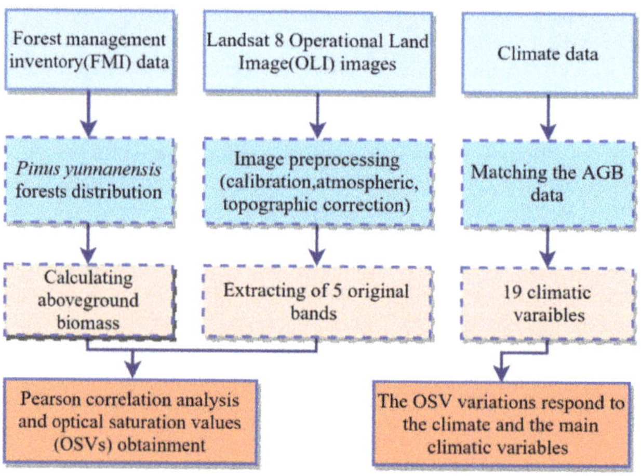

Figure 1. Methodological flowchart for estimating and analyzing OSV distribution patterns and variations.

2.1. Study Area

Yunnan Province, located in southwestern China, extends between latitudes 21°8′ N and 29°15′ N and longitudes 97°31′ E to 106°11′ E, encompassing a total area of 394,000 km^2 (Figure 2). It is in the region where the plateau and mountainous areas intersect, characterized by a complex and diverse topography including high mountains, canyons, hills, and plains. Its geographical location and complex terrain contribute to the high heterogeneity of climate, vegetation, and so on [26,32]. It exhibits a range of climatic types, including subtropical, temperate, and alpine climates. The altitude ranges from 76.4 m to 6740 m, while the precipitation varies between 500 mm and 2700 mm [18]. Meanwhile, the favorable climate in Yunnan provides optimal growth conditions for *Pinus yunnanensis* forests, enabling it to flourish in various ecological habitats, including subalpine forests, montane forests, and mixed evergreen–deciduous forests [33].

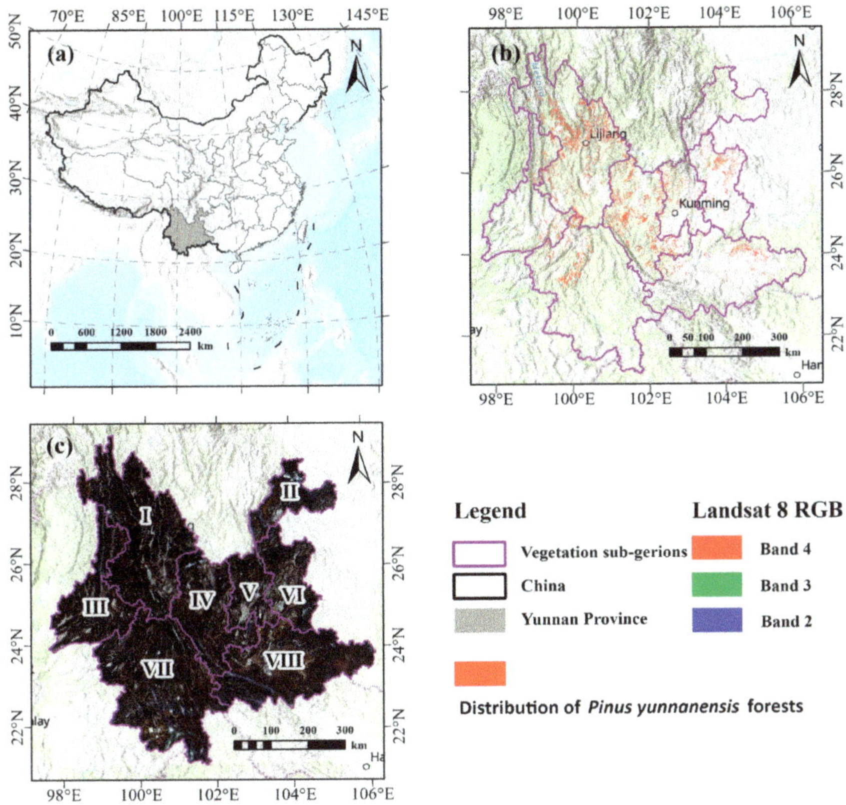

Figure 2. Overview of vegetation sub-regions and *Pinus yunnanensis* forests distribution: (**a**) illustrates the geographic location of Yunnan Province within China; (**b**) presents the spatial distribution of *Pinus yunnanensis* forests within Yunnan; and (**c**) shows Landsat 8 OLI imagery and eight sub-regions (I to VIII) within Yunnan.

2.2. Vegetation Sub-Regions

The Yunnan flora system was utilized to divide the Yunnan province into 8 sub-regions using ArcGIS 10.8, which was produced by combining the relationships between the distribution and phylogenetics of seed plant genera in 1983 [34,35]. This segmentation facilitated the alignment of climate and *Pinus yunnanensis* forest AGB data, enabling the derivation and comparative analysis of OSVs across sub-regions to investigate their variations (Figure 2).

2.3. Forest Aboveground Biomass Data

The distribution of *Pinus yunnanensis* forests was obtained from the 2016 FMI data for Yunnan province. The AGB per unit area of *Pinus yunnanensis* forests was calculated using the biomass conversion method based on a total of 175,511 selected sub-compartments [36]. The biomass conversion parameters are shown in Figure 3. The formula used for AGB calculation is as follows:

$$B = V \times SVD \times BEF \quad (1)$$

where B refers to the AGB for the sub-compartment (t/ha), V indicates the volume of storage per unit area in the sub-compartment (m^3/hm^2), SVD stands for the basic wood density (t/m^3), and BEF is the biomass conversion coefficient (dimensionless).

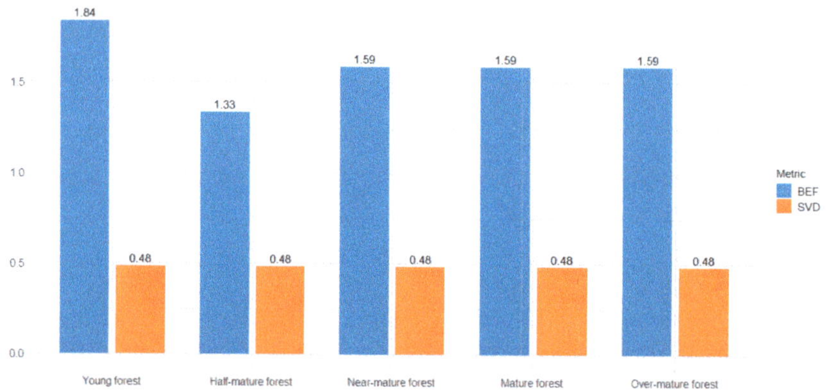

Figure 3. The parameters using the biomass conversion factor method.

All sub-compartments of *Pinus yunnanensis* forests were selected to estimate OSVs, providing a thorough estimation of the forest's spatial arrangement and AGB properties. The statistical metrics for each sub-region, including sample size, AGB range, mean AGB, and standard error, are detailed in Table 1. These values reveal significant variability in AGB across sub-regions. The standard error further underscores this heterogeneity, with sub-region I exhibiting marked variability (SE = 36.74 t/ha), while sub-region VI shows more consistency (SE = 18.33 t/ha). Such findings emphasize the need to account for regional differences in AGB and OSV estimations, given their sensitivity to local environmental dynamics.

Table 1. The statistical parameters of each sub-region for *Pinus yunnanensis* forest AGB.

Sub-Regions	n	AGB Range (t/ha)	Mean (t/ha)	SE (t/ha)
I	45,409	2.35–332.70	56.77	36.74
II	2478	1.64–373.09	39.68	18.49
III	11,085	3.05–433.93	64.35	35.45
IV	39,569	2.58–485.41	42.47	23.59
V	11,826	1.01–485.41	43.03	26.52
VI	16,547	3.56–264.20	37.13	18.33
VII	23,766	2.84–339.98	65.91	28.35
VIII	21,939	1.89–220.48	40.56	22.10

2.4. Remote Sensing Data

The original bands of Landsat 8 OLI data were used to derive the OSVs, with a spatial resolution of 30 m. The 29 Landsat 8 OLI images from 2016 in Yunnan were acquired from a website (http://www.gscloud.cn/ (accessed on 1 April 2023)). A substantial portion of the images had a cloud cover of less than 6.00%. Subsequent image preprocessing included radiometric calibration, atmospheric correction using FLAASH, topographic correction, and geographic alignment to ensure reliable and consistent land cover and surface property analysis, utilizing ENVI 5.3 software [37,38]. Finally, simultaneous Landsat 8 OLI images with the forest AGB data from 2016 were mosaicked to produce a seamless image.

2.5. Climate Data

The climate data were obtained from the WorldClim database (http://www.worldclim.org/ (accessed on 1 April 2023)), with a spatial resolution of 1 km × 1 km, containing 19 bio-climatic variables (Table 2). The climate data were georeferenced in ENVI 5.3 to align with the *Pinus yunnanensis* distribution and sub-regional boundaries and resampled to 30 × 30 m resolution for compatibility with the Landsat 8 imagery.

Table 2. The overview of climate variables.

Variables	Descriptions	Variables	Descriptions
AMT	Annual mean temperature (°C)	PRD	Precipitation in the driest quarter (mm)
MDR	Temperature diurnal range (°C)	PRS	Precipitation seasonality (mm)
ISO	Isothermality (%)	PWQ	Precipitation in the warmest quarter (mm)
MCQ	Mean temperature in the coldest quarter (°C)	PCQ	Precipitation in the coldest quarter (mm)
MTW	Temperature in the warmest month (°C)	PWM	Precipitation in the wettest month (mm)
MTC	Temperature in the coldest month (°C)	PDM	Precipitation in the driest month (mm)
TAR	Temperature annual range (°C)	TES	Temperature seasonality (°C)
MWQ	Mean temperature in the warmest quarter (°C)	PRW	Precipitation in the wettest quarter (mm)
MTD	Mean temperature in the driest quarter (°C)	ANP	Mean of annual precipitation (mm)
MTQ	Mean temperature in the wettest quarter (°C)		

2.6. Optical Saturation Values Obtainment

The study validated that the spherical model based on the semi-variance function provides highly accurate estimates for calculating OSVs both in Landsat TM [17] and Landsat 8 OLI imagery [18]. In this study, the spherical model was employed to estimate the OSVs using 5 bands, including Blue, Green, Red, NIR, and SWIR2, derived from Landsat 8 OLI imagery. The correlation between these bands and forest AGB was rigorously analyzed, with statistically significant correlations at the 0.01 level being selected. Subsequently, the OSVs were calculated across 8 sub-regions, allowing for a comprehensive analysis of OSV variations. The spherical model can be represented by the following equation:

$$y(AGB) = \begin{cases} c_0 + c(\frac{3AGB}{2BS} - \frac{AGB^3}{2BS^3})\ 0 \leq \mathrm{AGB} \leq \mathrm{BS} \\ c_0 + \mathrm{c}\ \mathrm{AGB} > \mathrm{BS} \end{cases} \quad (2)$$

where y(AGB) denotes the value of spectral reflectance; c_0 denotes the reflectance when biomass is zero; c stands for the rate of change in reflectance; $c_0 + c$ signifies the maximum or minimum reflectance value when biomass reaches saturation; BS corresponds to the spectral OSV for the specific bands. By setting $b_0 = c_0$, $b_1 = 3c/2BS$, $b_2 = -c/2BS^3$, $x = AGB$, the AGB value can be derived through least squares regression. Through the application of least squares fitting to Equation (3), the parameters for Equations (4) and (5) are obtained, enabling the calculation of OSVs by incorporating parameters b_1 and b_2 from Equation (6).

$$y = b_0 + b_1 x + b_2 x^3 \quad (3)$$

$$b_1 + b_2 = \frac{3c}{2\mathrm{OSV}} - \frac{\mathrm{C}}{2\mathrm{OSV}^3} \quad (4)$$

$$c = \frac{2}{3}\mathrm{OSV} \times b_1 \quad (5)$$

$$\mathrm{OSV} = \sqrt{-\frac{b_1}{3b_2}} \quad (6)$$

2.7. Optical Saturation Values' Variation Analyses

It was crucial to find the key variables affecting the OSV variations by clarifying the relationship between the OSVs and climatic variables. Canonical correlation analysis (CCA) is a statistical approach used to examine and quantify the linear relationships between two multivariate datasets [39,40]. Its advantages include revealing the correlations between multiple variables, dimensionality reduction, integration and interpretation of multiple datasets, and application in prediction and classification tasks [40,41]. Thus, CCA was applied to investigate the relationship between OSVs and climatic variables. The key factors influencing OSV variations were identified using the vegan package in R4.4.1 software as part of this study. Firstly, 19 climatic variables were standardized to ensure the variables were comparable across scales or units of measure, and we then eliminated the inequitable effects due to differences in the scales of variables [41]. Subsequently, standardized climatic variables and OSVs across 8 sub-regions underwent CCA, elucidating climate-responsive OSV variations and identifying key climatic determinants.

3. Results

3.1. Relationship between Forest Aboveground Biomass and Original Spectral Bands

Pearson correlation analysis was utilized between the *Pinus yunnanensis* forest AGB and five original bands, and the results show that all the original bands were significant at the 0.01 level. As shown in Figure 4, all the original bands were negative with the *Pinus yunnanensis* forest AGB in each sub-region, and the Red band showed the best performance, with all the absolute values of correlation coefficients greater than 0.2.

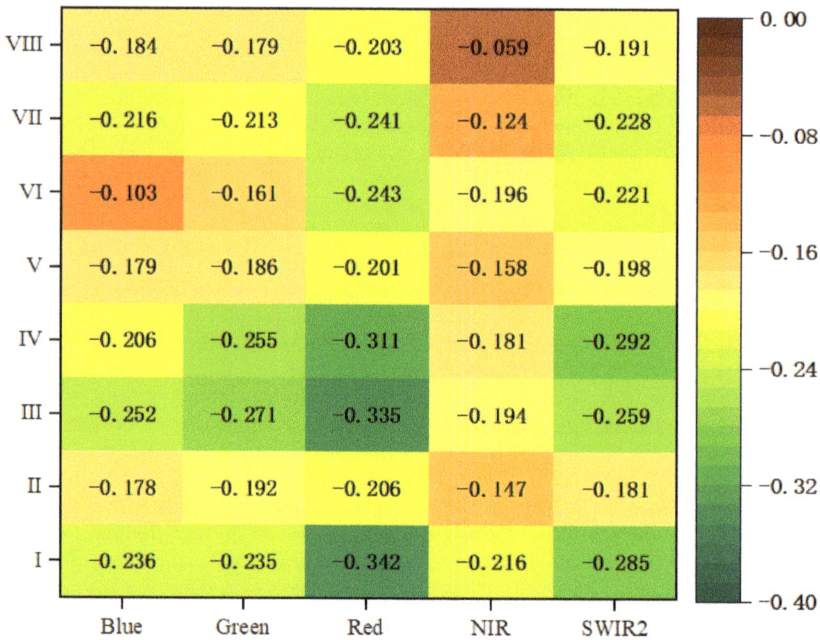

Figure 4. Relationship between *Pinus yunnanensis* forest AGB and the 5 original bands, with all original bands showing significance at the 0.01 level across the 8 sub-regions.

A significance test was applied to assess the correlation between *Pinus yunnanensis* forest AGB and the five original spectral bands, aiming to determine if there were notable differences in how each band contributed to the estimation of OSVs for forest AGB estimation. As shown in Figure 5, most bands were not statistically significant, with only a few exceptions (* $p < 0.05$, ** $p < 0.01$). This indicates that no single band significantly outperformed the others in estimating OSVs. Consequently, all five bands were incorporated into the final analysis to compute OSVs for each sub-region. In a more detailed sense, it illustrates that the correlations across the bands were quite similar, with minor variations. While the NIR and Red bands displayed slightly stronger negative correlations, these differences were not pronounced enough to prioritize one band over the others. Therefore, utilizing all five bands provided a more comprehensive and robust approach to estimating saturation values across the different regions.

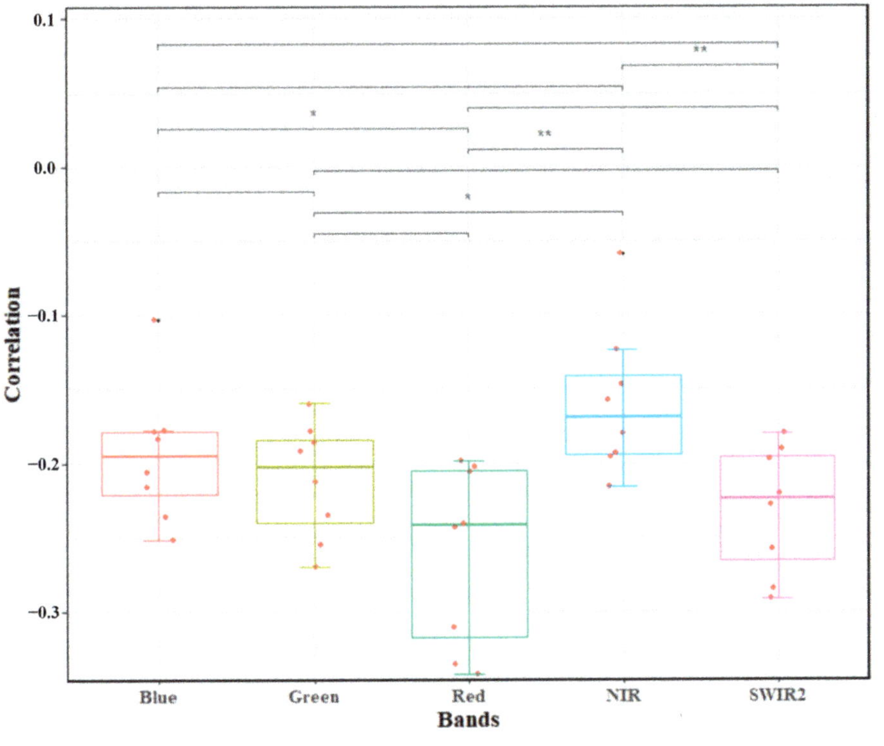

Figure 5. The significance test of the correlations between five original bands and *Pinus yunnanensis* forest AGB. Double asterisks (**) indicate significance at the 0.01 level ($p < 0.01$), while a single asterisk (*) denotes significance at the 0.05 level ($p < 0.05$).

3.2. OSV Variation Analysis

The OSVs in each vegetation sub-region displayed significant differences when using the same band (Figure 6), though similar OSVs were observed across sub-regions with different bands. Additionally, the OSVs for *Pinus yunnanensis* forests in Yunnan Province ranged from 104.42 t/ha to 209.11 t/ha. The highest OSV (209.11 t/ha) was recorded in sub-region VII using the Red band, while the lowest OSV (104.42 t/ha) was observed in sub-region VIII using the Green band. Then, most sub-regions, such as I, II, III, and V, were approximately 180 t/ha. Overall, the variations of OSV in *Pinus yunnanensis* forests were extremely noteworthy in Yunnan, showing a trend of lowest in the southeast and highest in the southwest.

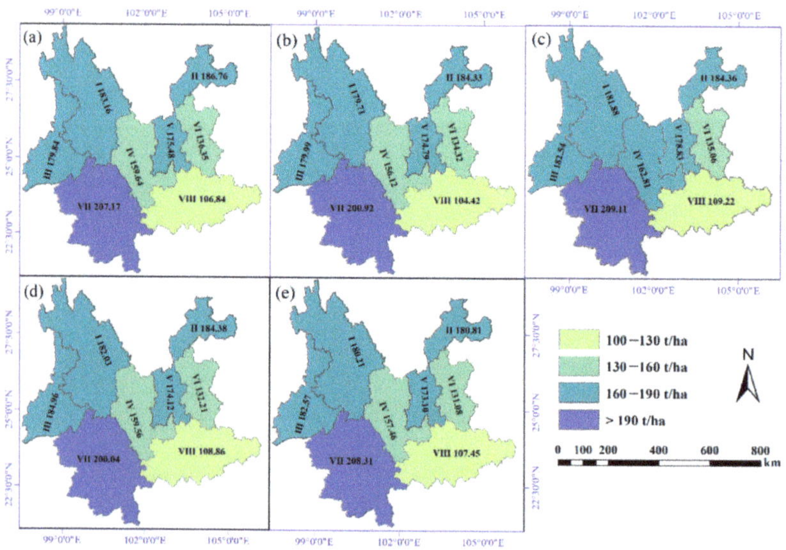

Figure 6. The OSVs derived from the 5 original bands in each vegetation sub-region; (**a**–**e**) correspond to the values obtained from the Blue, Green, Red, NIR, and SWIR2 bands.

3.3. *The OSV Variations Response to the Climate*

As shown in Figure 7, the OSV variations in response to climate were clarified through CCA. The results reveal that the first axis accounted for 88.3% of the OSV variations, while the second axis explained 4.9%, making a total of 93.2% of the OSV variations attributed to climatic factors. Additionally, annual mean temperature (AMT) contributed the most to the CCA, highlighting temperature as a key factor influencing OSV variations. The mean temperature during the wettest quarter (MTQ) had the largest impact on the first axis, whereas annual average precipitation (ANP) made the highest contribution on the second axis. These findings demonstrate that both temperature and precipitation were the primary factors driving OSV variations. It was observed that the first axis favored temperature changes, while the second axis was more influenced by precipitation. Furthermore, the OSV variations were notable in groups A, B, C, and D, with the order of OSVs being D > C > B > A. In the four groups, the area of group A was warmer but the driest, the areas of group B were warmer and drier, the areas of group C were colder and wetter, and the areas of group D were the warmest and the wettest. Therefore, these indicators show that the OSV variations were mainly affected by the temperature and humidity.

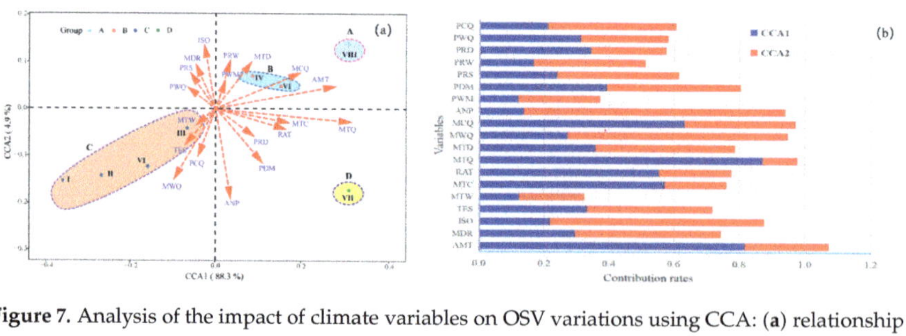

Figure 7. Analysis of the impact of climate variables on OSV variations using CCA: (**a**) relationship between OSVs and climate variables; (**b**) rates of contribution of climate variables to OSV variations.

4. Discussion

4.1. Original Bands

In this study, the original spectral bands exhibited a negative correlation with *Pinus yunnanensis* forests' AGB, indicating a heightened capacity for light absorption and scattering in the forest vegetation [42]. This phenomenon resulted in greater light absorption within areas of dense forest cover, leading to correspondingly lower reflectance values [10,43]. Conversely, sub-regions characterized by lower biomass, such as bare ground or grasslands, exhibited higher reflectance [44,45]. This resulted in substantial differences in reflectance among various sub-regions within the *Pinus yunnanensis* forests. This observation is consistent with previous findings [46], yet it provides further validation of the pronounced light reflectance variability inherent in highly heterogeneous landscapes like Yunnan. Areas of higher biomass, especially those dominated by tree species such as *Pinus yunnanensis*, display significantly enhanced light absorption due to their dense canopy structure, offering valuable contributions to the use of spectral imaging technology in such complex ecosystems [47]. These results underscore the critical need to incorporate spatial variability into modeling efforts, thereby enhancing the precision of AGB estimation in other similarly heterogeneous forest systems [48].

The Red band exhibited the strongest performance among the original spectral bands, which can be attributed to the fact that chlorophyll in the leaves efficiently absorbs red light for photosynthesis. Chlorophyll has a pronounced absorption peak in the red waveband, enabling plants to capture red light with exceptional efficiency for energy conversion [49,50]. This absorption of red light by chlorophyll not only drives photosynthesis, but also stimulates increased plant growth and biomass accumulation [51,52]. The dense canopies and high tree density in *Pinus yunnanensis* forests, characterized by an extensive leaf area index due to their numerous clustered leaves, further enhance this effect [53]. Consequently, a higher concentration of chlorophyll is required in regions with substantial AGB, absorbing more red light to fuel photosynthesis and promote the synthesis of organic compounds. This contributes to the greater AGB observed in *Pinus yunnanensis* forests in Yunnan, which, in turn, further reduces overall reflectance [54,55]. These findings offer novel empirical support for understanding how spectral characteristics, particularly red light absorption, directly influence biomass accumulation in high-biomass vegetation types such as *Pinus yunnanensis* [56]. This relationship holds significant potential for refining remote sensing models that address biomass saturation challenges in highly heterogeneous regions [57]. Moreover, our study suggests that Red band reflectance could serve as a robust indicator for biomass estimation in other dense forest ecosystems, providing valuable insights for enhancing global biomass monitoring efforts [58].

4.2. OSV Variations

In this study, the largest OSV was 208.31 t/ha in sub-region VII, while the smallest OSV was 107.45 t/ha in sub-region VIII, with a mean OSV of 165.11 t/ha when assessing OSVs derived from the SWIR2 band. Comparatively, the OSVs across different vegetation types varied between 100 t/ha and 159 t/ha in Zhejiang Province, with pine forests reaching an OSV of 159 t/ha [17]. Notably, the average OSV for *Pinus yunnanensis* forests in Yunnan surpassed that of the pine forests in Zhejiang, likely due to the mountainous plateau topography, complex stand structure, and heightened forest heterogeneity in Yunnan [59]. High forest heterogeneity and complex stand structures were identified as the primary contributors to the saturation issue [6]. In comparison, an associated study examining OSVs for oak forests in Yunnan reported a range from 104 t/hm^2 to 182 t/hm^2 [18]. The disparity in OSV ranges between these studies can largely be attributed to the intrinsic structural and physiological differences between coniferous forests, like *Pinus yunnanensis*, and broadleaf forests, such as oak. Coniferous species typically have denser canopies with narrower leaves, affecting light absorption and scattering, whereas broadleaf species, with their larger and more complex leaves, generally exhibit higher photosynthetic activity and greater biomass accumulation, which may contribute to distinct OSV patterns across these forest types [60,61]. This study is among the first to offer a quantitative analysis of how complex topography and forest heterogeneity influence OSV variations in *Pinus yunnanensis* forests. It reveals that OSVs are notably higher in highly heterogeneous and mountainous regions, providing important theoretical support for understanding the role of topography in biomass saturation. These findings suggest that similar effects of heterogeneity may be observed in other diverse forest ecosystems, emphasizing the need for region-specific OSV models that account for topographical complexity [62].

Moreover, it was observed that OSVs varied across different sub-regions, regardless of which original bands were used, indicating that OSV variation patterns exist inherently to some extent. A discernible trend emerged for *Pinus yunnanensis* forests in Yunnan, with the highest OSV recorded in sub-region VII, higher OSVs in sub-regions I, II, III, IV, and V, and lower OSVs in sub-region VI, with the lowest value in sub-region VIII. This pattern suggests that OSVs for *Pinus yunnanensis* forests in Yunnan are the highest in the southwestern regions and lowest in the southeastern areas. The systematic analysis of OSV differences across these sub-regions not only highlights the spatial variability of OSVs, but also provides novel insights into how OSVs respond to geographical and climatic influences. This finding underscores the necessity of developing predictive models that accurately capture the spatial variability of OSVs, particularly in heterogeneous environments like Yunnan [63]. Additionally, these results highlight the potential for designing targeted biomass management strategies that consider specific OSV variation patterns across diverse forest ecosystems.

4.3. The Key Climatic Variables Affecting OSV Variations

The OSV variations were highly responsive to climatic conditions, with 93.2% of the variation explained by climatic variables through CCA. Among these, the AMT made the greatest contribution on the first axis, indicating that temperature is one of the primary factors influencing OSV variations. Temperature affects the absorption of red and other visible light by influencing plant photosynthesis, which in turn impacts plant growth and contributes to a more complex stand structure [64,65]. Additionally, the MTQ and ANP made the highest contributions on the first and second axes, respectively. MTQ is crucial as it reflects the temperature conditions during the peak growing season, when precipitation is at its highest, significantly impacting physiological processes such as photosynthesis and respiration. Favorable temperatures enhance plant growth, nutrient uptake, and photosynthetic activity, ultimately increasing species heterogeneity [66,67]. ANP plays a vital role in determining water availability within an ecosystem [68]. Sufficient precipitation is essential for plant growth, as it provides the necessary water for photosynthesis and other metabolic processes. Conversely, inadequate rainfall leads to water stress, limiting plant productivity [69], and resulting in a lower leaf area index (LAI) and higher band reflectance, which, in turn, lowers biomass saturation values. On the other hand, higher ANP creates optimal conditions for plant growth, allowing for increased LAI and reduced band reflectance [70], which leads to higher saturation values. This comprehensive examination of the climatic influence on OSV variability provides profound insights into the climatic determinants of forest AGB and addresses the issue of saturation. By systematically examining the influence of key variables such as temperature and precipitation, this study advances our understanding of how climatic factors shape the distribution and OSVs of forest AGB, providing a robust framework for future climate-based biomass models [71]. Moreover, our findings suggest that the interplay between temperature and precipitation could serve as a key predictor for monitoring forest AGB in similarly forested regions worldwide, contributing to more accurate AGB estimations under varying climatic conditions.

The results indicate that OSV variations were primarily influenced by temperature and humidity, as demonstrated through CCA. This can be explained by the pivotal role temperature plays in regulating various biochemical and physiological processes in plants [72]. Temperature affects the efficiency of photosynthesis, the rate of metabolic reactions, and the overall growth and development of vegetation [73]. Similarly, humidity is closely tied to water availability, which is crucial for plant growth and survival [74]. Adequate humidity levels provide sufficient moisture for root absorption, facilitating the transport of nutrients and energy, promoting leaf development, and reducing band reflectance [75]. In contrast, insufficient humidity can lead to water stress and diminished plant productivity [76]. The climate in Yunnan, characterized by significant variability in temperature and humidity due to geographical location, elevation, and seasonality [77], has shaped the region's complex stand structure, diverse tree species composition, and heightened forest heterogeneity [9,78–80]. These findings enhance our understanding of the roles temperature and humidity play in shaping forest structural complexity and species diversity. This study's comprehensive examination of the relationship between climate and OSV variability advances the discourse on climate-driven processes in forest ecosystems, offering valuable theoretical and practical insights for improving AGB estimation models in heterogeneous environments [81]. Furthermore, by highlighting the intricate interrelationships between temperature, humidity, and the saturation problem, this research underscores the importance of developing integrative models that incorporate multiple climatic variables across diverse ecological contexts [82]. Numerous studies have demonstrated that the saturation problem and its variations are primarily driven by complex forest stand structures, diverse species composition, and high forest heterogeneity [6,10,83]. Accordingly, the OSV variations in *Pinus yunnanensis* forests were interpreted in response to climate and key climatic variables, with temperature and humidity identified as the primary factors influencing OSV variability. This study not only corroborates prior research on the significance of forest structure and heterogeneity in OSV variation, but also offers a comprehensive analysis of the climatic factors that govern these variations. By identifying the critical role of temperature and humidity, this work deepens our understanding of the intricate interplay between climatic conditions and saturation problems in forest AGB estimation using optical imagery.

4.4. Limitations and Future Research

This study leveraged Landsat 8 OLI imagery to quantify OSVs in *Pinus yunnanensis* forests across Yunnan. While this sensor offers valuable spectral data and an extensive historical archive, it has inherent limitations that may introduce uncertainties in estimating OSVs. The 30 m spatial resolution, although effective for large-scale forest monitoring, may fall short in capturing the fine-scale heterogeneity of forests, particularly in regions like Yunnan, where diverse forest structures and complex terrain are prevalent. This limitation can result in mixed pixels, where signals represent multiple land cover types, thereby diminishing the accuracy of OSV estimates [84]. Meanwhile, its lower spectral resolution compared to high-resolution sensors like WorldView-3 or hyperspectral platforms may result in the inadequate detection of subtle vegetation variations, potentially causing biomass overestimation or underestimation, especially in high-biomass areas susceptible to saturation effects [85]. Then, this study highlighted the limitations of using year-round data, which, while comprehensive, do not differentiate between the growing and non-growing seasons. Since vegetation productivity and spectral reflectance tend to peak during the growing season, this may impact OSV estimates [86]. We should explore seasonal data separation to better account for variations in vegetation dynamics across different times of the year in future research. Furthermore, other research found that the OSV was 192 t/ha using the Landsat 8 OLI, 247 t/ha using the Worldview-3, and 204 t/ha using Sentinel-2 MSI images [87], indicating that various remote sensing data had different saturation values [10]. Therefore, it is essential to investigate alternative remote sensing sources, including Sentinel-2, SPOT, MODIS, and QuickBird, for use in estimating forest AGB and OSV variations, particularly in areas of high forest heterogeneity.

Despite these sensor limitations, this study identified key climatic variables influencing OSV variations, with AMT emerging as the most significant factor using CCA. However, whether the inclusion of these climatic variables can enhance AGB estimation accuracy using Landsat 8 OLI remains an open question for future research. The observed OSV variations were also responsive to temperature and humidity, underscoring the need for further exploration into how OSV dynamics interact with climatic and environmental factors across diverse tree species and regions. It is crucial to explore the responses of the OSV variations using optical images of other tree species to climate and other environmental factors correspondingly.

5. Conclusions

This study analyzes how climatic factors influence OSV variations and identifies the key variables affecting these variations in *Pinus yunnanensis* forests using Landsat 8 OLI imagery. OSVs were analyzed across eight sub-regions of Yunnan Province, China, and CCA was used to determine the main climatic drivers. The results show that the Red band of Landsat 8 OLI had the strongest performance, with OSVs ranging from 104.42 t/ha to 209.11 t/ha, peaking in the southwest and being the lowest in the southeast. The CCA revealed that 93.2% of the OSV variation could be explained by climate, with temperature and humidity emerging as the most significant factors. The AMT was the highest-contributing variable of 19 climate variables, and the MTQ and ANP made the highest contributions in the first and second axes, respectively. Ultimately, it was found that the hydrothermal conditions were the main factors affecting the OSV variations at the general level, and the OSVs were larger in the warmer and wetter sub-regions.

Author Contributions: Y.W. participated in the collection of the data, conducted the data analysis, and wrote the draft of the paper; B.G., X.Z., H.L. (Hongbin Luo), Z.Y., H.L. (Huipeng Li) and K.S. helped with the data analysis, and constructed part of the graphs; L.W. and W.X. gave some suggestions and guidance; G.O. supervised and coordinated the research project, designed the experiment and revised the paper. All authors have read and agreed to the published version of the manuscript.

Funding: This research was jointly supported by the Ten Thousand Talent Plans for Young Top-notch Talents of Yunnan Province (YNWR-QNBJ-2018-184), the Scientific Research Fund Project of Yunnan Provincial Education Department (2023Y0732), and the Education Talent of Xingdian Talent Support Program of Yunnan Province, China.

Data Availability Statement: The original contributions presented in this study are included in the article, and further inquiries can be directed to the corresponding authors.

Acknowledgments: We would like to acknowledge all the people who have contributed to this paper.

Conflicts of Interest: The authors declare no conflicts of interest.

References

1. Schimel, D.S. Terrestrial ecosystems and the carbon cycle. *Glob. Chang. Biol.* **1995**, *1*, 77–91. [CrossRef]
2. Pearce, D.W. The economic value of forest ecosystems. *Ecosyst. Health* **2001**, *7*, 284–296. [CrossRef]
3. Mitchard, E.T. The tropical forest carbon cycle and climate change. *Nature* **2018**, *559*, 527–534. [CrossRef] [PubMed]
4. Keith, H.; Mackey, B.G.; Lindenmayer, D.B. Re-evaluation of forest biomass carbon stocks and lessons from the world's most carbon-dense forests. *Proc. Natl. Acad. Sci. USA* **2009**, *106*, 11635–11640. [CrossRef] [PubMed]
5. Sileshi, G.W. A critical review of forest biomass estimation models, common mistakes and corrective measures. *For. Ecol. Manag.* **2014**, *329*, 237–254. [CrossRef]
6. Lu, D.; Chen, Q.; Wang, G.; Liu, L.; Li, G.; Moran, E. A survey of remote sensing-based aboveground biomass estimation methods in forest ecosystems. *Int. J. Digital Earth* **2016**, *9*, 63–105. [CrossRef]
7. Tian, X.; Su, Z.; Chen, E.; Li, Z.; van der Tol, C.; Guo, J.; He, Q. Reprint of: Estimation of forest above-ground biomass using multi-parameter remote sensing data over a cold and arid area. *Int. J. Appl. Earth Obs. Geoinf.* **2012**, *17*, 102–110. [CrossRef]
8. Silveira, E.M.; Silva, S.H.G.; Acerbi-Junior, F.W.; Carvalho, M.C.; Carvalho, L.M.T.; Scolforo, J.R.S.; Wulder, M.A. Object-based random forest modelling of aboveground forest biomass outperforms a pixel-based approach in a heterogeneous and mountain tropical environment. *Int. J. Appl. Earth Obs. Geoinf.* **2019**, *78*, 175–188. [CrossRef]
9. Ou, G.; Li, C.; Lv, Y.; Wei, A.; Xiong, H.; Xu, H.; Wang, G. Improving aboveground biomass estimation of Pinus densata forests in Yunnan using Landsat 8 imagery by incorporating age dummy variable and method comparison. *Remote Sens.* **2019**, *11*, 738. [CrossRef]
10. Mutanga, O.; Masenyama, A.; Sibanda, M. Spectral saturation in the remote sensing of high-density vegetation traits: A systematic review of progress, challenges, and prospects. *ISPRS J. Photogramm. Remote Sens.* **2023**, *198*, 297–309. [CrossRef]
11. Allen, W.A.; Richardson, A.J. Interaction of light with a plant canopy. *JOSA* **1968**, *58*, 1023–1028. [CrossRef]

12. Di Bella, C.; Paruelo, J.; Becerra, J.; Bacour, C.; Baret, F. Effect of senescent leaves on NDVI-based estimates of APAR: Experimental and modelling evidences. *Int. J. Remote Sens.* **2004**, *25*, 5415–5427. [CrossRef]
13. Tucker, C.J. Red and photographic infrared linear combinations for monitoring vegetation. *Remote Sens. Environ.* **1979**, *8*, 127–150. [CrossRef]
14. Steininger, M. Satellite estimation of tropical secondary forest above-ground biomass: Data from Brazil and Bolivia. *Int. J. Remote Sens.* **2000**, *21*, 1139–1157. [CrossRef]
15. Mutanga, O.; Skidmore, A.K. Hyperspectral band depth analysis for a better estimation of grass biomass (*Cenchrus ciliaris*) measured under controlled laboratory conditions. *Int. J. Appl. Earth Obs. Geoinf.* **2004**, *5*, 87–96. [CrossRef]
16. Zhu, X.; Liu, D. Improving forest aboveground biomass estimation using seasonal Landsat NDVI time-series. *ISPRS J. Photogramm. Remote Sens.* **2015**, *102*, 222–231. [CrossRef]
17. Zhao, P.; Lu, D.; Wang, G.; Wu, C.; Huang, Y.; Yu, S. Examining spectral reflectance saturation in Landsat imagery and corresponding solutions to improve forest aboveground biomass estimation. *Remote Sens.* **2016**, *8*, 469. [CrossRef]
18. Wu, Y.; Ou, G.; Huang, T.; Zhang, X.; Liu, C.; Liu, Z.; Yu, Z.; Luo, H.; Lu, C.; Shi, K. Climate Interprets Saturation Value Variations Better Than Soil and Topography in Estimating Oak Forest Aboveground Biomass Using Landsat 8 OLI Imagery. *Remote Sens.* **2024**, *16*, 1338. [CrossRef]
19. Sturrock, R.; Frankel, S.; Brown, A.; Hennon, P.; Kliejunas, J.; Lewis, K.; Worrall, J.; Woods, A. Climate change and forest diseases. *Plant Pathol.* **2011**, *60*, 133–149. [CrossRef]
20. Xia, Y.; Hao, H.; Zanardo, G.; Deeks, A. Long term vibration monitoring of an RC slab: Temperature and humidity effect. *Eng. Struct.* **2006**, *28*, 441–452. [CrossRef]
21. Ainsworth, E.A.; Yendrek, C.R.; Sitch, S.; Collins, W.J.; Emberson, L.D. The effects of tropospheric ozone on net primary productivity and implications for climate change. *Annu. Rev. Plant Biol.* **2012**, *63*, 637–661. [CrossRef] [PubMed]
22. Gray, S.B.; Brady, S.M. Plant developmental responses to climate change. *Dev. Biol.* **2016**, *419*, 64–77. [CrossRef] [PubMed]
23. Ahanger, M.A.; Tomar, N.S.; Tittal, M.; Argal, S.; Agarwal, R. Plant growth under water/salt stress: ROS production; antioxidants and significance of added potassium under such conditions. *Physiol. Mol. Biol. Plants* **2017**, *23*, 731–744. [CrossRef] [PubMed]
24. Chelli, S.; Bricca, A.; Tsakalos, J.L.; Andreetta, A.; Bonari, G.; Campetella, G.; Carnicelli, S.; Cervellini, M.; Puletti, N.; Wellstein, C. Multiple drivers of functional diversity in temperate forest understories: Climate, soil, and forest structure effects. *Sci. Total Environ.* **2024**, *916*, 170258. [CrossRef] [PubMed]
25. Yang, Y.; Tian, K.; Hao, J.; Pei, S.; Yang, Y. Biodiversity and biodiversity conservation in Yunnan, China. *Biodivers. Conserv.* **2004**, *13*, 813–826. [CrossRef]
26. Li, X.; Walker, D. The plant geography of Yunnan Province, southwest China. *J. Biogeogr.* **1986**, *14*, 367–397.
27. Shen, J.; Li, Z.; Gao, C.; Li, S.; Huang, X.; Lang, X.; Su, J. Radial growth response of Pinus yunnanensis to rising temperature and drought stress on the Yunnan Plateau, southwestern China. *For. Ecol. Manage.* **2020**, *474*, 118357. [CrossRef]
28. Wu, Y.; Ou, G.; Lu, T.; Huang, T.; Zhang, X.; Liu, Z.; Yu, Z.; Guo, B.; Wang, E.; Feng, Z. Improving Aboveground Biomass Estimation in Lowland Tropical Forests across Aspect and Age Stratification: A Case Study in Xishuangbanna. *Remote Sens.* **2024**, *16*, 1276. [CrossRef]
29. Felipe-Lucia, M.R.; Soliveres, S.; Penone, C.; Manning, P.; van der Plas, F.; Boch, S.; Prati, D.; Ammer, C.; Schall, P.; Gossner, M.M. Multiple forest attributes underpin the supply of multiple ecosystem services. *Nature Commun.* **2018**, *9*, 4839. [CrossRef]
30. Swanson, M.E.; Franklin, J.F.; Beschta, R.L.; Crisafulli, C.M.; DellaSala, D.A.; Hutto, R.L.; Lindenmayer, D.B.; Swanson, F.J. The forgotten stage of forest succession: Early-successional ecosystems on forest sites. *Front. Ecol. Environ.* **2011**, *9*, 117–125. [CrossRef]
31. Duivenvoorden, J.E. Tree species composition and rain forest-environment relationships in the middle Caquetá area, Colombia, NW Amazonia. *Vegetation* **1995**, *120*, 91–113. [CrossRef]
32. Peng, J.; Yang, Y.; Liu, Y.; Du, Y.; Meersmans, J.; Qiu, S. Linking ecosystem services and circuit theory to identify ecological security patterns. *Sci. Total Environ.* **2018**, *644*, 781–790. [CrossRef]
33. Zhang, Y. Water and Carbon Balances of Deciduous and Evergreen Broadleaf Trees from a Subtropical Cloud Forest in Southwest China. Ph.D. Thesis, University of Miami, Coral Gables, FL, USA, 2012.
34. Li, R.; Kraft, N.J.; Yang, J.; Wang, Y. A phylogenetically informed delineation of floristic regions within a biodiversity hotspot in Yunnan, China. *Sci. Rep.* **2015**, *5*, 9396. [CrossRef] [PubMed]
35. Fang, P.; Ou, G.; Li, R.; Wang, L.; Xu, W.; Dai, Q.; Huang, X. Regionalized classification of stand tree species in mountainous forests by fusing advanced classifiers and ecological niche model. *GISci. Remote Sens.* **2023**, *60*, 2211881. [CrossRef]
36. Xu, H.; Zhang, Z.; Ou, G.; Shi, H. *A Study on Estimation and Distribution for Forest Biomass and Carbon Storage in Yunnan Province*; Yunnan Science and Technology Press: Kunming, China, 2019.
37. Kaufman, Y.J.; Tanre, D. Strategy for direct and indirect methods for correcting the aerosol effect on remote sensing: From AVHRR to EOS-MODIS. *Remote Sens. Environ.* **1996**, *55*, 65–79. [CrossRef]
38. Young, N.E.; Anderson, R.S.; Chignell, S.M.; Vorster, A.G.; Lawrence, R.; Evangelista, P.H. A survival guide to Landsat preprocessing. *Ecology* **2017**, *98*, 920–932. [CrossRef]
39. Thompson, B. Canonical correlation analysis. In *Handbook of Applied Multivariate Statistics and Mathematical Modeling*; Elsevier: Amsterdam, The Netherlands, 2000.
40. Andrew, G.; Arora, R.; Bilmes, J.; Livescu, K. Deep canonical correlation analysis. In Proceedings of the International Conference on Machine Learning, Atlanta, GA, USA, 16–21 June 2013; pp. 1247–1255.

41. Hardoon, D.R.; Szedmak, S.; Shawe-Taylor, J. Canonical correlation analysis: An overview with application to learning methods. *Neural Comput.* **2004**, *16*, 2639–2664. [CrossRef]
42. Ollinger, S.V. Sources of variability in canopy reflectance and the convergent properties of plants. *New Phytol.* **2011**, *189*, 375–394. [CrossRef]
43. Gao, X.; Huete, A.R.; Ni, W.; Miura, T. Optical–biophysical relationships of vegetation spectra without background contamination. *Remote Sens. Environ.* **2000**, *74*, 609–620. [CrossRef]
44. Jin, Y.; Yang, X.; Qiu, J.; Li, J.; Gao, T.; Wu, Q.; Zhao, F.; Ma, H.; Yu, H.; Xu, B. Remote sensing-based biomass estimation and its spatio-temporal variations in temperate grassland, Northern China. *Remote Sens.* **2014**, *6*, 1496–1513. [CrossRef]
45. Hütt, C.; Bolten, A.; Hüging, H.; Bareth, G. Uav lidar metrics for monitoring crop height, biomass and nitrogen uptake: A case study on a winter wheat field trial. *PFG–J. Photogramm. Remote Sens. Geoinf. Sci.* **2023**, *91*, 65–76. [CrossRef]
46. Chave, J.; Réjou-Méchain, M.; Búrquez, A.; Chidumayo, E.; Colgan, M.S.; Delitti, W.B.; Duque, A.; Eid, T.; Fearnside, P.M.; Goodman, R.C. Improved allometric models to estimate the aboveground biomass of tropical trees. *Global Chang. Biol.* **2014**, *20*, 3177–3190. [CrossRef] [PubMed]
47. Schimel, D.; Pavlick, R.; Fisher, J.B.; Asner, G.P.; Saatchi, S.; Townsend, P.; Miller, C.; Frankenberg, C.; Hibbard, K.; Cox, P. Observing terrestrial ecosystems and the carbon cycle from space. *Glob. Chang. Biol.* **2015**, *21*, 1762–1776. [CrossRef] [PubMed]
48. Gitelson, A.A.; Peng, Y.; Arkebauer, T.J.; Schepers, J. Relationships between gross primary production, green LAI, and canopy chlorophyll content in maize: Implications for remote sensing of primary production. *Remote Sens. Environ.* **2014**, *144*, 65–72. [CrossRef]
49. Huete, A.; Liu, H.; Batchily, K.; Van Leeuwen, W. A comparison of vegetation indices over a global set of TM images for EOS-MODIS. *Remote Sens. Environ.* **1997**, *59*, 440–451. [CrossRef]
50. Terashima, I.; Fujita, T.; Inoue, T.; Chow, W.S.; Oguchi, R. Green light drives leaf photosynthesis more efficiently than red light in strong white light: Revisiting the enigmatic question of why leaves are green. *Plant Cell Physiol.* **2009**, *50*, 684–697. [CrossRef]
51. Zhu, L.; Bloomfield, K.J.; Hocart, C.H.; Egerton, J.J.; O'Sullivan, O.S.; Penillard, A.; Weerasinghe, L.K.; Atkin, O.K. Plasticity of photosynthetic heat tolerance in plants adapted to thermally contrasting biomes. *Plant Cell Environ.* **2018**, *41*, 1251–1262. [CrossRef]
52. Moriwaki, T.; Falcioni, R.; Tanaka, F.A.O.; Cardoso, K.A.K.; Souza, L.; Benedito, E.; Nanni, M.R.; Bonato, C.M.; Antunes, W.C. Nitrogen-improved photosynthesis quantum yield is driven by increased thylakoid density, enhancing green light absorption. *Plant Sci.* **2019**, *278*, 1–11. [CrossRef]
53. Lin, Q.; Huang, H.; Chen, L.; Wang, J.; Huang, K.; Liu, Y. Using the 3D model RAPID to invert the shoot dieback ratio of vertically heterogeneous Yunnan pine forests to detect beetle damage. *Remote Sens. Environ.* **2021**, *260*, 112475. [CrossRef]
54. Chen, L.; Wang, Y.; Ren, C.; Zhang, B.; Wang, Z. Optimal combination of predictors and algorithms for forest above-ground biomass mapping from Sentinel and SRTM data. *Remote Sens.* **2019**, *11*, 414. [CrossRef]
55. Tu, J.; Li, Z.; Xu, Y.; Sun, J.; Zhao, X.; Huang, Z. Sapling root ecological stoichiometric characteristics of dominant species in three forests in Yunnan, China, and their relationship with soil physicochemical factors. *Appl. Veg. Sci.* **2022**, *33*, e13166. [CrossRef]
56. Ali, I.; Greifeneder, F.; Stamenkovic, J.; Neumann, M.; Notarnicola, C. Review of machine learning approaches for biomass and soil moisture retrievals from remote sensing data. *Remote Sens.* **2015**, *7*, 16398–16421. [CrossRef]
57. Houghton, R.; Lawrence, K.; Hackler, J.; Brown, S. The spatial distribution of forest biomass in the Brazilian Amazon: A comparison of estimates. *Glob. Chang. Biol.* **2001**, *7*, 731–746. [CrossRef]
58. Kobayashi, H.; Dye, D.G. Atmospheric conditions for monitoring the long-term vegetation dynamics in the Amazon using normalized difference vegetation index. *Remote Sens. Environ.* **2005**, *97*, 519–525. [CrossRef]
59. Liu, S.; An, N.; Yang, J.; Dong, S.; Wang, C.; Yin, Y. Prediction of soil organic matter variability associated with different land use types in mountainous landscape in southwestern Yunnan province, China. *Catena* **2015**, *133*, 137–144. [CrossRef]
60. Asner, G.P.; Scurlock, J.M.; Hicke, J.A. Global synthesis of leaf area index observations: Implications for ecological and remote sensing studies. *Global Ecol. Biogeogr.* **2003**, *12*, 191–205. [CrossRef]
61. Niinemets, Ü. Responses of forest trees to single and multiple environmental stresses from seedlings to mature plants: Past stress history, stress interactions, tolerance and acclimation. *For. Ecol. Manag.* **2010**, *260*, 1623–1639. [CrossRef]
62. Saatchi, S.S.; Harris, N.L.; Brown, S.; Lefsky, M.; Mitchard, E.T.; Salas, W.; Zutta, B.R.; Buermann, W.; Lewis, S.L.; Hagen, S. Benchmark map of forest carbon stocks in tropical regions across three continents. *Proc. Natl. Acad. Sci. USA* **2011**, *108*, 9899–9904. [CrossRef]
63. Harris, N.L.; Brown, S.; Hagen, S.C.; Saatchi, S.S.; Petrova, S.; Salas, W.; Hansen, M.C.; Potapov, P.V.; Lotsch, A. Baseline map of carbon emissions from deforestation in tropical regions. *Science* **2012**, *336*, 1573–1576. [CrossRef]
64. Ahmed, H.A.; Tong, Y.-X.; Yang, Q.-C. Optimal control of environmental conditions affecting lettuce plant growth in a controlled environment with artificial lighting: A review. *S. Afr. J. Bot.* **2020**, *130*, 75–89. [CrossRef]
65. Lichtenthaler, H.K.; Rinderle, U. The role of chlorophyll fluorescence in the detection of stress conditions in plants. *Crit Rev. Anal. Chem.* **1988**, *19*, S29–S85. [CrossRef]
66. Usoltsev, V.A.; Zukow, W.; Osmirko, A.A.; Tsepordey, I.S.; Chasovskikh, V.P. Additive biomass models for Quercus spp. single-trees sensitive to temperature and precipitation in Eurasia. *Ecol. Questi.* **2019**, *30*, 29–40. [CrossRef]
67. Hatfield, J.L.; Prueger, J.H. Temperature extremes: Effect on plant growth and development. *Weather Clim. Extrem.* **2015**, *10*, 4–10. [CrossRef]

68. Margalef, O.; Sardans, J.; Fernández-Martínez, M.; Molowny-Horas, R.; Janssens, I.; Ciais, P.; Goll, D.; Richter, A.; Obersteiner, M.; Asensio, D. Global patterns of phosphatase activity in natural soils. *Sci. Rep.* **2017**, *7*, 1337. [CrossRef] [PubMed]
69. Ali, M.; Talukder, M. Increasing water productivity in crop production—A synthesis. *Agric. Water Manag.* **2008**, *95*, 1201–1213. [CrossRef]
70. Han, T.; Ren, H.; Wang, J.; Lu, H.; Song, G.; Chazdon, R.L. Variations of leaf eco-physiological traits in relation to environmental factors during forest succession. *Ecol. Indic.* **2020**, *117*, 106511. [CrossRef]
71. Baccini, A.; Goetz, S.; Walker, W.; Laporte, N.; Sun, M.; Sulla-Menashe, D.; Hackler, J.; Beck, P.; Dubayah, R.; Friedl, M. Estimated carbon dioxide emissions from tropical deforestation improved by carbon-density maps. *Nat. Clim. Chang.* **2012**, *2*, 182–185. [CrossRef]
72. Cohen, I.; Zandalinas, S.I.; Huck, C.; Fritschi, F.B.; Mittler, R. Meta-analysis of drought and heat stress combination impact on crop yield and yield components. *Physiol. Plant.* **2021**, *171*, 66–76. [CrossRef]
73. Tripathy, B.C.; Brown, C.S.; Levine, H.G.; Krikorian, A.D. Growth and photosynthetic responses of wheat plants grown in space. *Plant Physiol.* **1996**, *110*, 801–806. [CrossRef]
74. Passioura, J.B. Drought and drought tolerance. *Plant Growth Regul.* **1996**, *20*, 79–83. [CrossRef]
75. Grossiord, C.; Buckley, T.N.; Cernusak, L.A.; Novick, K.A.; Poulter, B.; Siegwolf, R.T.; Sperry, J.S.; McDowell, N.G. Plant responses to rising vapor pressure deficit. *New Phytol.* **2020**, *226*, 1550–1566. [CrossRef] [PubMed]
76. Blum, A. Effective use of water (EUW) and not water-use efficiency (WUE) is the target of crop yield improvement under drought stress. *Field Crops Res.* **2009**, *112*, 119–123. [CrossRef]
77. Rangwala, I.; Miller, J.R. Climate change in mountains: A review of elevation-dependent warming and its possible causes. *Clim. Chang.* **2012**, *114*, 527–547. [CrossRef]
78. Jin, Y.; Liu, C.; Qian, S.S.; Luo, Y.; Zhou, R.; Tang, J.; Bao, W. Large-scale patterns of understory biomass and its allocation across China's forests. *Sci. Total Environ.* **2022**, *804*, 150169. [CrossRef] [PubMed]
79. Zheng, Z.; Feng, Z.; Cao, M.; Li, Z.; Zhang, J. Forest Structure and Biomass of a Tropical Seasonal Rain Forest in Xishuangbanna, Southwest China. *Biotropica J. Biol. Conserv.* **2006**, *38*, 318–327.
80. Meng, L.-Z.; Martin, K.; Weigel, A.; Liu, J.-X. Impact of rubber plantation on carabid beetle communities and species distribution in a changing tropical landscape (southern Yunnan, China). *J. Insect Conserv.* **2012**, *16*, 423–432. [CrossRef]
81. Lawrence, D.; Vandecar, K. Effects of tropical deforestation on climate and agriculture. *Nat. Clim. Chang.* **2015**, *5*, 27–36. [CrossRef]
82. Nemani, R.R.; Keeling, C.D.; Hashimoto, H.; Jolly, W.M.; Piper, S.C.; Tucker, C.J.; Myneni, R.B.; Running, S.W. Climate-driven increases in global terrestrial net primary production from 1982 to 1999. *Science* **2003**, *300*, 1560–1563. [CrossRef]
83. Gleason, C.J.; Im, J. A review of remote sensing of forest biomass and biofuel: Options for small-area applications. *GISci. Remote Sens.* **2011**, *48*, 141–170. [CrossRef]
84. Huang, T.; Ou, G.; Wu, Y.; Zhang, X.; Liu, Z.; Xu, H.; Xu, X.; Wang, Z.; Xu, C. Estimating the Aboveground Biomass of Various Forest Types with High Heterogeneity at the Provincial Scale Based on Multi-Source Data. *Remote Sens.* **2023**, *15*, 3550. [CrossRef]
85. Choudhury, M.A.M.; Marcheggiani, E.; Galli, A.; Modica, G.; Somers, B. Mapping the urban atmospheric carbon stock by lidar and worldview-3 data. *Forests* **2021**, *12*, 692. [CrossRef]
86. Hardisky, M.A.; Daiber, F.C.; Roman, C.T.; Klemas, V. Remote sensing of biomass and annual net aerial primary productivity of a salt marsh. *Remote Sens. Environ.* **1984**, *16*, 91–106. [CrossRef]
87. Jha, N.; Tripathi, N.K.; Barbier, N.; Virdis, S.G.; Chanthorn, W.; Viennois, G.; Brockelman, W.Y.; Nathalang, A.; Tongsima, S.; Sasaki, N. The real potential of current passive satellite data to map aboveground biomass in tropical forests. *Remote Sens. Ecol. Conserv.* **2021**, *7*, 504–520. [CrossRef]

land

MDPI

Article

Soil-Based Emissions and Context-Specific Climate Change Planning to Support the United Nations (UN) Sustainable Development Goal (SDG) on Climate Action: A Case Study of Georgia (USA)

Davis G. Nelson [1], Elena A. Mikhailova [1,*], Hamdi A. Zurqani [2,3], Lili Lin [4], Zhenbang Hao [5], Christopher J. Post [1], Mark A. Schlautman [6] and George B. Shepherd [7]

[1] Department of Forestry and Environmental Conservation, Clemson University, Clemson, SC 29634, USA; dgn@clemson.edu (D.G.N.); cpost@clemson.edu (C.J.P.)
[2] University of Arkansas Division of Agriculture, Arkansas Forest Resources Center, University of Arkansas System, Monticello, AR 71656, USA; Zurqani@uamont.edu
[3] College of Forestry, Agriculture, and Natural Resources, University of Arkansas at Monticello, Monticello, AR 71656, USA
[4] Department of Biological Science and Biotechnology, Minnan Normal University, Zhangzhou 363000, China; lll2639@mnnu.edu.cn
[5] Department of Electronic Information, Zhangzhou Institute of Technology, Zhangzhou 363000, China; haozhenbang@126.com
[6] Department of Environmental Engineering and Earth Sciences, Clemson University, Anderson, SC 29625, USA; mschlau@clemson.edu
[7] School of Law, Emory University, Atlanta, GA 30322, USA; gshep@law.emory.edu
* Correspondence: eleanam@clemson.edu

Abstract: Soil-based emissions from land conversions are often overlooked in climate planning. The objectives of this study were to use quantitative data on soil-based greenhouse gas (GHG) emissions for the state of Georgia (GA) (USA) to examine context-specific (temporal, biophysical, economic, and social) climate planning and legal options to deal with these emissions. Currently, 30% of the land in GA has experienced anthropogenic land degradation (LD) primarily due to agriculture (64%). All seven soil orders were subject to various degrees of anthropogenic LD. Increases in overall LD between 2001 and 2021 indicate a lack of land degradation neutrality (LDN) in GA. Besides agricultural LD, there was also LD caused by increased development through urbanization, with 15,197.1 km^2 developed, causing midpoint losses of 1.2×10^{11} kg of total soil carbon (TSC) with a corresponding midpoint social cost from carbon dioxide (CO_2) emissions (SC-CO_2) of USD \$20.4B (where B = billion = 10^9, \$ = U.S. dollars (USD)). Most developments occurred in the Metro Atlanta and Coastal Economic Development Regions, which indicates reverse climate change adaptation (RCCA). Soil consumption from developments is an important issue because it limits future soil or forest carbon (C) sequestration potential in these areas. Soil-based emissions should be included in GA's carbon footprint. Understanding the geospatial and temporal context of land conversion decisions, as well as the social and economic costs, could be used to create incentives for land management that limit soil-based GHG emissions in a local context with implications for relevant United Nations (UN) initiatives.

Keywords: carbon; damage; decarbonization; gas; greenhouse; land use

1. Introduction

Climate change affects many facets of society (e.g., social, economic, etc.) and requires context-specific climate planning (Figure 1). In general, *"context is defined as the interrelated conditions in which something exists or occurs"* [1]. For example, soil-based emissions occur because of economic activity and biophysical changes from land conversions leading to damages, which can be described in terms of societal costs (e.g., social costs of CO_2 emissions, SC-CO_2 [2]) (Figure 1). Most states in the United States of America (USA) do not have climate change preparation and adaptation plans, and the very few states that have plans often overlook soil-based emissions (https://www.georgetownclimate.org/adaptation/plans.html (accessed on 15 March 2024)) [3]. The omission of soil-based emissions can lead to the underestimation of a state's carbon footprint and can harm overall climate change mitigation efforts. Understanding emissions patterns provides important geospatial and temporal information about land and soil consumption patterns. Soil consumption from developments is an important issue because it limits future soil or forest C sequestration potential in these areas. Understanding the context of land conversion decisions, as well as the social and economic costs, could be used to create mechanisms to incentivize land use decisions that limit soil-based greenhouse gas (GHG) emissions within a local context [4].

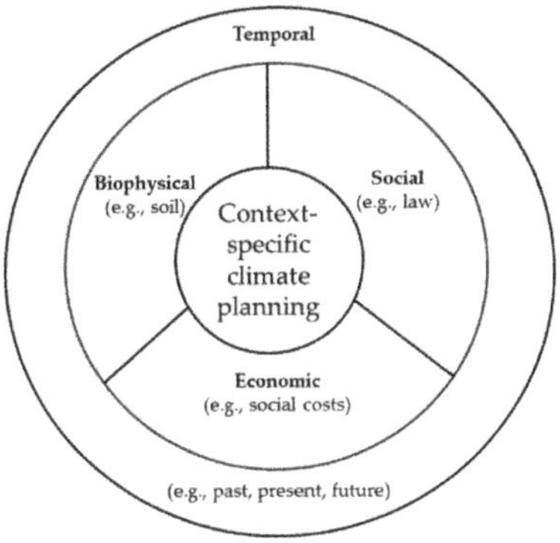

Figure 1. Examples of different contexts which can be used in context-specific climate planning.

The state of Georgia (GA) currently does not have climate change preparation and adaptation plans. However, GA conducted research for the state's climate action using an approach developed specifically for GA, which is called "Drawdown GA" (https://www.drawdownga.org/carbon-reduction-visualizer/ (accessed on 16 August 2024)) [5–7]. This GHG reduction framework evaluates the baseline emissions for GA and identifies and examines possible emission reduction options [6]. This research evaluated 20 "high impact" carbon reduction strategies but assumed there was no change in soil-based C sequestration [7]. A companion interactive website, "Drawdown GA," further explores the impact of the proposed reductions and includes the concept of "Climate Smart Agriculture," which can use agronomic conservation practices to increase C sequestration, as well as the assumption that land is only a sink of GHG emissions. Soil-based emissions because of land conversions are not accounted for in the "Drawdown GA" framework [5]. Brown et al. (2021) [7] stress the importance of "place" and "context" in developing climate action plans. In addition to these GA research initiatives, on July 14, 2023, the EPA granted the state of GA $3 million ($, USD) to develop its first-ever climate plan as a part of the United States (US) Congressional Inflation Reduction Act (IRA) of 2022 [8–10].

The omission of soil-based emissions from land conversions in GA climate change preparation and adaptation plans can lead to the underestimation of emissions from the state as well as an inability to reduce future potential emissions through sustainable land and soil management. The soil-based emissions potential from land conversions is dependent on many factors, including the inherent soil quality of soil types found in GA (Figure 2). The inherent SQ of GA is dominated by strongly weathered Ultisols (77%) with lower SQ status (Figure 2) [11]. All other soil orders have limited presence in the state (Table 1). Citizens of GA have selected the State Soil as Tifton (soil order: Ultisols) because it is among the most extensive soils in the state [12]. Soils of GA provide a wide range of ecosystem services (ES) (cultural, provisioning, regulation/maintenance) to the economic development regions for GA (Figure 2, Table S1) [11]. Regulating ES provided by GA soils is particularly relevant to the state's climate change planning, especially concerning soil C storage and the potential for C loss because of disturbance, which results in the oxidation of soil organic matter (SOM) and subsequent release of carbon dioxide gas (CO_2) [11]. The knowledge of soil C stocks is relevant to the GA climate change plan; however, it should be noted that this stock was already significantly depleted during the state's land and soil use history [13]. As of 2021, the remaining estimated total mid-point storage for TSC for GA was 1.4×10^{12} kg C with an estimated total mid-point monetary value of \$244.2B (i.e., \$244.2B billion U.S. dollars (USD), where B = billion = 10^9 in social costs of C (SC-CO_2) [2] (Table 1)). From these total estimates, SOC was 60% of the total value (1.3×10^{12} kg C, \$220.4B), and SIC represented 40% of the total value (1.4×10^{11} kg C, \$23.8B). We have previously reported that the state of GA ranked 15th for SOC [14], 34th for SIC [15], and 26th for TSC [16] for the SC-CO_2 values among the 48 contiguous US states. Soil orders with the highest midpoint monetary value and storage for TSC were Ultisols (8.0×10^{11} kg C, \$135.1B), Histosols (2.2×10^{11} kg C, \$37.2B), and Entisols (1.6×10^{11} kg C, \$26.9B) (Table 1). Counties in GA having the highest midpoint TSC values included Charlton (1.1×10^{11} kg C, \$18.6B), Ware (1.1×10^{11} kg C, \$18.6B), and Clinch (3.5×10^{10} kg C, \$5.9B) (Table 1). Estimated values of the remaining TSC storage and its social costs estimate the potential C footprint of GA soils upon disturbance with its variability (minimum, midpoint, and maximum) reported in the Supplemental Materials (Table S2).

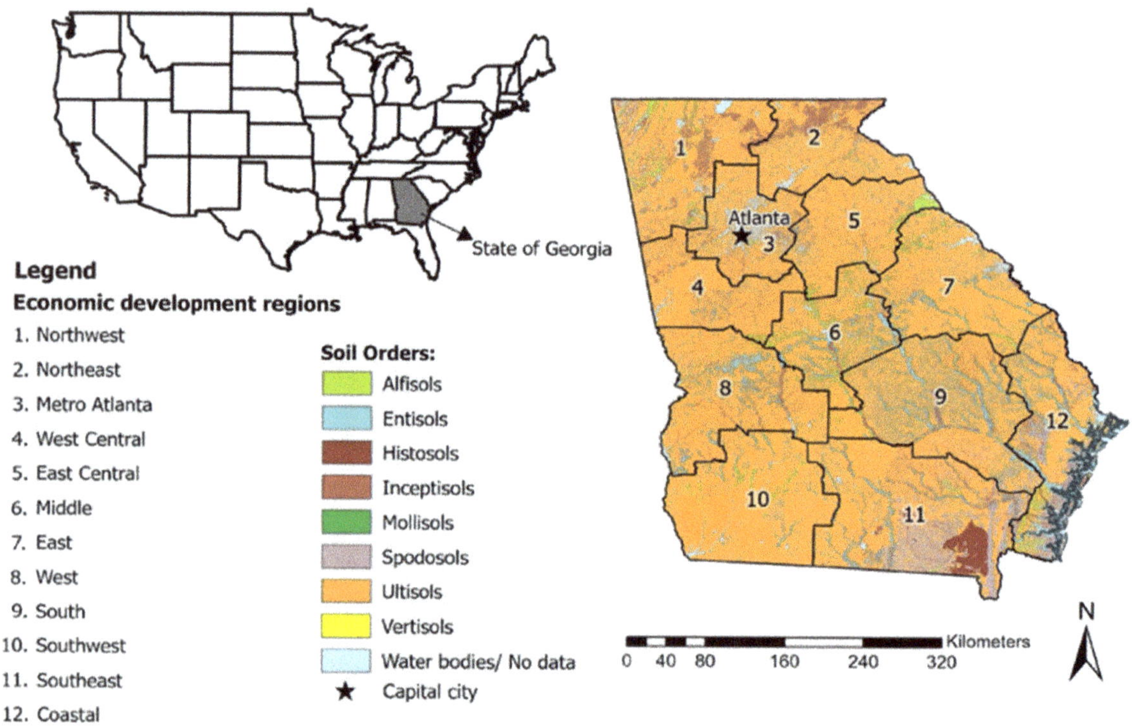

Figure 2. State of Georgia (GA) (USA) soil map (30° 35′ N to 35° N; 80° 50′ W to 85° 36′ W) derived from the SSURGO soils database [17] with boundaries of economic development regions [18]. The inherent soil quality (soil suitability) of GA is dominated by strongly weathered Ultisols (77%) with lower inherent soil quality status.

Table 1. Distribution of inherent soil quality and soil carbon regulating ecosystem services in the state of Georgia (GA) (USA) by soil order (photos courtesy of USDA/NRCS [19]).

Soil Regulating Ecosystem Services in the State of Georgia (USA)						
Degree of Weathering and Soil Development (Inherent Soil Quality)						
Slight (17%)			**Moderate (3%)**		**Strong (80%)**	
Entisols 8%	**Inceptisols** 8%	**Histosols** 1%	**Alfisols** 3%	**Mollisols** <1%	**Spodosols** 3%	**Ultisols** 77%

Midpoint storage and social cost of soil organic carbon (SOC): 1.3 × 10^{12} kg C, $220.4B						
9.9 × 10^{10} kg	1.0 × 10^{11} kg	2.2 × 10^{11} kg	2.8 × 10^{10} kg	3.8 × 10^{7} kg	6.1 × 10^{10} kg	8.0 × 10^{11} kg
$16.8B	$17.0B	$36.6B	$4.7B	$0.6B	$10.2B	$135.1B
8%	8%	17%	2%	<1%	5%	61%
Midpoint storage and social cost of soil inorganic carbon (SIC): 1.4 × 10^{11} kg C, $23.8B						
6.0 × 10^{10} kg	5.8 × 10^{10} kg	3.7 × 10^{9} kg	1.6 × 10^{10} kg	3.2 × 10^{7} kg	3.0 × 10^{9} kg	0
$10.2B	$9.8B	$0.6B	$2.7B	$5.4M	$0.5B	$0
43%	41%	3%	11%	<1%	2%	0%
Midpoint storage and social cost of total soil carbon (TSC): 1.4 × 10^{12} kg C, $244.2B						
1.6 × 10^{11} kg	1.6 × 10^{11} kg	2.2 × 10^{11} kg	4.4 × 10^{10} kg	7.0 × 10^{7} kg	6.4 × 10^{10} kg	8.0 × 10^{11} kg
$26.9B	$26.8B	$37.2B	$7.4B	$12M	$10.7B	$135.1B
11%	11%	15%	3%	<1%	4%	55%
Sensitivity to climate change						
Low	Low	High	High	High	Low	Low
SOC and SIC sequestration (recarbonization) potential						
Low	Low	Low	Low	Low	Low	Low

Note: Entisols, Inceptisols, Alfisols, Mollisols, Spodosols, and Ultisols are mineral soils. Histosols are mostly organic soils. M = million = 10^6; B = billion = 10^9; $ = U.S. dollars (USD). See Supplementary Materials Table S2 for minimum and maximum values.

Previously, Brown et al. (2021) [7] claimed that GA lands serve as a sink for GHG emissions with only the potential to increase GHG sequestration and not as a contributor to emissions. However, this assumption does not consider past, current, and future soil-based emissions from land conversions. This study hypothesizes that it is possible to quantitatively evaluate soil-based greenhouse gas (GHG) emissions for the state of GA within temporal, biophysical, economic, and social contexts to enable climate planning and to help use either existing or new legal strategies to limit and provide greater responsibility and accountability for damages from these emissions.

The primary objectives of this study were to use satellite-based remote sensing (Multi-Resolution Land Characteristics Consortium (MRLC) [20]) and soil spatial databases (Soil Survey Geographic Database (SSURGO) [17,21], and the State Soil Geographic Database, (STATSGO) [22]) to examine soil-based emissions from land conversions in GA in biophysical, economic, and social contexts. Sub-objectives include: (1) quantifying the total area of anthropogenic land degradation (LD) and potential land for nature-based solutions (NBS) disaggregated by type of LD and soil type; (2) quantifying the total area of past (prior to and through 2021) and recent (2001-2021) developments in GA by soil type; (3) quantifying the total soil C loss as a result of the past and recent developments based on information provided by Guo et al. (2006) [23]; (4) estimating the social cost of C [2] loss from past and recent developments; (5) presenting results in both tabular and spatial formats (e.g., maps) to identify emission hotspots for climate change planning and (6) discussing the use of existing or novel legal strategies to provide greater accountability for emission damages in GA and worldwide.

2. Materials and Methods

This study used an accounting framework (Table S3) to examine soil-based emissions in biophysical, economic, and social contexts and their temporal changes in GA. The biophysical analyses involved calculating soil organic, inorganic, and total carbon stocks (kg), the area (km^2), and proportion (%) of LD and LDN by soil type, land cover type, and administrative unit (state, county). Land cover analysis using classified satellite remote sensing data (30-m) provided by the Multi-Resolution Land Characteristics Consortium (MRLC) [20] allowed the evaluation of LD based on NLCD class legend and descriptions between 2001 and 2021 (Figure S1). To combine the land cover data with high-resolution soil spatial layers (SSURGO) [21], the data were converted to vector from the original raster format and then unioned with the soil data. This combined dataset of land cover and soil information was analyzed using ArcGIS Pro 2.6 [24] to examine the soil types associated with various land cover types and changes.

The economic analysis focused on calculating monetary damages from LD within various administrative units. The soil spatial analysis was combined with values provided by Guo et al. (2006) [23] to estimate soil C contents for SOC, SIC, and TSC (kg m^{-2}) by administrative unit (e.g., county) and soil order (Table S3, Table S4). For the economic analysis, these calculated soil contents were used to calculate the C that was likely lost through CO_2 emissions as the social cost of carbon (SC-CO_2) (Table S4) in monetary terms. These calculations relied on the EPA SC-CO_2 valuation of $46 per metric ton of CO_2, which serves as a damage estimate from CO_2 release but underestimates the true costs and impacts associated with climate change damage [2]. Monetary values ($ m^{-2}) were determined for each area using Equation (1), while totals were calculated by summing within the polygon boundary (with SC = soil carbon and a metric tonne equal to 1 megagram (Mg) or 1000 kilograms (kg)):

$$\frac{\$\,\mathrm{USD}}{\mathrm{m}^2} = \left(\mathrm{SC\,Content}, \frac{\mathrm{kg}}{\mathrm{m}^2}\right) \times \frac{1\,\mathrm{Mg}}{10^3\,\mathrm{kg}} \times \frac{44\,\mathrm{Mg\,CO_2}}{12\,\mathrm{Mg\,SC}} \times \frac{\$46\,\mathrm{USD}}{\mathrm{Mg\,CO_2}} \quad (1)$$

As a calculation example for areas with Alfisols soil order, Guo et al. (2006) [23] provided a midpoint estimate of 7.5 kg m^{-2} for SOC content (2-m soil depth; Table S4). This reported soil content is then used in Equation (1) to calculate an area-normalized SOC value of $1.27 m^{-2} (Table S4). The SOC content with its area-normalized value for that area is subsequently multiplied by the area of Alfisols within GA (3699.0 km^2) to create an SOC stock estimate of 2.8×10^{10} kg and a $4.7B monetary value of SC-CO_2.

3. Results

3.1. Biophysical Context

3.1.1. Total Area of Anthropogenic Land Degradation (LD) and Potential Land for Nature-Based Solutions (NBS) Disaggregated by Type of LD and Soil Type

As of 2021, almost 30% of the land in GA experienced anthropogenic land degradation (LD) primarily due to agriculture (64%) (Figure 3, Table 2, Figure S2). All seven soil orders were subject to varying degrees of anthropogenic soil degradation and LD: Ultisols (35%), Mollisols (30%), Alfisols (20%), Entisols (13%), Spodosols (13%), Inceptisols (13%), and Histosols (0.1%). Increases in soil and LD between 2001 and 2021 (+3.7%) indicated a lack of land degradation neutrality (LDN) in GA (Table 3). Almost 44% of GA is covered by mixed, deciduous, and evergreen forests, which are primarily found in soils classified as Ultisols because of their high susceptibility to soil erosion and leaching [12]. Land degradation in GA has a long history since European settlers changed the natural soils in GA with various agricultural uses (e.g., corn, cotton, tobacco, soybeans, etc.), and these soils now require modern technologies to support their fertility status because the underlying material of these soils has low native fertility [25]. Only 8.8% of the land in GA is potentially available for NBS (Figures S3 and S4), the availability of which is further complicated by the high amount of private land ownership (90.3%) in the state [26]. Most of the potential land for NBS belonged to the soil order of Ultisols (79.5%) with inherently low SQ and high susceptibility to erosion (Table 3).

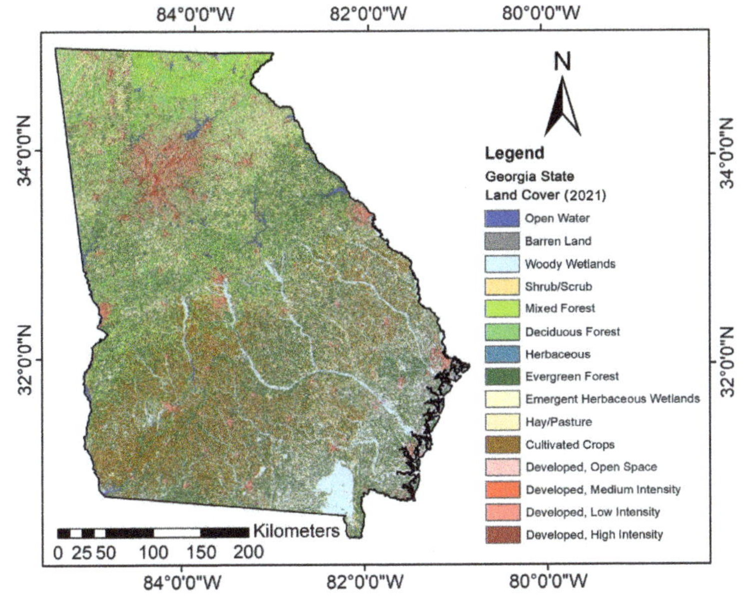

Figure 3. State of Georgia (GA) (USA) 2021 land cover map (30° 35′ N to 35° N; 80° 50′ W to 85° 36′ W) (using data from Multi-Resolution Land Characteristics Consortium (MRLC) [20]).

Table 2. Land use/land cover (LULC) by soil order in Georgia (GA) (USA) in 2021.

NLCD Land Cover Classes (LULC), Soil Health Continuum	2021 Total Area by LULC (km^2)	Degree of Weathering and Soil Development (Inherent Soil Quality)						
		Slight			Moderate		Strong	
		Entisols	Inceptisols	Histosols	Alfisols	Mollisols	Spodosols	Ultisols
		2021 Area by Soil Order (km^2)						
Woody wetlands	24073.8	5009.2	4143.2	1515.0	662.3	0.3	857.7	11886.1
Shrub/Scrub	6051.0	408.2	176.4	0.4	174.3	0.1	604.7	4686.8
Mixed forest	9544.6	610.0	1175.9	0.1	279.3	0.6	11.1	7467.8
Deciduous forest	18518.2	879.4	2795.8	0.0	503.6	0.4	6.1	14332.9
Herbaceous	6579.4	487.6	182.0	1.5	173.9	0.2	348.8	5385.5
Evergreen forest	35692.0	1824.6	1321.5	4.1	1124.0	0.5	2414.6	29002.7
Emergent herbaceous wetlands	2561.1	1561.6	143.3	31.3	30.8	0.0	78.6	715.6
Hay/Pasture	10192.7	451.6	534.2	0.1	376.0	0.5	96.0	8734.3
Cultivated crops	17879.5	447.1	207.8	0.0	65.1	0.2	83.7	17075.6
Developed, open space	7992.8	369.0	410.9	0.9	168.6	0.0	235.6	6807.8
Developed, low intensity	4629.3	192.1	162.9	0.2	85.2	0.0	133.5	4055.3
Developed, medium intensity	1899.9	96.3	74.7	0.1	34.5	0.0	49.4	1644.8
Developed, high intensity	675.0	39.3	26.0	0.0	12.7	0.0	14.3	582.8
Barren land	309.4	55.3	13.4	0.0	8.9	0.0	14.0	217.8
Totals	**146,598.9**	**12,431.4**	**11,367.9**	**1553.8**	**3699.0**	**2.8**	**4948.0**	**112,596.0**

Note: Entisols, Inceptisols, Alfisols, Mollisols, Spodosols, and Ultisols are mineral soils. Histosols are most often organic soils.

Table 3. Anthropogenic land degradation status and potential land for nature-based solutions by soil order for the state of Georgia (GA) in the United States of America (USA) in 2021. Percent changes in area from 2001 to 2021 are shown in parentheses. Reported values have been rounded; therefore, calculated sums and percentages may exhibit minor discrepancies.

Soil Order	Total Area		Anthropogenically Degraded Land	Types of Anthropogenic Degradation			Potential Land for Nature-Based Solutions
				Barren	Developed	Agriculture	
	(km^2)	(%)	(km^2)	(km^2)	(km^2)	(km^2)	(km^2)
Slightly Weathered Soils							
	25,353	**17.3**	**3082 (+5.9)**	**68 (−3.4)**	**1372 (+24.7)**	**1641 (−5.7)**	**1325 (+5.3)**
Entisols	12,431	8.5	1651 (+5.9)	55 (−0.2)	697 (+22.2)	899 (−3.8)	951 (+9.2)
Inceptisols	11,368	7.8	1430 (+5.8)	13 (−14.7)	674 (+27.4)	742 (−7.9)	372 (−3.9)
Histosols	1554	1.1	1 (+22.9)	0 (0)	1 (+20.4)	0 (0)	2 (+151.6)
Moderately Weathered Soils							
	3701	**2.5**	**752 (+2.4)**	**9 (−4.4)**	**301 (+25.1)**	**442 (−8.8)**	**357 (+9.2)**
Alfisols	3699	2.5	751 (+2.4)	9 (−4.5)	301 (+25.1)	441 (−8.8)	357 (+9.2)
Mollisols	3	0	1(+6.1)	0 (0)	0 (0)	1 (+5.0)	0 (0)
Strongly Weathered Soils							
	117,544	**80.2**	**39,745 (+3.3)**	**232 (−10.6)**	**13,524 (+26.6)**	**25,990 (−5.3)**	**11,258 (+19.8)**
Spodosols	4948	3.4	626 (+27.6)	14 (+93.2)	433 (+16.7)	180 (+59.5)	967 (+24.0)
Ultisols	112,596	76.8	39,118 (+3.3)	218 (−13.5)	13,091 (+26.9)	25,810 (−5.5)	10,290 (+19.4)
All Soils							
Totals	**146,599**	**100.0**	**43,579 (+3.7)**	**309 (−8.9)**	**15,197 (+26.4)**	**28,072 (−5.3)**	**12,940 (+17.8)**

Note: Entisols, Inceptisols, Mollisols, Spodosols, Ultisols, and Alfisols are mineral soils. Histosols are mostly organic soils. Anthropogenically degraded land was calculated as a sum of degraded land from agriculture (hay/pasture, and cultivated crops), from development (developed, open space; developed, low intensity; developed, medium intensity; developed, high intensity), and barren land. Developed land includes categories: developed, open space; developed, low intensity; developed, medium intensity; developed, high intensity. Agriculture includes categories: hay/pasture; and cultivated crops. Potential land for nature-based solutions (NBS) is limited to barren land, shrub/scrub, and herbaceous land cover classes, to provide potential land areas without impacting current land uses. Change in the area was calculated as follows: ((2021 Area − 2001 Area) / 2001 Area) × 100%.

3.1.2. Biophysical Losses and Damages to Ecosystem Services

Anthropogenic LD is a dynamic process that causes various damages to ecosystem services (ES) (cultural, regulation/maintenance, provisioning), which need to be quantified both spatially and temporally. Table 4 shows anthropogenic LD trends by soil type in GA from 2001 to 2021, which demonstrates an increase in developments in place of more land-conserving LULC (e.g., mixed forest, deciduous forest, etc.). As an example, developments within the state of GA caused loss and damage (L&D) to regulating ES because of the loss of land that could potentially be used for soil carbon (C) sequestration with a sum of 15,197.1 km^2 of land area converted to developments before and through 2021 (Table 3). The largest area losses from developments were found in Gwinnett (741.7 km^2), Cobb (577.9 km^2), and Fulton (441.2 km^2) counties (Table S5). Between 2001 and 2021, new developments led to a total of 3564.9 km^2 of land being converted to developments (Table S5). The areas with the largest losses from development were found in Gwinnett (208.8 km^2), Fulton (149.1 km^2), and Henry (117.0 km^2) counties (Table S5). Most developments occurred adjacent to the Atlanta urban area and came at the expense of cultivated and forest areas (Figure 4). This analysis determined that between 2001 and 2021, land developments mainly occurred near existing urban and coastal areas. Georgia is dominated by the soil order of Ultisols, which have inherently low C sequestration potential. Projected urbanization increases will likely cause a future reduction in land available for C sequestration. Another example of L&D is from soil carbon (C) loss and the associated emissions from the land conversion process to create developments in GA (USA) (Figure 5). These losses that occurred before and up to 2021 resulted in an estimated midpoint total of 1.2×10^{11} kg of C losses (Table S6). The largest soil C losses were seen in Gwinnett

(5.8×10^9 kg C), Cobb (4.3×10^9 kg C), and Fulton (3.4×10^9 kg C) counties (Table S5). All these counties are in proximity to the urban center of Atlanta. New development activity between 2001 and 2021 caused a total of 6.5×10^{10} kg in C losses (Table S7). The highest soil C losses were seen in Gwinnett (5.3×10^9 kg C), Fulton (3.4×10^9 kg C), and Chatham (2.2×10^9 kg C) counties (Figure 5, Table S5).

Table 4. Land use/land cover (LULC) change (%) by soil order in Georgia (USA) from 2001 to 2021.

NLCD Land Cover Classes (LULC), Soil Health Continuum	Change in Area, 2001–2021 (%)	Degree of Weathering and Soil Development (Inherent Soil Quality)						
		Slight			Moderate		Strong	
		Entisols	Inceptisols	Histosols	Alfisols	Mollisols	Spodosols	Ultisols
		Change in Area, 2001–2021 (%)						
Woody wetlands	0.7	1.1	0.8	1.3	0.9	1.9	2.3	0.2
Shrub/Scrub	55.9	24.4	3.9	44.6	45.0	400.0	97.4	58.5
Mixed forest	−4.0	−9.2	0.2	−32.9	−1.9	−3.0	−44.5	−4.1
Deciduous forest	−13.2	−15.5	−3.8	−51.0	−12.0	−7.9	−83.3	−14.6
Herbaceous	−2.7	0.1	−9.6	218.9	−12.0	100.0	−25.3	−0.4
Evergreen forest	−2.3	−0.8	−2.4	−18.6	1.0	−27.2	−11.6	−1.7
Emergent herbaceous wetlands	14.8	1.7	22.6	−28.2	41.6	52.9	33.4	57.4
Hay/Pasture	−11.3	−8.0	−11.5	29.8	−11.2	−6.6	121.6	−12.1
Cultivated crops	−1.5	1.0	2.7	71.4	7.8	42.6	20.7	−1.8
Developed, open space	5.0	−0.1	9.8	6.1	4.3	−28.9	−5.6	5.4
Developed, low intensity	37.7	38.2	39.2	73.6	42.7	−6.3	35.6	37.6
Developed, medium intensity	154.4	129.2	171.9	126.0	159.1	1000.0	192.3	154.2
Developed, high intensity	122.6	99.2	144.9	150.0	118.0	0.0	143.1	123.1
Barren land	−8.9	−0.2	−14.7	216.7	−4.5	0.0	93.2	−13.5

Note: Entisols, Inceptisols, Alfisols, Mollisols, Spodosols, and Ultisols are mineral soils. Histosols are most often organic soils. Change in the area was calculated as follows: ((2021 LULC Area − 2001 LULC Area) / 2001 LULC Area) * 100%.

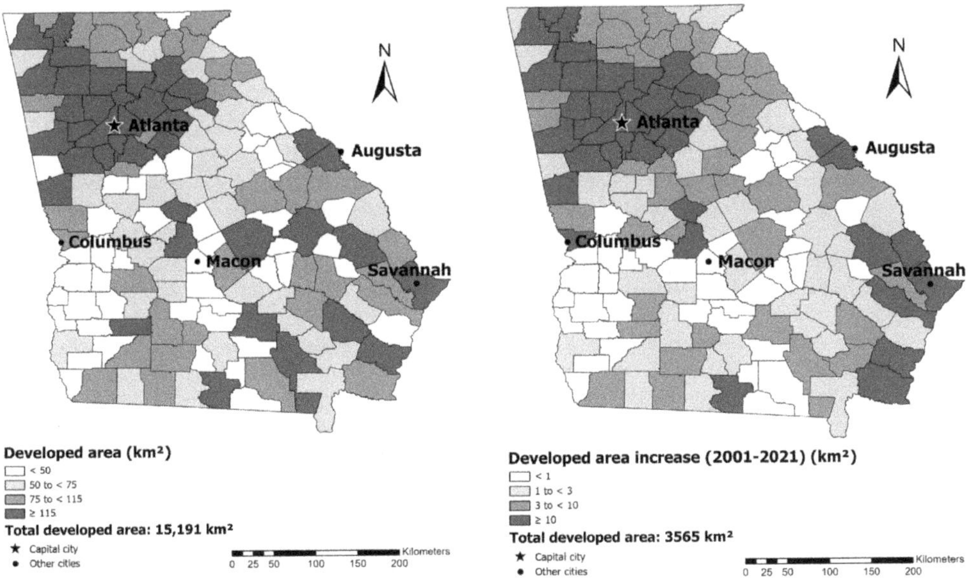

(a) (b)

Figure 4. Loss and damage (L&D) because of loss of land for potential soil carbon (C) sequestration from (**a**) past developments (prior and through 2021), and (**b**) land developments that occurred in time interval (2001–2021) for Georgia (GA) (USA).

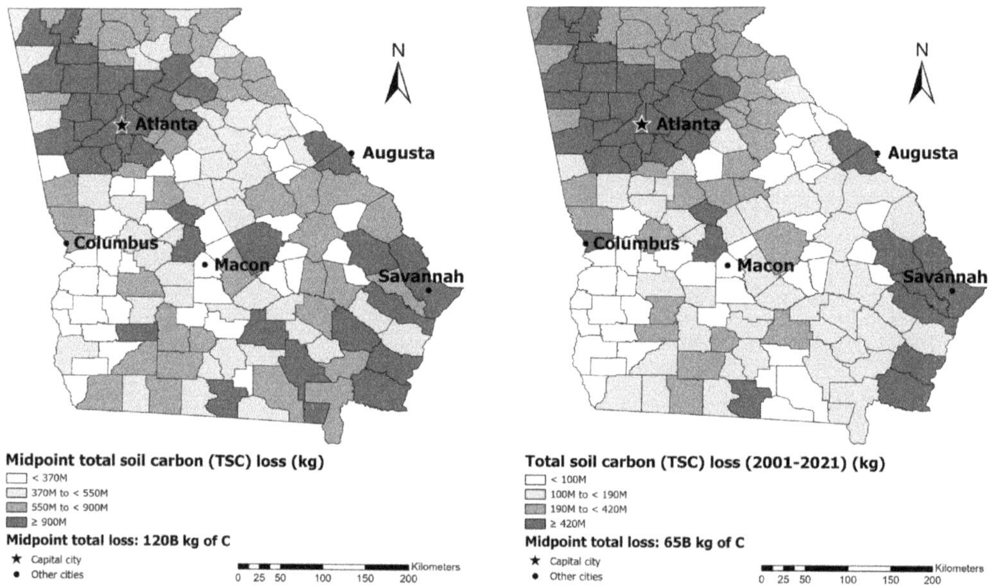

(a) (b)

Figure 5. Loss and damage (L&D) because of soil carbon (C) loss with associated emissions from (**a**) past land developments (through 2021), and (**b**) more recent land developments (2001–2021) in Georgia (GA) (USA). Note: M = million = 10^6; B = billion = 10^9.

3.2. Economic Context

3.2.1. Anthropogenic Land Degradation (LD) as a Proxy for Economic Development

Anthropogenic LD in GA is closely associated with past and current economic activities in the state, with a high degree of spatial variability between counties and economic development regions (Table 5, Figure 6). There were 68 counties below the value of 29.7% anthropogenic LD for the state as a whole and 91 counties at or above this value. Southwest (48.5%) and Metro Atlanta (42.5%) economic development regions had the highest proportions of anthropogenic LD in the state (Table 5) compared to 29.7% of anthropogenic LD for the whole state. Table 5 shows anthropogenic LD status in 2021 but likely does not account for historical anthropogenic LD or most inherent LD. Metro Atlanta (+18.3%) and Coastal (+11.0%) economic development regions experienced the highest increase in development between 2001 and 2021. Among GA counties, Cobb (71.8%), Gwinnett (71.6%), and Clayton (65.4%) had the highest anthropogenic LD proportions. All three counties are in the Metro Atlanta economic development region. Counties with the lowest anthropogenic LD were McIntosh (5.1%) (in the Coastal economic region), Charlton (5.2%), and Clinch (6.5%) (both in the Southeast economic development region).

It should be noted that 29.7% of the current level of anthropogenic LD does not account for historic anthropogenic LD, where as much as 95% of the forests in GA were removed by the 1920s for agriculture and had to be subsequently reforested after agricultural uses collapsed [27]. The rapid development of the area in and around Atlanta has led to the rapid loss and fragmentation of forests [28] in concert with the increase in LD. This loss of forest habitat has likely caused a range of ecological and habitat damages [28], as well as loss of above-ground C with forest removal. A study by Obata et al. (2020) [29] examined forest disturbance between 1987 and 2016 and found that 29.2% of the state was disturbed, noting the difference between forestry cycles and urbanization.

Table 5. Land degradation (LD) status and trends in economic development regions of Georgia (GA) (USA). Percent changes in area from 2001 to 2021 are shown in parentheses. Reported values have been rounded; therefore, calculated sums and percentages may exhibit minor discrepancies. This table shows anthropogenic LD status in 2021 but likely does not account for historical anthropogenic LD as well as most inherent LD.

Georgia Economic Development Regions	Land Degradation (through 2021)	
	Area (Change from 2001 to 2021) (km^2, %)	Proportion from Total Region Area (%)
Metro Atlanta	3665.5 (+18.3)	42.5
Northwest	3728.4 (+3.3)	28.5
Southeast	5162.5 (+4.3)	30.9
Coastal	2299.6 (+11.0)	23.1
Northeast	2606.8 (+7.8)	22.0
South	4700.9 (+0.1)	25.6
East Central	3045.0 (+1.9)	25.8
East	3054.1 (-0.7)	31.9
Southwest	7116.2 (+0.9)	48.5
Middle	2143.2 (+4.5)	27.7
West Central	2345.5 (+0.3)	23.7
West	3711.0 (+1.0)	25.9
Overall State Total	**43,578.7 (+3.7)**	**29.7 (State)**

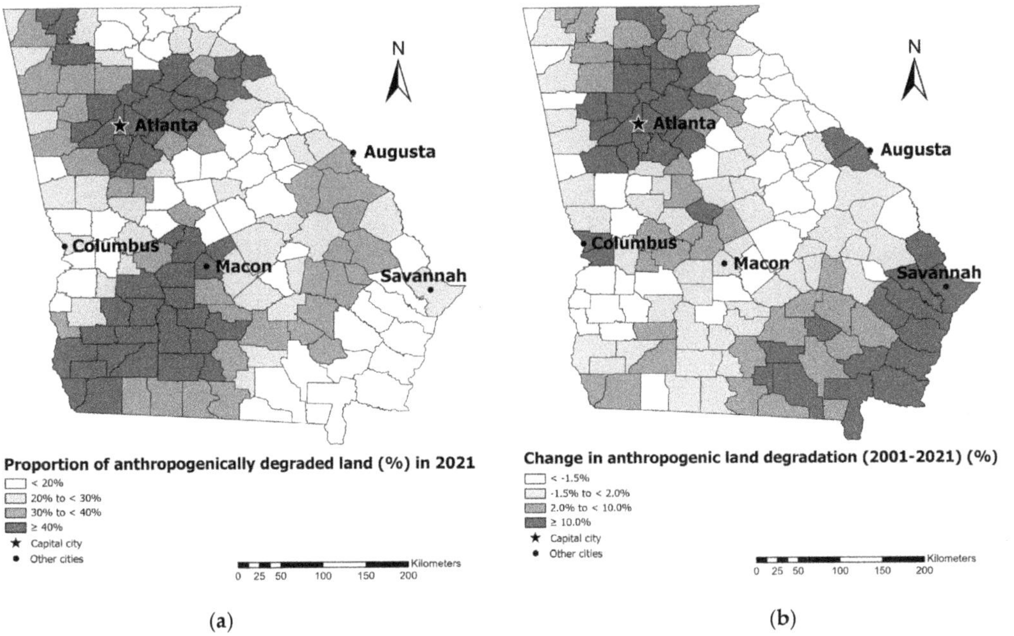

Figure 6. Maps of (**a**) the proportion (%) by county of anthropogenically degraded land in 2021, (**b**) more recent land degradation (%) between 2001 and 2021 in Georgia (GA) (USA). Land subject to anthropogenic degradation was calculated as a sum of developed land (developed, open space; developed, high intensity; developed, medium intensity; developed, low intensity), agriculture (cultivated crops, and hay/pasture), and barren land.

3.2.2. Global Social Cost of Soil-Based Emissions Associated with Economic Development

Loss and damage from land conversions associated with developments extend beyond the boundaries of GA, which can be quantified as the "realized" social costs of soil carbon (C) (SC-CO_2) [2] released from land conversion because of soil organic matter decomposition and other disturbances [11]. The SC-CO_2 is a fixed, non-market-based value intended to monetize damages to society from a metric tonne of CO_2 emissions [30], which is often used only for government purposes. The SC-CO_2 from land conversions to developments before and into 2021 within the state of GA (USA) results in a total midpoint value of $20.4B SC-$CO_2$ (Table S6). The highest social costs were found in Gwinnett ($983.8M), Cobb ($732.7M), and Fulton ($580.4M) counties, which are all located in the Metro Atlanta economic development region. From 2001 to 2021, new developments caused $11.0B in SC-$CO_2$. The highest costs were found in Gwinnett ($904.2M) and Fulton ($582.7M), which are also located in the Metro Atlanta region and Chatham ($375.9M) in the Coastal economic development region (Figure 7). For economic development regions, the highest "past" and "recent" SC-CO_2 values were associated with Metro Atlanta (past: $3.9B; recent: $3.1B), Northwest ($2.3B; $1.7B), and Coastal ($1.8B; $1.3B) economic development regions (Table 6). It should be pointed out that SC-CO_2 values are calculated from the developed areas and soil types within these areas, and values of SC-CO_2 can be quite different even if the developed areas are the same in size because of the variability in soil C content between soil types (Table 6). For example, even though the Coastal economic development region has a smaller area (km^2) than the Northeast economic development region, it has more SC-CO_2 from developments because of the higher soil C content in soils for these developed areas (Table 6). Potential solutions to negative externalities associated with damages from developments can include market-based payments in proportion to these damages [31].

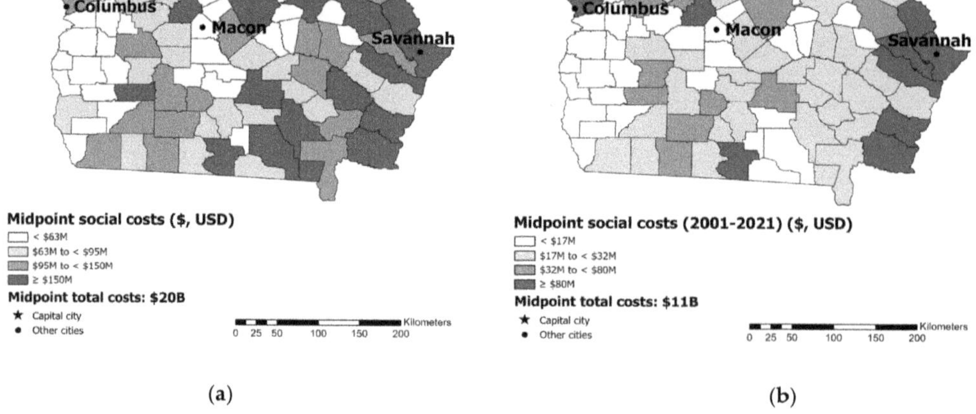

Figure 7. Loss and damage (L&D) can be measured as "realized" social costs of soil carbon (C) (SC-CO_2) from (**a**) past developments (prior and through 2021), and (**b**) recent land developments in the state of Georgia (GA) (USA) from 2001 to 2021. Note: M = million = 10^6, B = billion = 10^9, $ = U.S. dollars (USD).

Table 6. Past and recent loss and damages (L&D) from developments by economic development regions, Georgia (GA) (USA).

Georgia Economic Development Regions	Past Developments (through 2021)			Recent Developments (2001-2021)		
	Area (km^2)	Midpoint TSC loss (kg)	Midpoint SC-CO_2 ($, USD)	Area (km^2)	Midpoint TSC loss (kg)	Midpoint SC-CO_2 ($, USD)
Metro Atlanta	3025.6	2.3×10^{10}	$3.9B	871.1	1.8×10^{10}	$3.1B
Northwest	1634.0	1.4×10^{10}	$2.3B	433.1	1.0×10^{10}	$1.7B
Southeast	1257.1	1.1×10^{10}	$1.8B	193.3	2.8×10^{9}	$478.1M
Coastal	1120.7	1.1×10^{10}	$1.8B	349.8	7.4×10^{9}	$1.3B
Northeast	1345.1	1.0×10^{10}	$1.7B	352.5	5.8×10^{9}	$980.4M
South	1160.8	9.0×10^{9}	$1.5B	138.8	2.0×10^{9}	$335.1M
East Central	1133.8	8.4×10^{9}	$1.4B	321.5	5.0×10^{9}	$849.3M
East	915.0	7.0×10^{9}	$1.2B	160.5	2.3×10^{9}	$382.9M
Southwest	980.1	7.2×10^{9}	$1.2B	137.2	1.9×10^{9}	$316.8M
Middle	829.5	6.5×10^{9}	$1.1B	215.4	3.1×10^{9}	$522.1M
West Central	880.6	6.5×10^{9}	$1.1B	221.3	3.5×10^{9}	$595.3M
West	908.6	7.0×10^{9}	$1.2B	170.4	2.8×10^{9}	$467.1M
Overall State Total	**15,190.8**	$\mathbf{1.2 \times 10^{11}}$	**$20.4B**	**3564.9**	$\mathbf{6.5 \times 10^{10}}$	**$11.0B**

Note: TSC = total soil carbon; SC-CO_2 = social costs of carbon dioxide emissions; M = million = 10^6; B = billion = 10^9; $ = U.S. dollars (USD).

3.2.3. Reverse Climate Change Adaptation (RCCA) Linked to Economic Development

Increased GHG emissions in GA and worldwide have contributed to sea level rise which threatens the GA coast. Eleven out of GA's 159 counties are potentially impacted by projected sea level rise (Figure 8, Table 7). All these 11 counties experienced increases in development between 2001 and 2021 (Table 7), resulting in loss of land for C sequestration and RCCA.

Sea level rise (3.2 cm per decade) is already occurring along the GA coast and appears to be 30% higher than global averages [32,33] because of the regional land subsidence of the GA coastal plain [34]. Future flood modeling predicts increases in coastal flooding for GA, with an increase to more than 8 days of flooding within approximately 125 events each year by 2060 [32]. The largest economic impact from flooding caused by extreme weather events is seen in the GA coastal communities because coastal buildings and infrastructure have high risks from sea level rise and proximity to storm surges [35]. The economic impact of these extreme weather events is exacerbated by the 300% increase in coastal property value between 1980 and 2000 [36], which has likely continued to increase to the present day. While a recent study has found some reduced property value (3.1%) for homes at high flood risk [37], this is nearly insignificant when looking at the overall value of coastal property subject to risk from these natural disasters. Therefore, there are few negative incentives for coastal development (e.g., increased insurance cost or even uninsurability) that could serve to disincentivize development that caused LD and the associated social costs while also putting additional homeowners at risk from climate-change-related extreme weather events. Government-supported insurance for coastal developments may further aggravate the problem of RCCA by providing incentives that result in increased development in these hazard-prone areas [38]. An unanswered question is what the economic cost would be of relocation of property and people who currently live in coastal areas, which will not be habitable because of the increasing disaster risk. It should be noted that flooding risk in GA is not only limited to the coastal areas, but other parts of the state as well. Ferguson and Ashley (2017) [39] conducted a spatiotemporal analysis of residential flood risk in the Atlanta metropolitan area and concluded that an increase in developments contributes to greater flood risks.

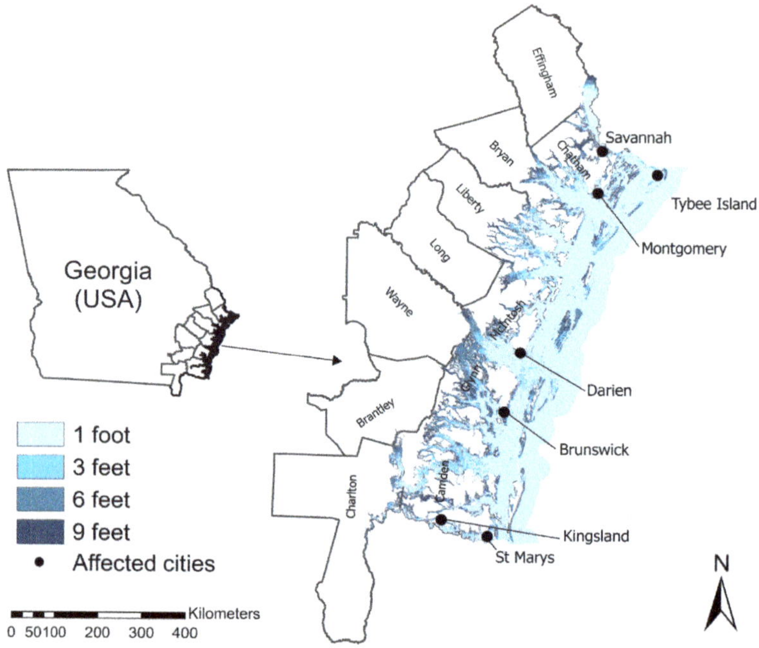

Figure 8. Projections of future sea rise and land loss due to climate change in some coastal counties of the state of Georgia (GA) (USA).

Table 7. Selected county area changes in the developed area (2001-2021) and county area loss (%) due to sea rise in the state of Georgia (GA) (USA) (based on original ArcGIS Pro 2.6 [24] analysis of data from the National Oceanic and Atmospheric Administration (NOAA) [40]).

Counties (Affected by Sea Rise)	Change in Developed Area (2001-2021) (km^2, %)	County Area Loss due to Sea Rise (%)			
		1 foot	3 feet	6 feet	9 feet
Brantley	8.3 (+14.5)	0.0	0.0	0.1	0.3
Bryan	20.2 (+31.5)	0.0	0.0	14.6	18.7
Camden	21.0 (+22.7)	21.4	28.8	34.4	44.6
Charlton	5.0 (+9.3)	1.0	1.9	3.9	6.2
Chatham	81.2 (+55.8)	37.3	43.2	50.4	60.3
Effingham	35.3 (+46.9)	2.7	3.2	3.8	4.4
Glynn	27.0 (+25.0)	27.7	35.1	48.1	63.0
Liberty	21.1 (+25.4)	14.0	17.9	21.8	26.3
Long	9.2 (+20.1)	0.0	0.0	0.0	0.1
McIntosh	3.8 (+9.2)	34.2	39.6	46.5	54.7
Wayne	10.2 (+9.3)	0.0	0.0	0.3	0.7

Note: 11 out of Georgia's (USA) 159 counties are potentially affected by the projected sea level rise. 1 foot = 0.3048 meters.

3.3. Social Context

3.3.1. Significance of the Results for Georgia's Soil Health Legislation

Georgia passed a soil health-related Bill No. 391-1-6 (https://rules.sos.ga.gov/gac/391-1-6?urlRedirected=yes&data=admin&lookingfor=391-1-6 (accessed on 8 July 2024) [41]. This bill, "Georgia Conservation Tax Credit Program," provides income tax credits for land that is accepted into the program in return for agreeing to a permanent conservation of various types of lands. This includes the protection of streams, lakes, wetlands or rivers, wildlife habitats, cultural sites, and lands used for outdoor recreation. Additionally, this program can be used to protect prime forestry or agricultural lands, with a stipulation requiring the use of forestry or agricultural best management practices (BMPs). For agricultural lands, the program requires the use of the Georgia Soil and Water Commission BMPs [42], which encourage but do not require the use of conservation tillage, which can improve soil organic matter (SOM), as well as various practices that can reduce soil erosion. While this program can protect land from development, it is not specifically focused on improving soil health and C sequestration. Another program in GA also incentivizes maintaining forest, agricultural, and environmentally sensitive areas by reducing the tax rate on lands in the Conservation Use Valuation Assessment (CUVA) system [43]. This reduces the property taxes for land that is maintained in the specified land uses (forest, agricultural, and environmentally sensitive areas) for ten years, with financial penalties for breaking the conservation agreement. This CUVA program has been extensively used in north Georgia [44] and likely serves to reduce some development but does not require soil conservation or improvement. While these types of programs may reduce LD and the associated social cost of emissions, they may not be the most efficient approach because they do not directly target incentives to prevent LD or GHG emissions that land use conversions cause (e.g., soil-based emissions from land development), which vary based on the type of disturbance and soil type.

3.3.2. Importance of the Results for Georgia's Climate Change

Despite the ongoing impacts of climate change on GA, there are no completed state-led climate change preparation and adaptation plans (https://www.georgetownclimate.org/adaptation/plans.html (accessed on 8 August 2024) [3]. Georgia has been experiencing a variety of impacts from climate change: rising atmospheric temperatures [45,46] and precipitation, more severe floods [46] and droughts, sea level rise [32,33] and sinking coastline [34], and many others [47]. Droughts often decrease soil available water, making soils drier in most of GA, which can result in reduced agricultural output and forest cover [47]. Our study shows that GA had an increase (+13.3%) in developments associated with impervious surfaces, which will only aggravate the problem of flooding and urban heat islands [48].

Results of our study show that the state of GA experienced LD and associated GHG emissions, which can be mapped and quantified using geospatial techniques. These GHG emissions go beyond the state of GA boundaries and should be accounted for (e.g., "polluter-pays-principle" [49], etc.) in the global loss and damage (L&D) accounting. Increases in developments in coastal counties of GA between 2001 and 2021 most likely indicate reverse climate change adaptation with potentially detrimental consequences to property and human life (Table 7). These coastal developments should not be further incentivized by providing federal assistance to support property insurance or repair and recovery of properties in coastal areas at high risk from future climate risks.

3.4. Temporal Context

Temporal context plays an important role in climate change planning. Our study likely underestimates soil-based emissions from ***past*** land conversions because of the history of deforestation and subsequent agriculture that was prevalent in GA and the Southeast region of the US in the 19th century, which has now largely returned to forest cover in many areas. This LD caused gullies due to soil erosion, with the most extreme example being evident at Providence Canyon State Park, also known as Georgia's "Little Grand Canyon," where historic LD (Figure 9a) is still evident after reforestation (Figure 9b) [50]. Government intervention was necessary for land restoration and the resulting economic hardship caused by failed agricultural activity [51], and it is unknown if the reforestation efforts were sufficient to compensate for the past GHG emissions caused by the combination of deforestation and agriculture.

The ***present*** context is demonstrated by our study, which documents patterns of ongoing LD and development very often following historic patterns of past LD and development ("business as usual"). Technological advances, including the advent of soil spatial data and satellite-based LULC analysis, allow us to quantify spatial patterns and trends in emissions. Also, the realization that GHG emissions contribute to climate change damages with real social costs necessitates new methods to understand and track the physical and economic impacts of anthropogenic land conversions.

In terms of ***future*** context, our study offers a methodology to quantify and understand how the impact of land management decisions can be modeled so that governments can provide incentives that minimize LD and emissions by optimizing the decision-making process. Future advances in high-resolution remote sensing technologies, combined with automated artificial intelligence classification and prediction techniques, will track land conversions and predict future potential LULC changes [52] that cause LD and soil GHG emissions. This may help optimize land planning by providing the opportunity to disincentivize developments with significant negative impacts.

(a) (b)

Figure 9. Degraded land caused by human activity in the Providence Canyon State Park, Stewart County, Georgia (USA) (https://gastateparks.org/ProvidenceCanyon) (accessed on 16 August 2024) [53]: (**a**) an oblique photo of eroded soil in 1937 from the Library of Congress (control number: 2017775702) [54], and (**b**) a modern aerial photo showing these erosion features (from 2023 National Agriculture Imagery Program (NAIP) aerial photography [55]).

3.5. Legal Context

In the legal literature, an important distinction exists between the "law as written" and the "law in context." The distinction is between the law as written in a statute or court decision, on the one hand, and the law as actually applied in an actual society, on the other [56]. According to Selznick (2003) [56]: *"[T]he phrase 'law in context' points to the many ways legal norms and institutions are conditioned by culture and social organization."* That is, *"[w]e see … how much the authority and self-confidence of legal institutions depend on underlying realities of class and power; how legal rules fit into broader contexts of custom and morality"*. Indeed a whole movement has developed that recognizes the importance of understanding the law as it exists within a society's specific economic and moral context. The founding principle of the Law and Society movement, with its Law and Society Association, is that identical written laws could lead to fundamentally different outcomes in societies with different conditions. This contextual approach has become commonplace in the teaching and application of law [57].

Legal scholars have recognized that the societal context is vitally important to whether a state or country will successfully address climate change. According to Osofsky (2003) [58], first, the success of a state's efforts to control climate change will depend on pressures from above and below: *"vertical pressures from 'above' (international negotiations for the post-2012 regime) and 'below' (state and local efforts)"*. An important influence from above would be the United Nations. Success will also depend on horizontal influences, *"namely climate change litigation and executive policy, as well as advocacy efforts by a range of nongovernmental actors"* [58]. For example, studies of household-level adaptation to flooding indicate the profound impacts of the context of prior experience with flooding, "*the influence of the media or the behavior of others, and demographic factors such as age or education*" [59].

Whether our results will prove fruitful in contributing to appropriate measures to address climate change in GA will require both understanding the local context in which such efforts will occur and also working to influence that context [60]. Georgia has traditionally been a conservative state with powerful agricultural and business interests. Efforts to address climate change in GA must recognize this context and propose measures that align with it. First, the context can be altered by distributing the information from our paper that land disturbance can be a major source of GHG; a main reason why GA has not taken measures to address this danger is that most people do not know about it.

Second, as much as possible, measures to address excessive land disturbance and climate change should be crafted to align contextually with existing interests and preferences. In the contexts of other states, it might be sufficient to motivate change by invoking the dangers of increases in sea levels. In these other states, reports or agreements from the UN or other groups might prove persuasive. In contrast, in a proudly independent state such as GA, change in the state's context might be presented as promoting business and agricultural interests: limiting land disturbance not only reduces climate change but also reduces soil erosion.

4. Significance of Results in a Broader Context

The methodology used by the UNCCD leverages Trends.Earth 2.1.17 cloud-based software to evaluate LD by using a range of datasets at various spatial resolutions to evaluate the SDG 15.3.1 sub-indicators of land productivity (LP), land cover (LC) change and SOC stocks [61]. A recent study evaluated this methodology to identify LD in the European Union and found that degraded land that was identified using the Trends.Earth software had poor agreement with low C and highly eroded soils, which is an alternative way to identify areas with LD [62]. Recommendations for improving the accuracy of LD analysis included identifying a baseline of prior LD status, increasing the spatial resolution of datasets to better contextualize changes in LC and LP, and incorporating local-scale datasets, including soil surveys, to better understand the relationship between LC change and LD [62]. The present study for the state of GA is innovative because it uses higher resolution (30 m) classified imagery than the UNCCD standard method to more accurately identify LC to developments. Zimba et al. (2024) [63] found that using 30-m Landsat-derived land cover maps was much more accurate, particularly in the development and cropping categories, than the standard land cover maps used in the Trends.Earth analysis, which has a spatial resolution of 300 m. This study is also innovative in its use of high-resolution soil survey maps to link LC change to soil types to better link soil capacity and soil C contents to land conversions for LD analysis. High-resolution soil survey data (e.g., SSURGO) is likely the most accurate way to account for landscape soil C, because it is unrealistic to obtain the density and coverage of soil samples necessary to accurately characterize impacts for land cover change over large areas [64]. There is a realization that higher-resolution data are needed to properly understand LD [65]; however, this study and related studies do not incorporate detailed soil spatial data to help contextualize LD. Econometric models can link practices to CO_2 emissions on a global scale [66]; however, they do not typically provide the spatial information necessary to understand the local context and impact of land use change. Also, we account for TSC because there are potential CO_2 emissions from both SOC and SIC from land development.

This study's relevance to multiple UN initiatives includes Sustainable Development Goals (SDGs), adopted in 2015 [67], and many other UN initiatives (e.g., UN Convention to Combat Desertification [68,69]; UN Convention on Biological Diversity [70]; UN Kunming-Montreal Global Biodiversity Framework [71]; Ramsar Convention on Wetlands [72–75], etc.) because GA is a state in the contiguous US. The UN suggests disaggregating indicators whenever possible. Therefore, this study linked soil and land use relationships to UN SDGs to better direct land management to meet UN SDGs, because country-level analysis can easily mask differences within states and regions [67]. This study's results are essential for UN initiatives and goals for the following reasons:

- There was an overall decrease in cultivated crops (−1.5%) and hay/pasture (−11.3%) between 2001 and 2021 in GA (Table 5). This may indicate a reduction in available farmlands overall, as well as the conversion of hay/pasture to more destructive land uses. (Relevant for *UN SDG 2: Zero Hunger*);
- For the state of GA, this study found a spatial link between high soil-based GHG emissions areas and likely vulnerability to climate change. The projected GA land losses from expected sea level rise (Table 7) will impact several highly populated areas as well as areas with high-value real estate, causing potential human displacement and damage to infrastructure and buildings. Table 7 also shows an increase in developments in the GA counties impacted by the rising sea level, which can be an indication of reverse climate change adaptation (RCCA). (Relevant for *UN SDG 11: Sustainable Cities and Communities*);
- Land conversions that occurred across all of the seven soil orders found in GA were caused by land development at the expense of mixed (−4.0%), deciduous (−13.2%), and evergreen (−2.3%) forests (Table 5). Land development occurred on soils with high agricultural productivity (e.g., Alfisols and Mollisols), while C-rich Histolsols were also developed in place of mixed forest (−32.9%), deciduous forest (−51.0%) and evergreen forest (−18.6%) (Table 5). This shows that C-sequestering and productive soils were impacted by land development. (Relevant for *UN SDG 12: Responsible Consumption and Production*);
- No climate change plans for GA's preparation and adaptation have been completed (https://www.georgetownclimate.org/adaptation/plans.html (accessed on 8 August 2024) [3]. The state of GA was awarded a $3 million noncompetitive planning grant to develop a climate action plan, which intends to generate a Priority Climate Action Plan (PCAP) by March 1, 2024, a Comprehensive Climate Action Plan (CCAP) by June 30, 2025, and a status report, due at the close of the 4-year grant period, which ends in 2027 [8,9]. Data from this study that estimates soil-based GHG emissions from land developments can support the development of a future plan. This study's quantitative soil-based GHG emissions estimates are from both past and recent land conversions and the resulting monetary social C cost (SC-CO_2) values. Also, this research quantified the area no longer available for C sequestration in GA. Prior to and before 2021, GA lost an area of 15,197.1 km^2 to developments with a midpoint of 1.2×10^{11} of total soil carbon (TSC) losses and midpoint values of $20.4B (where B = billion = 10^9, $ = U.S. dollars (USD)) in SC-CO_2. "New" land developments (3564.9 km^2) that occurred from 2001 to 2021 likely caused a loss of midpoint 6.5×10^{10} kg of TSC, causing a midpoint of $11.0B SC-$CO_2$. There is very little land (8.8% of total land area) available for nature-based C sequestration (e.g., 0.2% barren land, 4.1% shrub/scrub, 4.5% herbaceous) (Table 2). Georgia's soils typically have low inherent potential for C sequestration because they are dominated by low-fertility and highly leached Ultisols. Projected levels of sea level rise and expected urbanization will likely reduce land availability for C sequestration further. (Addressing *UN SDG 13: Climate Action*);
- Nearly 30% of GA's land area has had anthropogenic LD, mainly due to agriculture (64%) before and through 2021. All seven soil orders received varying degrees of anthropogenic LD: Ultisols (35%), Inceptisols (13%), Mollisols (30%), Entisols (13%), Spodosols (13%), Alfisols (20%), and Histosols (0.1%). Recent trends (2001-2021) showed a +3.7% increase in anthropogenic LD and an increase of +26.4% in the developed type of LD in the state, which was not balanced by the potential NBS land. Development has resulted in a reduction of soil resources because of LULC change between 2001 and 2021 for nearly all 159 counties and 12 economic development regions in GA (Table 3, Table S5). There were decreases in the total areas of deciduous (−13.2%), mixed (−4.0%), and evergreen (−2.3%) forests, hay/pasture (−11.3%), herbaceous (−2.7%) land covers needed for atmospheric pollution reduction and

C sequestration (Table 5). (Addressing *UN SDG 15: Life on Land; UN Convention to Combat Desertification; UN Convention on Biological Diversity; UN Kunming-Montreal Global Biodiversity Framework*);

- At the international level, there is renewed attention on preserving ecosystem resilience and integrity, as shown by the agreement from the UN's fifteenth meeting of the conference of the parties (COP 15), which adopted the UN Kunming-Montreal Global Biodiversity Framework [71]. This framework includes the goal (Goal A) of maintaining, enhancing, and restoring the resilience, connectivity, and integrity of all ecosystems and includes the target (Target 11) to both restore as well as maintain and enhance ecosystem functions and services (e.g., air, water, soil health, and regulation of climate). This study shows that GA did not reach LDN between 2001 and 2021, with developments occurring in all soil orders, including the agriculturally important soil orders of Alfisols and Mollisols and the C-rich Histosols soil order. The creation of these new developments likely decreased biodiversity through the loss of pedodiversity (soil diversity). This study's techniques can guide decision-making by providing methods to create the best possible data, which supports Target 21, which focuses on the importance of data development to support equitable governance. (Relevant to *UN Kunming-Montreal Global Biodiversity Framework*).
- The Ramsar Convention on Wetlands was adopted in 1971 with a focus on the conservation of wetlands, especially as they relate to habitat for waterfowl [72–74]. The United States joined the Ramsar Convention on Wetlands in 1986 and currently has 41 designated Ramsar sites that contain critical wetlands areas, including the Okefenokee National Wildlife Refuge (designated as a wetland of international importance), which is both in the states of GA and Florida (FL) [75]. As part of the agreement, the United States supports the Wetlands for the Future (WFF) initiative, which funds training and is focused on wetland management and conservation as part of the development process [72]. Key to the Ramsar Convention is the concept of the "wise use" of wetlands to maintain the "ecological character (of wetlands) . . . within the context of sustainable development." This convention also obligates the US to work to conserve all wetlands, including those wetlands outside of the designated sites [73]. Initiatives from the U.S. government in 1989 and 1993 promoted the concept of no net loss of wetlands [73]. A more recent resolution from the Ramsar Convention noted the significance of some wetland types for C storage in relation to climate change [74]. Our study leverages satellite change analysis, combined with soil spatial databases, to identify changes in LULC related to wetlands (e.g., emergent herbaceous wetlands) and related soil types (e.g., Histosols) in GA. One way to evaluate if wetland areas had no net loss is to use satellite land cover data over time. Table 5 shows this analysis for the state of GA between 2001 and 2021, which indicates that there was a net gain in overall wetlands. However, further analysis reveals that there was a loss within Histosols (-28.2%) in wetland areas, which indicates the loss of C-rich soils and wetlands to development or other LULC conversions. Future analysis should use soil spatial data to help understand and disaggregate LULC analysis to quantify wetland change. Changes in wetland areas that contain Histosols can have a much larger impact on soil C emissions because of their much higher C contents compared to other soil types. Histosols account for only 1% of soils in GA, but these soils are a significant source ("hotspot") of SOC (17% of the total SOC of GA) and TSC (15% of the total TSC of GA) (Table 1). There was an overall reduction in Histosols in the state of GA of -28.2% between 2001 and 2021 (Table 5). In this period of time, 0.2 km^2 of Histosols were converted to developments, which resulted in the loss of 29.4M kg of TSC and corresponding SC-CO_2 in the amount of $5.0M USD. This type of analysis can aid the recent resolution from a Ramsar Convention to quantify wetland changes in C storage in relation to climate change. In addition, in the case of GA (USA), the counties that contain the Okefenokee National Wildlife Refuge [74] saw reductions between 2001 and 2021 in the amount of emergent herbaceous wetlands: Charlton (-9.5%), Clinch

(-3.2%), and Ware (-16.7%). Furthermore, this analysis can provide additional details concerning which soils were impacted at the county level and showed a reduction in Histosols within the emergent herbaceous wetlands LULC: Charlton (-10.49%), Clinch (-77.46%), and Ware (-17.48%). Histosols are C-rich soils commonly associated with wetlands. There were also large increases in development in these three counties: Charlton (+9.3%), Clinch (+4.1%), and Ware (+9.9%), which is likely related to the wetland and Histosols identified losses. Methods used in our study can also estimate the CO_2 release and SC-CO_2 associated with the development of Histosols in wetlands, which demonstrates why it is important to preserve wetlands and protect these C-rich soils. (Relevant to *Ramsar Convention on Wetlands*).

- The Revised World Soil Charter, which was endorsed by member states of the Food and Agriculture Organization (FAO), provides guidelines to ensure that "soils are managed sustainably and that degraded soils are rehabilitated or restored" [76]. This Charter calls for the limiting of soil degradation to preserve soil ecosystem services and support LDN. Our study shows that the state of GA has experienced an increase in both LD and soil degradation, as indicated in Table 4, with an overall +3.7% increase in LD between 2001 and 2021. Land and soil degradation occurred across all soil types during this study period primarily due to the rise in developments. The state of GA was not LD neutral, as indicated by the data in Table 4. This case study in GA is an important contribution to the ongoing research on climate governance [77], which should include soil governance as well [78–80]. (Relevant to *The Revised World Soil Charter*).

5. Conclusions

This study reveals that the current assumption that land and soil in GA only serve as a GHG emissions sink is unlikely to be true. Our results show that the state of GA (USA) experienced significant historical and present LD and soil degradation, which are (were) accompanied by GHG emissions with corresponding social costs, which are (were) not accounted for in the business activities responsible for these LD and soil degradation. This research most likely underestimated the GHG emissions because only areas subjected to developments were considered, and it was not possible to calculate emissions from agriculture and other non-development-related land cover changes. This is an important finding because past and future land use decisions have impacted and will impact soil-based GHG emissions. Also, the social costs calculated likely underestimate the true impact of soil-based emissions, both because agricultural and other land use conversions were not considered and because the social costs were calculated using standard methods that are based on fixed, non-market values. Study limitations include potential errors associated with satellite image classification. However, the resolution of remote sensing data used for this study (30 m) is higher than the land cover data used for most LD analyses. Also, our study compared land cover data over twenty years (2001 to 2021), which only documented the overall change and not the yearly change and other LULC cycles that occurred within those dates (e.g., forestry and agricultural practices). Future studies could leverage yearly or even more fine-scaled land cover data to more quickly identify critical changes and relationships.

Despite the limitations of our study, the results are useful for future climate planning efforts in GA (USA) because they provide a technique to monitor LULC changes and to account for soil-based emissions from the rapid development of GA. The analysis has several direct benefits for the planning process, including spatial analysis of soil type-specific emissions from developments, which also show development patterns, including areas that are susceptible to sea level rise leading to reverse climate change adaptation. Also, soil-based emissions were driven by developments linked to high-value economic development regions, including Metro Atlanta and Coastal economic development regions. Considering that the state of GA is in the process of using context-specific climate change planning, these soil-based emissions can also be examined using a context-specific framework, which could be used to develop regulatory or incentive-based methods to account for the past soil-based emissions and reduce future soil-based emissions tied to LD and soil degradation. Our study demonstrated how soil-based emissions can be understood in temporal, biophysical, economic, and social contexts, which can be used for existing or novel legal strategies to provide greater responsibility and accountability for emissions damages in the state of GA (USA) and worldwide. Potential solutions to negative externalities associated with damages from developments can include market-based payments in proportion to damages from LD and developments. To achieve success in GA, new policies must recognize GA's specific context. Accordingly, policies should be crafted to appeal not just to the broad worldwide virtues of addressing climate change but also to the benefits of such policies to GA itself.

Supplementary Materials: The following supporting information can be downloaded at: https://www.mdpi.com/article/10.3390/land13101669/s1, Table S1: Soil diversity (pedodiversity) is expressed as taxonomic diversity at the level of soil order in the state of Georgia (GA) (USA) [11]; Table S2: Distribution of soil carbon regulating ecosystem services in the state of Georgia (GA) (USA) by soil order; Table S3: An overview of the accounting framework used by this study (adapted from Groshans et al. (2019) [15]) for the state of Georgia (GA) (USA); Table S4: Area-normalized content (kg m^{-2}) and monetary values (\$ m^{-2}) of soil organic carbon (SOC), soil inorganic carbon (SIC), and total soil carbon (TSC = SOC + SIC) by soil order using data developed by Guo et al. (2006) [23] for the upper 2-m of soil and an avoided social cost of carbon (SC-CO_2) of \$46 per metric ton of CO_2, applicable for 2025 (2007 U.S. dollars with an average discount rate of 3% [2]); Table S5: Anthropogenic land degradation status and potential land for nature-based solutions in the state of Georgia (GA) in the contiguous United States of America (USA) in 2021. Percent changes in area from 2001 to 2021 are shown in parentheses. Reported values have been rounded; therefore, calculated sums and percentages may exhibit minor discrepancies. This table shows the anthropogenic land degradation status in 2021 but most likely does not account for historical anthropogenic land degradation as well as most of the inherent land degradation; Table S6: Developed land and potential for realized social costs of carbon (C) due to complete loss of total soil carbon (TSC) of developed land by soil order in the state of Georgia (GA) (USA) prior to and through 2021; Table S7: Increases in developed land and potential for realized social costs of carbon (C) due to complete loss of total soil carbon (TSC) of developed land by soil order in the state of Georgia (GA) (USA) from 2001 to 2021; Figure S1: High-resolution aerial photos showing examples of land classes (LULC) which were used to determine anthropogenically degraded land (LD) in the state of Georgia (GA) (USA) by assuming that degraded lands are represented by the land classes (LULC) for agriculture (hay/pasture, and cultivated crops), development (developed, open space; developed, low intensity; developed, medium intensity; developed, high intensity) and barren lands. Representative examples were located using a land cover map of Georgia for 2021 (based on data from the Multi-Resolution Land Characteristics Consortium (MRLC) with detailed descriptions of the land classes [20]); Figure S2: Maps of (a) anthropogenically degraded land in 2021 (km^2) and (b) more recent land degradation (km^2) between 2001 and 2021 in Georgia (GA) (USA). Land subject to anthropogenic degradation was calculated as a sum of developed land (developed, open space; developed, high intensity; developed, medium intensity; developed, low intensity), agriculture (cultivated crops, and hay/pasture), and barren land; Figure S3: Maps of (a) potential land area (km^2) for nature-based solutions (NBS) in 2021 and (b) change in potential land area (km^2) for nature-based solutions (NBS) between 2001 and 2021 in Georgia (GA) (USA). Potential land for NBS is limited to barren land, shrub/scrub, and herbaceous land cover classes, to provide potential land areas without impacting current land uses; Figure S4: Maps of (a) the proportion of potential nature-based solutions (NBS) land over the total land area (%) in 2021 and (b) the change in potential land (%) for nature-based solutions (NBS) between 2001 and 2021 in Georgia (GA) (USA). Potential land for NBS is limited to barren land, shrub/scrub, and herbaceous land cover classes, to provide potential land areas without impacting current land uses.

Author Contributions: Conceptualization, E.A.M.; methodology, E.A.M., M.A.S. and H.A.Z.; formal analysis, E.A.M. and D.G.N.; writing—original draft preparation, E.A.M.; writing—review and editing, E.A.M., C.J.P., M.A.S. and G.B.S.; visualization, H.A.Z., L.L. and Z.H. All authors have read and agreed to the published version of the manuscript.

Funding: This research received no external funding.

Data Availability Statement: Data are contained within the article and supplementary materials.

Acknowledgments: We would like to thank the reviewers for their constructive comments and suggestions.

Conflicts of Interest: The authors declare no conflict of interest.

Glossary

B	Billion
CCA	Climate Change Adaptation
CO_2	Carbon dioxide
EPA	Environmental Protection Agency
FAO	Food and Agriculture Organization
GA	Georgia
GHG	Greenhouse gases
LD	Land degradation
LDN	Land degradation neutrality
L&D	Loss and damage
LULC	Land use/land cover
M	Million
MRLC	Multi-Resolution Land Characteristics Consortium
N	North
NBS	Nature-based solutions
NLCD	National Land Cover Database
NOAA	National Oceanic and Atmospheric Administration
NRCS	Natural Resources Conservation Service
RCCA	Reverse climate change adaptation
SC-CO_2	Social cost of carbon emissions
SDGs	Sustainable Development Goals
SIC	Soil inorganic carbon
SOC	Soil organic carbon
SSURGO	Soil Survey Geographic Database
STATSGO	State Soil Geographic Database
TSC	Total soil carbon
UN	United Nations
UNCCD	United Nations Convention to Combat Desertification
USDA	United States Department of Agriculture
W	West

References

1. Merriam-Webster Dictionary. Available online: https://www.merriam-webster.com/dictionary/context (accessed on 15 April 2024).
2. EPA—United States Environmental Protection Agency. 2016. The Social Cost of Carbon. EPA Fact Sheet. Available online: https://19january2017snapshot.epa.gov/climatechange/social-cost-carbon_.html (accessed on 10 April 2024).
3. Georgetown Law. Georgetown Climate Center. State Adaptation Progress Tracker. Available online: https://www.georgetownclimate.org/adaptation/plans.html (accessed on 9 March 2024).
4. Newman, D.H.; Brooks, T.A.; Dangerfield Jr, C.W. Conservation use valuation and land protection in Georgia. *For. Policy Econ.* **2000**, *1*(3-4), 257–266. [CrossRef]
5. "Drawdown Georgia". Available online: https://www.drawdownga.org/carbon-reduction-visualizer/ (accessed on 15 April 2024).
6. Brown, M.A.; Beasley, B.; Atalay, F.; Cobb, K.M.; Dwiveldi, P.; Hubbs, J.; Iwaniek, D.M.; Mani, S.; Matisoff, D.; Mohan, J.E.; Mullen, J. Translating a global emission-reduction framework for subnational climate action: A case study from the state of Georgia. *Environ. Manag.* **2021**, *67*(2), 205–227. [CrossRef] [PubMed]
7. Brown, M.A.; Dwivedi, P.; Mani, S.; Matisoff, D.; Mohan, J.E.; Mullen, J.; Oxman, M.; Rodgers, M.; Simmons, R.; Beasley, B.; Polepeddi, L. A framework for localizing global climate solutions and their carbon reduction potential. *Proc. Natl. Acad. Sci. USA* **2021**, *118*(31), p.e2100008118. [CrossRef]

8. Jones, E. Georgia is making its first-ever climate change plan. The Current GA. 2023. Available online: https://thecurrentga.org/2023/07/11/georgia-is-making-its-first-ever-climate-change-plan/#:~:text=For%20the%20first%20time,%20Georgia%20is%20working%20on,state%20to%20make%20a%20climate%20pollution%20reduction%20plan (accessed on 10 April 2024).
9. Kann, D. Georgia is creating its first-ever climate plan. Here's what to expect. *Atlanta Journal-Constitution*. 2023. Available online: https://www.ajc.com/news/georgia-is-creating-its-first-ever-climate-plan-heres-what-to-expect/RKDANAY5DBCJLHYCUXVYCUDB5Q/#:~:text=Georgia%20is%20developing%20its%20first-ever%20climate%20plan%20with,will%20also%20be%20the%20first%20of%20its%20kind (accessed on 10 April 2024).
10. H.R.5376 - Inflation Reduction Act of 2022, Title VI, Subtitle A, Sec 60102. 2022. Available online: https://www.congress.gov/bill/117th-congress/house-bill/5376 (accessed on 30 August 2023).
11. Brady, N.C.; Weil, R.R. *The Nature and Properties of Soils*, 13th ed.; Pearson Education: London, UK, 2002.
12. Soil Science Society of America. n.d. USDA. Tifton – Georgia State Soil. Available online: https://www.soils4teachers.org/files/s4t/k12outreach/ga-state-soil-booklet.pdf (accessed on 16 April 2024).
13. Sutter, P.S. What gullies mean: Georgia's "Little Grand Canyon" and Southern environmental history. *J. South. Hist.* **2010**, *76*(3), 579–616.
14. Mikhailova, E.A.; Groshans, G.R.; Post, C.J.; Schlautman, M.A.; Post, G.C. Valuation of soil organic carbon stocks in the contiguous United States based on the avoided social cost of carbon emissions. *Resources* **2019**, *8*, 153. [CrossRef]
15. Groshans, G.R.; Mikhailova, E.A.; Post, C.J.; Schlautman, M.A.; Zhang, L. Determining the value of soil inorganic carbon stocks in the contiguous United States based on the avoided social cost of carbon emissions. *Resources* **2019**, *8*, 119. [CrossRef]
16. Mikhailova, E.A.; Groshans, G.R.; Post, C.J.; Schlautman, M.A.; Post, C.J. Valuation of total soil carbon stocks in the contiguous United States based on the avoided social cost of carbon emissions. *Resources* **2019**, *8*, 157. [CrossRef]
17. Soil Survey Staff, Natural Resources Conservation Service, United States Department of Agriculture. n.d.a. *Soil Survey Geographic (SSURGO) Database*. Available online: https://nrcs.app.box.com/v/soils (accessed on 16 August 2024).
18. The United States Census Bureau. TIGER/Line Boundary Shapefiles. 2018. Available online: https://www.census.gov/geographies/mapping-files/time-series/geo/tiger-line-file.2018.html (accessed on 8 August 2024).
19. Soil Survey Staff, Natural Resources Conservation Service, United States Department of Agriculture. Photos of Soil Orders. Available online: https://www.nrcs.usda.gov/wps/portal/nrcs/detail/soils/edu/?cid=nrcs142p2_053588 (accessed on 9 April 2024).
20. Multi-Resolution Land Characteristics Consortium—MRLC. Available online: https://www.mrlc.gov/ (accessed on 10 April 2024).
21. Sharma, R.; Levi, M.R.; Ricker, M.C.; Thompson, A.; King, E.G.; Robertson, K. 2024. Scaling of soil organic carbon in space and time in the Southern Coastal Plain, USA. *Sci. Total Environ.* **2024**, *933*, 173060. [CrossRef]
22. Soil Survey Staff, Natural Resources Conservation Service, United States Department of Agriculture. U.S. General Soil Map (STATSGO2). Available online: https://sdmdataaccess.sc.egov.usda.gov (accessed on 9 April 2024).
23. Guo, Y.; Amundson, R.; Gong, P.; Yu, Q. Quantity and spatial variability of soil carbon in the conterminous United States. *Soil Sci. Soc. Am. J.* **2006**, *70*, 590–600. [CrossRef]
24. ESRI (Environmental Systems Research Institute). ArcGIS Pro 2.6. Available online: https://pro.arcgis.com/en/pro-app/2.6/get-started/whats-new-in-arcgis-pro.htm (accessed on 1 July 2024).
25. Daniels, R.B. 1987. Soil erosion and degradation in the southern Piedmont of the USA. In *Land Transformation in Agriculture*; Wolman, M.G., Fournier, F.G.A., Eds.; John Wiley & Sons Ltd., Hoboken, New Jersey, USA, pp. 407–428.
26. *U.S. Bureau of the Census. Statistical Abstract of the United States: 1991*; U.S. Bureau of the Census: Washington, DC, USA, 1991; p. 201. Available online: https://www.census.gov/library/publications/1991/compendia/statab/111ed.html (accessed on 4 July 2024).
27. Pinder, J.E.; Rea, T.E.; Funsch, D.E. Deforestation, reforestation and forest fragmentation on the upper coastal plain of South Carolina and Georgia. *Am. Midl. Nat.* **1999**, *142*(2), 213–228. [CrossRef]
28. Miller, M.D. The impacts of Atlanta's urban sprawl on forest cover and fragmentation. *Appl. Geogr.* **2012**, *34*, 171–179. [CrossRef]
29. Obata, S.; Bettinger, P.; Cieszewski, C.J.; Lowe III, R.C. Mapping forest disturbances between 1987–2016 using all available time series Landsat TM/ETM+ imagery: Developing a reliable methodology for Georgia, United States. *Forests* **2020**, *11*(3), 335. [CrossRef]
30. Rennert, K.; Errickson, F.; Prest, B.C.; Rennels, L.; Newell, R.G.; Pizer, W.; Kingdon, C.; Wingenroth, J.; Cooke, R.; Parthum, B.; Smith, D. Comprehensive evidence implies a higher social cost of CO_2. *Nature* **2022**, *610*(7933), 687–692. [CrossRef] [PubMed]
31. Varadarajan, R. Market exchanges, negative externalities and sustainability. *J. Macromarketing* **2020**, *40*(3), 309–318. [CrossRef]
32. Evans, J.M.; Gambill, J.; McDowell, R.J.; Prichard, P.W.; Hopkinson, C.S. 2016. Tybee Island Sea Level Rise Adaptation Plan. Georgia Sea Grant. Available online: https://resilienttybee.com/wp-content/uploads/2022/05/Tybee_SLR_Adaptation_Plan.pdf (accessed on 1 October 2024).
33. Gambill, J.; Russell, M.; Spratt, K.; Whitehead, J.; Alfonso, M.; Hopkinson, C.; Evans, J.M. St. Mary's Flood Resiliency Project. Georgia Sea Grant. 2017. Available online: https://cms3.revize.com/revize/stmarysnew/docs/StMarys_Plan_03_06_2017_JME.pdf (accessed on 1 October 2024).
34. Davis, G.H. Land subsidence and sea level rise on the Atlantic Coastal Plain of the United States. *Environ. Geol. Water Sci.* **1987**, *10*, 67–80. [CrossRef]

35. Binita, K.C.; Shepherd, J.M.; Gaither, C.J. Climate change vulnerability assessment in Georgia. *Appl. Geogr.* **2015**, *62*, 62–74. [CrossRef]
36. Burkett, V.; et al. 2001. Potential Consequences of Climate Variability and Change for the Southeastern United States. In Climate change Impacts in the United States: Potential Consequences of Climate Variability and Change, edited by National Assessment Team, U.S. Global Research Program, 137–64. Cambridge: Cambridge University Press.
37. Beck, J.; Lin, M. The impact of sea level rise on real estate prices in coastal Georgia. *Rev. Reg. Stud.* **2020**, *50*(1), 43–52. [CrossRef]
38. Mikhailova, E.A.; Lin, L.; Hao, Z.; Zurqani, H.A.; Post, C.J.; Schlautman, M.A.; Post, G.C.; Shepherd, G.B.; Kolarik, S.J. Enhancing the definitions of climate-change loss and damage based on land conversion in Florida, USA. *Urban Science* **2023**, *7*(2), 40. [CrossRef]
39. Ferguson, A.P.; Ashley, W.S. Spatiotemporal analysis of residential flood exposure in the Atlanta, Georgia metropolitan area. *Nat. Hazards* **2017**, *87*, 989–1016. [CrossRef]
40. National Oceanic and Atmospheric Administration (NOAA). Climate.gov. Available online: https://www.climate.gov/maps-data (accessed on 16 August 2024).
41. Rules and Regulations of the State of Georgia. Subject 391-1-6 Georgia Conservation Tax Credit Program. Available online: https://rules.sos.ga.gov/gac/391-1-6?urlRedirected=yes&data=admin&lookingfor=391-1-6 (accessed on 16 August 2024).
42. Georgia Soil & Water Conservation Commission. 2013. Best Management Practices for Georgia Agriculture. Second edition. Available online: https://gaswcc.georgia.gov/sites/gaswcc.georgia.gov/files/2013AgManual_Intro+Ch1.pdf (accessed on 16 August 2024).
43. Georgia (USA) Department of Revenue. Conservation Use Land Values. Conservation Use Valuation Assessment (CUVA). Available online: https://dor.georgia.gov/local-government-services/digest-compliance/conservation-use-land-values (accessed on 16 August 2024).
44. Chhetri, S.G.; Li, Y.; Hepinstall-Cymerman, J.; Siry, J.; Gordon, J. Factors influencing forestland enrollment in Georgia's preferential property tax programs. *For. Policy Econ.* **2024**, *160*, 103126. [CrossRef]
45. Tebaldi, C.; Adams-Smith, D.; Heller, N. 2012. The Heat is On: U.S. Temperature Trends. Climate Central. Available online: https://ccimgs.s3.amazonaws.com/HeatIsOnReport.pdf (accessed on 16 August 2024).
46. National Oceanic and Atmospheric Administration (NOAA) National Centers for Environmental Information. 2022. Georgia State Climate Summary 2022. NOAA Technical Report NESDIS 150-GA. NOAA/NESDIS, Silver Spring, MD. Available online: https://statesummaries.ncics.org/downloads/Georgia-StateClimateSummary2022.pdf (accessed on 16 August 2024).
47. EPA – United States Environmental Protection Agency. What climate change means for Georgia. EPA 430-F-16-012. August 2016. Available online: https://19january2017snapshot.epa.gov/sites/production/files/2016-09/documents/climate-change-ga.pdf/ (accessed 16 August 2024).
48. Zhou, Y.; Shepherd, J.M. Atlanta's urban heat island under extreme heat conditions and potential mitigation strategies. *Nat. Hazards* **2010**, *52*, 639–668. [CrossRef]
49. Khan, M.R. Polluter-pays-principle: The cardinal instrument for addressing climate change. *Laws* **2015**, *4*, 638–653. [CrossRef]
50. Daniels, R.B. 1987. Soil erosion and degradation in the southern Piedmont of the USA. Land transformation in agriculture, pp. 407–428.
51. Alderman, D.H.; Alderman, D.G.S. Kudzu: A tale of two vines. *Southern Cultures* **2001**, *7*(3), 49–64. [CrossRef]
52. Ebenezer, P.A.; Manohar, S. Land use/land cover change classification and prediction using deep learning approaches. *Signal, Image Video P.* **2024**, *18*(1), 223–232. [CrossRef]
53. Providence Canyon State Park, Stewart County, Georgia (USA) (https://gastateparks.org/ProvidenceCanyon) (accessed on 16 August 2024).
54. Library of Congress (control number: 2017775702). Available online: https://www.loc.gov/item/2017775702/ (accessed on 16 August 2024).
55. National Agriculture Imagery Program (NAIP) aerial photography for 2023. Available online: https://naip-usdaonline.hub.arcgis.com/ (accessed on 16 August 2024).
56. Selznick, P. 'Law in context' revisited. *J. Law Soc.* **2003**, *30*(2), 177–186. [CrossRef]
57. Havemann, P. Law in context-taking context seriously. *Waikato L. Rev.* **1995**, *3*, 137.
58. Osofsky, H.M. Climate change legislation in context. *Nw. L. Rev. Colloquy* **2007**, *102*, 245.
59. Wilson, R. Adaptation is context specific. *Nat. Clim. Change* **2022**, *12*, 8–14. [CrossRef]
60. Noll, B.; Filatova, T.; Need, A.; Taberna, A. Contextualizing cross-national patterns in household climate change adaptation. *Nat. Clim. Change* **2022**, *12*, 30–35. [CrossRef]
61. Conservation International. Trends.Earth. 2022. Available online at: http://trends.earth (accessed on 16 August 2024).
62. Schillaci, C.; Jones, A.; Vieira, D.; Munafò, M.; Montanarella, L. Evaluation of the United Nations Sustainable Development Goal 15.3. 1 indicator of land degradation in the European Union. *Land Degrad. Dev.* **2023**, *34*(1), 250–268. [CrossRef]
63. Zimba, H.M.; Banda, K.E.; Mbewe, S.; Nyambe, I.A. Integrated use of the CA–Markov model and the Trends. Earth module to enhance the assessment of land cover degradation. *Environ. Syst. Res.* **2024**, *13*(1), 25. [CrossRef]
64. Bradford, M.A.; Eash, L.; Polussa, A.; Jevon, F.V.; Kuebbing, S.E.; Hammac, W.A.; Rosenzweig, S.; Oldfield, E.E. Testing the feasibility of quantifying change in agricultural soil carbon stocks through empirical sampling. *Geoderma* **2023**, *440*, 116719. [CrossRef]

65. Gianoli, F.; Weynants, M.; Cherlet, M. Land degradation in the European Union—Where does the evidence converge? *Land Degrad. Dev.* **2023**, *34*(8), 2256–2275. [CrossRef]
66. Asumadu-Sarkodie, S.; Owusu, P.A. The causal nexus between carbon dioxide emissions and agricultural ecosystem—an econometric approach. *Environ. Sci. Pollut. Res.* **2017**, *24*, 1608–1618. [CrossRef] [PubMed]
67. United Nations (UN). Transforming Our World: The 2030 Agenda for Sustainable Development; The Resolution Adopted by the General Assembly on 25 September 2015; United Nations: New York, NY, USA, 2015; 35p.
68. United Nations Convention to Combat Desertification. Available online: https://www.unccd.int/convention/overview (accessed on 8 August 2024).
69. United Nations (UN) Convention to Combat Desertification. Decision 3/COP.12 (COP 12, Ankara, 2015). Integration of the Sustainable Development Goals and Target into the Implementation of the United Nations Convention to Combat Desertification and the Intergovernmental Working Group Report on Land Degradation Neutrality. Available online: https://www.unccd.int/official-documentscop-12-ankara-2015/3cop12 (accessed on 16 August 2024).
70. UN. *Convention on Biological Diversity*. 1992. Treaty Collection. Available online: https://www.cbd.int/doc/legal/cbd-en.pdf (accessed on 16 August 2024).
71. UN. *Convention on Biological Diversity*. 2022. Kunming-Montreal Global Biodiversity Framework. Available online: https://www.cbd.int/article/cop15-final-text-kunming-montreal-gbf-221222 (accessed on 16 August 2024).
72. Ramsar Convention on Wetlands. Available online: https://www.ramsar.org/ (accessed on 16 August 2024).
73. Podolsky, M.J. US wetlands policy, legislation, and case law as applied to the wise use concept of the Ramsar convention. *Case W. Res. L. Rev.* **2001**, *52*, 627.
74. Ramsar COP13 Resolution XIII.13. 13th Meeting of the Conference of the Contracting Parties to the Ramsar Convention on Wetlands. "Wetlands for a Sustainable Urban Future", Dubai, United Arab Emirates, 21-29 October 2018. Resolution XIII.13. Restoration of degraded peatlands to mitigate and adapt to climate change and enhance biodiversity and disaster risk reduction. Available online: https://www.ramsar.org/sites/default/files/documents/library/xiii.13_peatland_restoration_e.pdf (accessed on 16 August 2024).
75. Ramsar Sites Information Service. Annotated List of Wetlands of International Importance: United States of America. Available online: https://rsis.ramsar.org/sites/default/files/rsiswp_search/exports/Ramsar-Sites-annotated-summary-United-States-of-America.pdf (accessed on 16 August 2024).
76. Food and Agriculture Organization (FAO). Revised World Soil Charter. Available online: https://www.fao.org/fileadmin/user_upload/GSP/docs/ITPS_Pillars/annexVII_WSC.pdf (accessed on 16 August 2024).
77. Sridhar, A.; Dubash, N.K.; Averchenkova, A.; Higham, C.; Rumble, O.; Gilder, A. Climate Governance Functions: Towards Context-specific Climate Laws. Center for Policy Research, The Grantham Research Institute at London School of Economics and Climate Legal. October 14, 2022. Available online: https://cprindia.org/wp-content/uploads/2022/10/Climate-Governance-Functions_17_Nov_22.pdf (accessed on 10 July 2024).
78. Ruppel, O.C. Overview of international soil law. *Soil Security* **2022**, *6*, 100056. [CrossRef]
79. Fox, S.J. Soil governance and private property. *Utah Law Rev.* **2024**, *1*, 1–57. [CrossRef]
80. Bodle, R. International soil governance. *Soil Security* **2022**, *6*, 100037. [CrossRef]

land

MDPI

Article

Spatiotemporal Variation and Driving Factors of Carbon Sequestration Rate in Terrestrial Ecosystems of Ningxia, China

Yi Zhang [1], Chunxiao Cheng [1], Zhihui Wang [1,2,*], Hongxin Hai [1,3] and Lulu Miao [1,4]

1 Key Laboratory of Soil and Water Conservation on the Loess Plateau of Ministry of Water Resources, Yellow River Institute of Hydraulic Research, Yellow River Conservancy Commission, Zhengzhou 450003, China; zy1607299084@163.com (Y.Z.); ccx02@126.com (C.C.); z202210151100@stu.ncwu.edu.cn (H.H.); 212204020083@home.hpu.edu.cn (L.M.)

2 School of Social and Environmental Sustainability, University of Glasgow, Dumfries DG1 4UL, UK

3 College of Surveying and Geo-Informatics, North China University of Water Resources and Electric Power, Zhengzhou 450046, China

4 School of Surveying and Land Information Engineering, Henan Polytechnic University, Jiaozuo 454003, China

* Correspondence: wzh8588@aliyun.com; Tel.: +86-13353806578

Abstract: This study investigates the spatiotemporal variation characteristics and influencing factors of an ecosystem's carbon sequestration rate (CSR) in the Ningxia region from 2001 to 2023, providing scientific evidence for assessing the regional carbon sequestration capacity and formulating carbon neutrality policies. Based on ground observation data and multimodal datasets, the optimal machine learning model (EXT) was used to invert a 30 m high-resolution vegetation and soil carbon density dataset for Ningxia from 2000 to 2023. Annual variation analysis and geographical detector methods were employed to assess the spatiotemporal distribution characteristics of the CSR from 2001 to 2023 and identify the primary influencing factors. The results show that from 2001 to 2023, the CSR of the Ningxia ecosystem exhibits a spatial distribution pattern characterized by higher values in the south and lower values in the north, with a mean value of 21.95 $gC{\cdot}m^{-2}$, and an overall fluctuating increasing trend, with an annual growth rate of 0.53 $gC{\cdot}m^{-2}\ a^{-1}$. Significant differences in the CSR exist across different ecological regions. In terms of land use types, the ranking of carbon sequestration capacity is forest > farmland > grassland > barren, while the ranking of the carbon sequestration enhancement capacity is farmland > forest > grassland > barren. Among land use change types, the carbon sequestration enhancement capacity significantly increased when grassland was converted to forest or shrubland, farmland to forest–grassland, and bare land to forest–grassland, with increases of 42.9%, 9.2%, and 34.6%, respectively. The NDVI is the primary driver of CSR spatiotemporal variation, while the interaction between the Enhanced Vegetation Index (EVI) and soil bulk density has a more significant explanatory power for CSR spatial differentiation. This study shows that ecological restoration projects, such as the conversion of cropland to forest (or grassland) and protective farmland measures, play a significant role in enhancing the carbon sequestration capacity in Ningxia.

Keywords: carbon sequestration rate; spatio-temporal; Ningxia

1. Introduction

The carbon sequestration rate (CSR) is an important indicator for measuring the ecosystem carbon sink function [1], which has attracted increasing attention, particularly in the context of intensifying global climate change. It is estimated that global terrestrial ecosystems sequester approximately 2.4 billion tons of carbon per year, absorbing around

30% of the CO_2 emissions caused by human activities [2], thus playing a crucial role in combating climate change and maintaining ecosystem health. However, the differences in natural geographic conditions and human activities lead to significant spatial heterogeneity in the regional CSR, particularly in Ningxia, a semi-arid region with a fragile ecological environment, low vegetation cover, and a high susceptibility to climate conditions and human activities [3]. Therefore, CSR estimates based on fixed values are unable to accurately reflect the carbon sequestration status in different regions, thereby increasing the uncertainty of the results.

In recent years, with the implementation of a series of ecological restoration projects, vegetation conditions in Ningxia have improved; however, systematic research on the spatiotemporal variation characteristics of its CSR is still lacking [4]. Therefore, an in-depth study of the CSR in the Ningxia region will provide scientific support for more accurate carbon sequestration accounting and the implementation of ecological protection and restoration projects. Currently, methods for estimating CSR mainly include traditional approaches based on ground-based measurements and process models or atmospheric inversion methods based on remote sensing data [5]. Traditional methods primarily rely on long-term observational data from ground-based sampling points. While they can provide relatively accurate CSR information, their application on large scales is limited by the number and spatial distribution of the sampling points [6]. With technological advancements, methods such as the eddy covariance technique [7] and ecosystem process models (e.g., CENTURY and Biome-BGC models) [8] have gradually become important tools for assessing the CSR. In recent years, the application of remote sensing technology has provided new perspectives for the large-scale monitoring and assessment of ecosystem carbon stocks [9,10]. For example, Feng et al. (2013) used remote sensing technology to quantitatively assess carbon sink changes from 2000 to 2008 in the Grain for Green project on the Loess Plateau, finding that the total carbon sink in the region was approximately 96.1 Tg [11].

Significant progress has been made in CSR research both domestically and internationally. For example, Lin demonstrated that forests have a higher CSR potential due to their rich biomass [12]. Mitsch, by comparing the CSR of different types of wetlands (e.g., marshes, peatlands, and mangroves), concluded that wetlands, particularly peatlands, have a high carbon sequestration capacity [13]. Lal and Aubrey, among others, have demonstrated the importance of soil carbon sequestration in agricultural lands for global climate change and food security [14,15]. Building on this, CSR research has gradually focused on the synergistic effects of various influencing factors. For instance, Mekonnen et al. demonstrated that climate change is a key factor affecting the CSR [16]. Moisa et al. indicated that human activities, such as land use changes and urbanization, significantly impact the CSR, with large-scale land reclamation and vegetation destruction reducing the carbon storage capacity of ecosystems [17]. Topographic factors (elevation, slope, and aspect), soil factors (the soil organic matter content, fertility, and pH), and vegetation types have all been shown to significantly affect the CSR [18–20]. Bu et al.'s study indicated that from 2000 to 2015, Ningxia's wetland restoration project led to an increase of 204,900 tons of carbon storage [2]. Although previous studies have provided some theoretical support for Ningxia's carbon sequestration capacity, most existing research focuses on the effects of single ecosystem types or specific factors, lacking a spatiotemporal dynamic analysis of the CSR across the entire terrestrial ecosystem at the regional scale under the influence of multiple factors. Research has mainly focused on regional-scale studies, with less attention given to the spatial heterogeneity of the CSR within the region, and limited studies on the current CSR status in Ningxia.

Therefore, this study proposes to utilize extensive ground-based data, combined with long-term remote sensing data, topographic and soil data, and machine learning techniques to generate vegetation and soil carbon density datasets for the period 2000–2023. The dataset will be further used to derive a 30 m high-resolution CSR dataset for 2001–2023, which will be analyzed for spatial heterogeneity and dynamic characteristics across different ecological regions of Ningxia. This will reveal the evolution of the CSR in Ningxia and provide scientific support for achieving the region's dual carbon goals. At the same time, this study will assess the impact of different land use types on the regional CSR, providing decision-making references for optimizing the land use structure and the rational layout of vegetation restoration projects. This will contribute to the promotion of ecological environment construction in Ningxia and support the achievement of carbon neutrality goals, offering practical value and scientific support for regional sustainable development.

2. Materials and Methods

2.1. Study Area

The Ningxia Hui Autonomous Region (104°17′–107°39′ E, 35°14′–39°23′ N) is located in the upper reaches of the Yellow River in northwestern China, covering an area of approximately 66,400 square km(Figure 1). The region has a continental semi-humid, semi-arid climate, characterized by dryness and low precipitation. The annual average temperature ranges from 6.9 °C to 11.5 °C, increasing from south to north. Precipitation is mainly concentrated in the summer, with annual rainfall decreasing from south to north. The southern part receives 350–600 mm of rainfall, while the northern part receives only about 200 mm. The terrain is complex and diverse, with elevations ranging from 1000 to 3500 m. Based on topography, landforms, and soil erosion characteristics, the region can be divided into seven typical ecological zones: Helan Mountain water erosion zone, Potential wind erosion area of Yinchuan Plain zone, Hill platform arid grassland geomantic erosion staggered zone, Liupan Mountain water erosion zone, Water erosion area of Loess hilly and gully residual tableland zone, Loess hilly and gully water erosion zone, and Loess hilly and gully water wind erosion staggered zone. Due to severe climate aridity, soil erosion, and desertification, which have led to significant vegetation degradation, vegetation restoration and the enhancement of the region's carbon sink function are of great significance for achieving ecological restoration and carbon neutrality goals in Ningxia.

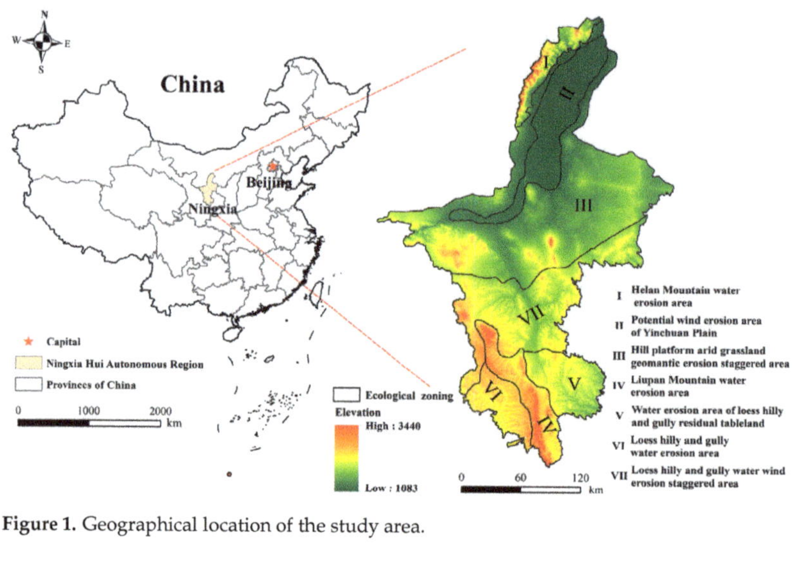

Figure 1. Geographical location of the study area.

2.2. Data Sources and Preprocessing

2.2.1. Field Survey Data

In 2019, Xu [21] compiled a carbon density dataset for China's terrestrial ecosystems, sourced from 1,036 relevant studies and the author's field survey plot data. The data were collected between 2004 and 2014 and cover major ecosystems such as forests, grasslands, farmlands, wetlands, and shrub. The shared dataset includes vegetation total biomass (aboveground and belowground) carbon density (VTCD) data and soil organic carbon density (SOCD) data for depths of 0–100 cm. This study extracted all available plot data within Ningxia Hui Autonomous Region from this dataset, including 92 vegetation total biomass density plots and 67 soil carbon density plots(Figure 2). From 28 November to 4 December 2023, a VTCD and SOCD survey and field sampling were conducted in Ningxia Autonomous Region. Considering factors such as the proportion of area under restoration measures, accessibility, plot spacing (5–10 km), and a minimum restoration area of 10,000 m^2, 200 survey plots (30 m × 30 m) were selected. Within each plot, three sub-plots with high, medium, and low coverages (50 cm × 50 cm) were selected. All herbaceous vegetation aboveground and belowground biomass within each sub-plot was harvested, dried, weighed, and used to calculate VTCD. Information on tree species and total number of each tree and shrub species was recorded. Diameter at breast height (DBH), height, and crown width were measured using a tape measure and laser hypsometer. The allometric equations for tree and shrub species developed by previous studies [22] were used to estimate aboveground and belowground biomass, which were multiplied by an empirical coefficient to calculate VTCD. Three sampling points were evenly selected within each plot. Soil samples were collected to a depth of 100 cm using a soil auger, and the samples were sent to the laboratory for analysis. Soil organic carbon content was measured and converted into SOCD density.

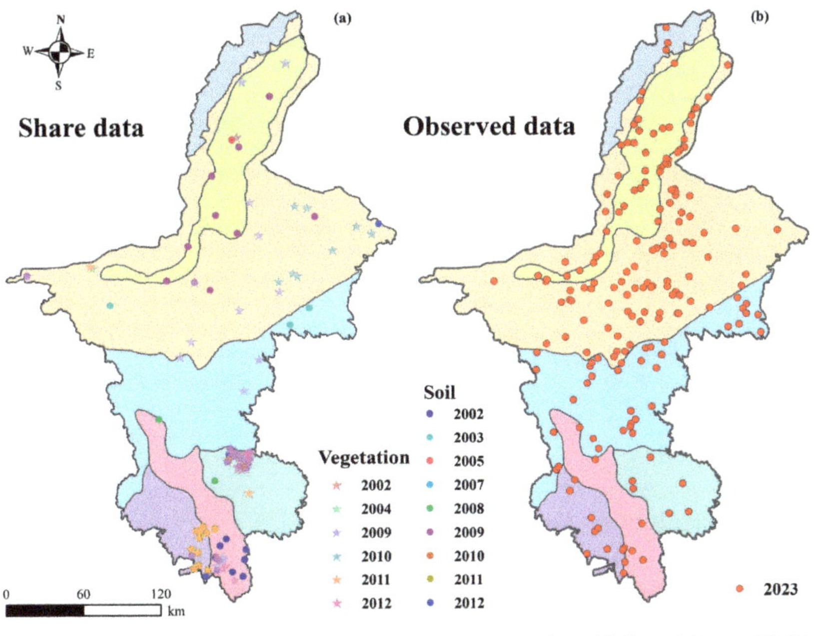

Figure 2. Spatial distribution of carbon density measurement data: (**a**) Shared dataset SOCD and VTCD Field Survey data; (**b**) Vegetation and soil manual measurement sampling points.

2.2.2. Multimodal Datasets

The detailed dataset is shown in Table 1. In this study, the Landsat SR (https://developers.google.com/earth-engine/datasets/catalog/landsat (accessed on 5 May 2024)) product was used to obtain surface reflectance, covering six bands: blue, green, red, near-infrared, shortwave infrared 1, and shortwave infrared 2. Based on these band data, several vegetation indices were calculated, including the Normalized Difference Vegetation Index (NDVI) [23], Enhanced Vegetation Index (EVI) [24], Normalized Difference Moisture Index (NDMI) [25], and Ratio Vegetation Index (RVI) by using GIS (ArcMap 10.8) software. [26]. For meteorological data, rainfall data were provided by the NCEP product from the National Centers for Environmental Prediction (https://www.noaa.gov/jetstream/ncep (accessed on 10 May 2024)), while temperature data were obtained from the ERA5 product of the European Center for Medium-Range Weather Forecasts (https://www.ecmwf.int/en/forecasts/dataset/ecmwf-reanalysis-v5 (accessed on 9 May 2024)). For topographic features, the AW3D30 (https://www.eorc.jaxa.jp/ALOS/en/dataset/aw3d30/aw3d30_e.htm (accessed on 6 May 2024)) product was used to obtain the Digital Elevation Model (DEM), from which slope aspect, slope, topographic index, and soil erosion slope-length factor were extracted. The Terrace dataset (https://figshare.com/articles/dataset/unet_T_d_1_89_LP_tif/17121941 (accessed on 11 May 2024)), based on the distribution map of the Loess Plateau (TDMLP), was developed by the Institute of Geography, Chinese Academy of Sciences (1.89 m) [27]. Additionally, soil type and property data were obtained from the SoilGrids 250m product (https://soilgrids.org/ (accessed on 10 May 2024)). Land use data were sourced from the Yellow River Conservancy Commission's YRCC_LPLC product (https://doi.org/10.5281/zenodo.10225564 (accessed on 11 May 2024)) [28].

Table 1. Detailed information about raster dataset used in this study.

Data Subcategory	Data Name	Product	Spatial Resolution	Temporal Duration
Topography	DEM	AW3D30	30 m	2010
	Slope	-	30 m	2010
	Aspect	-	30 m	2010
	Terrace	TDMLP	1.89 m	2021
	LS	-	30 m	2010
	Topographic index	-	30 m	2010
Climate	Precipitation	NCEP	0.25°	2000–2023
	Temperature	ERA5	0.25°	2000–2023
Vegetation index	Surface reflectance	Landsat SR	30 m	2000–2023
	NDVI	-	30 m	2000–2023
	EVI	-	30 m	2000–2023
	NDMI	-	30 m	2000–2023
	RVI	-	30 m	2000–2023
Soil properties	Bulk density	SoilGrids	250 m	2021
	Clay content	SoilGrids	250 m	2021
	Sand	SoilGrids	250 m	2021
	Slit	SoilGrids	250 m	2021

2.3. Remote Sensing Inversion Methods for Carbon Density

Taking into account the influence of climate, soil properties, topography, terraces, and vegetation on the spatial–temporal distribution of carbon density [29,30], a total of 16 features were used to estimate VTCD and SOCD, including DEM, slope, aspect, temperature, precipitation, terrace, LS, topographic index, bulk density, clay content, sand, silt, NDVI, EVI, NDMI, and RVI. This study used six machine learning algorithms—Random Forest (RF) [31], Artificial Neural Network (ANN) [32], Support Vector Machine

(SVM) [33], Light Gradient-Boosting Machine (Light GBM) [34], eXtreme Gradient Boosting (XGBoost) [35], and Extra Tree Regressor (EXT)—to construct vegetation and soil carbon density inversion models. Various spread values were tested until the mean square error or the number of neurons met the required criteria, optimizing the network parameters to determine the best carbon storage model for each algorithm.

2.4. Carbon Sequestration Rate Calculation

The carbon sequestration rate refers to the change in carbon storage over a unit of time, typically calculated on an annual basis. This study uses the "next year minus previous year" method, based on the 2000–2023 carbon storage dataset, to calculate the carbon sequestration rate in Ningxia from 2001 to 2023. The specific calculation formula is as follows:

$$CSR = \frac{C_{t+1} - C_t}{(t+1) - t} \quad (1)$$

where C_{t+1} and C_t represent the carbon storage in the following year and the previous year, respectively. The CSR is the change in carbon storage over a unit of time (year).

2.5. Trend Slope Analysis

Combining the Ningxia CSR data from 2000 to 2023, the trend slope analysis method was used to calculate the long-term trend of carbon sequestration rate changes in Ningxia. The formula for calculating the linear slope is as follows:

$$Slope = \frac{n \times \sum_{i=1}^{n} i \times X_i - \sum_{i=1}^{n} i \sum_{i=1}^{n} X_i}{n \times \sum_{i=1}^{n} i^2 - (\sum_{i=1}^{n} i)^2} \quad (2)$$

where X_i is the carbon sequestration rate for year i and n is the total number of years. When the *slope* > 0, the CSR shows an increasing trend; when the *slope* < 0, the CSR shows a decreasing trend.

2.6. Geographical Detector

This study employed factor detectors and interaction detectors to reveal the relationships between the carbon sequestration rate of the Ningxia ecosystem and its key driving factors [36]. The factor detector measures the explanatory power of each influencing factor on the spatial differentiation of ecosystem carbon storage by calculating the q-value. The formula is as follows:

$$q = 1 - \frac{\sum_{c=1}^{L} N_c D_c}{ND} \quad (3)$$

where q represents the explanatory power of factor X; $C = 1, 2, 3 \ldots L$ are the partitions of explanatory variable X, N_c and N represent the number of sub-zones and the total number of units in the whole region, respectively, and D_c and D represent the variance of Y values in the sub-zones and the whole region, respectively. The q-value ranges from [−1,1], with higher values indicating stronger explanatory power for the spatial differentiation of CSR.

The interaction detector is used to analyze whether there is an interaction between different driving factors and assess how this interaction affects the spatial differentiation of CSR. Specifically, the interaction detector compares the q-values of individual factors X_1 and X_2 with their interaction q-value $q(X_1 \cap X_2)$ to determine the relationship between the driving factors. It indicates that the combined effect of the two factors on CSR is greater than the sum of their individual effects, implying a synergistic effect. This study used the geographical detector to calculate the explanatory power of each factor and their interactions on the long-term average CSR, thereby analyzing the main influencing factors of the spatial changes in CSR in Ningxia from 2000 to 2023.

3. Results

3.1. Carbon Density Inversion Result Accuracy Verification

The field observation dataset was randomly divided into a training set (2/3) and a test set (1/3), and prediction models for the VTCD and SOCD were constructed. The best carbon density model was selected by comparing the performance of different machine learning algorithms in the carbon density prediction. The results for vegetation and soil training sets are shown in Figure 3a,b, respectively. All models exhibited a good fitting performance, with the coefficient of determination (R^2) for VTCD prediction generally above 0.8, and the root mean square error (RMSE) controlled at a low level. The comparison results for the test set are shown in Figure 3c,d. In the VTCD test set (Figure 3c), the Extreme Random Tree (EXT) model exhibited the best prediction performance (R^2 = 0.71, RMSE = 0.56 Kg C m^{-2}, MSE = 0.31 Kg C m^{-2}), followed by the XGBoost model, while the Artificial Neural Network (ANN) showed relatively lower prediction accuracy. For the SOCD test set (Figure 3d), the EXT model also performed excellently (R^2 = 0.70, RMSE = 1.61 Kg C m^{-2}, MSE = 2.6 Kg C m^{-2}), followed by the Support Vector Regression (SVR) model, while the XGBoost model showed the poorest prediction performance. Considering both the model performance indicators and prediction stability, the EXT model showed the best performance for both the VTCD and SOCD, thus it was chosen as the optimal carbon density model for Ningxia.

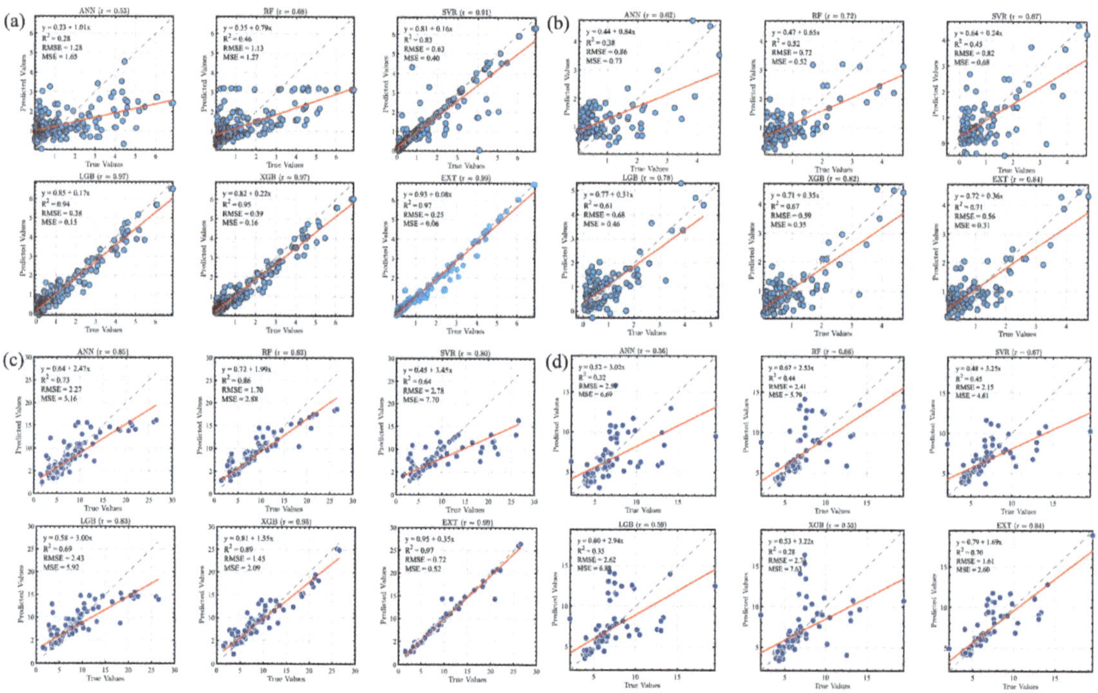

Figure 3. Accuracy evaluation of VTCD (**a**,**c**) and SOCD (**b**,**d**) inversion results based on the validation sample dataset.

Based on the accuracy metrics of the validation sample dataset, the EXT model was selected to perform the pixel-by-pixel inversion of the vegetation and soil carbon density across Ningxia, generating the 2000–2023 vegetation and soil carbon density dataset. The final inversion dataset was re-extracted using all sample data and compared with measured values to perform accuracy validation (Figure 4). From the scatter plot distribution, a good linear relationship between predicted and measured values is observed, with the points closely aligned to the 1:1 line, indicating high reliability of the model's predictions.

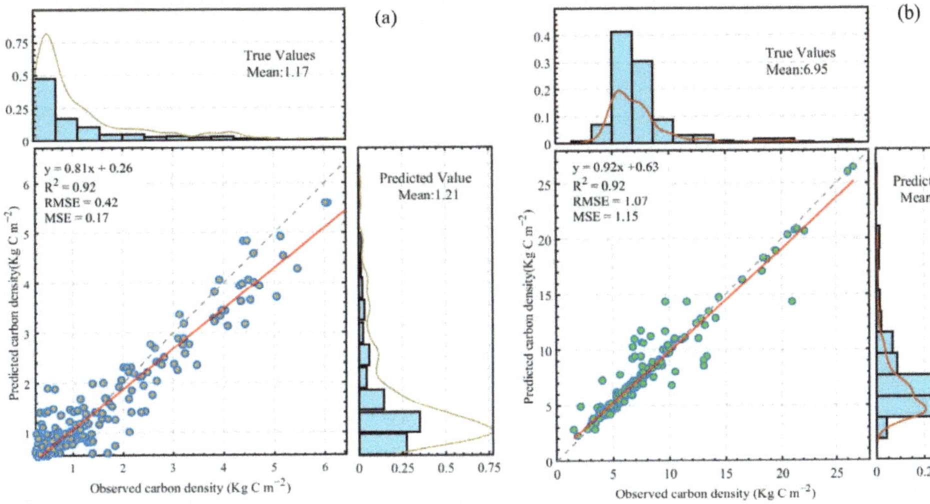

Figure 4. Accuracy validation of the EXT model prediction results based on the entire sample dataset: (**a**) VTCD and (**b**) SOCD.

The EXT model was used to perform the pixel-by-pixel inversion of the vegetation and soil carbon density in the Ningxia region, generating a carbon density dataset from 2000 to 2023. The final inversion dataset was then re-extracted using all sample data, and compared with measured values to generate accuracy validation scatter plots for the vegetation and soil carbon density (Figure 4a,b). From the scatter plot distribution, the predicted values show a strong linear relationship with the measured values, with points tightly clustered around the 1:1 line, indicating high prediction accuracy. Both models exhibit high coefficients of determination (R^2 = 0.92) and a low root mean square error (RMSE) and mean square error (MSE), further confirming the reliability of the EXT model in the carbon density prediction.

3.2. Spatial Variation Characteristics of CSR in Ningxia

The average CSR values for Ningxia in 2001, 2011, and 2023 were 13.37, 22.59, and 25.41 gC·m^{-2}, respectively. As shown in Figure 5a–c, the carbon sequestration rates in the southern and northwestern regions of Ningxia have significantly increased. The percentage of areas with CSR < 4 gC·m^{-2} in Ningxia was 8.8%, 3%, and 1.8% in 2001, 2011, and 2023, showing a decreasing trend year by year. In contrast, the percentage of areas with CSR > 40 gC·m^{-2} increased from 2% in 2001 to 10.6% in 2011 and 17.6% in 2023. From 2001 to 2023, the average CSR for the Ningxia ecosystem exhibited a spatial distribution pattern of decreasing values from south to north, as shown in Figure 5d, with the average CSR of 21.95 gC·m^{-2}. Low CSR values are primarily concentrated in the northern region of the Helan Mountain water erosion area and the western region of the Hill platform arid grassland geomantic erosion staggered area. High CSR values are mainly distributed in the Liupan Mountain water erosion zone, the water erosion area of Loess hilly and gully residual tableland, and the southern region of the Loess hilly and gully water erosion zone. As shown in Figure 5e, the Liupan Mountain water erosion zone has the highest CSR, with an average value of 46.51 gC·m^{-2}, followed by the Loess hilly and gully water erosion zone, with an average CSR of 37.92 gC·m^{-2}. The Helan Mountain water erosion zone has the lowest average CSR, at 11.34 gC·m^{-2}.

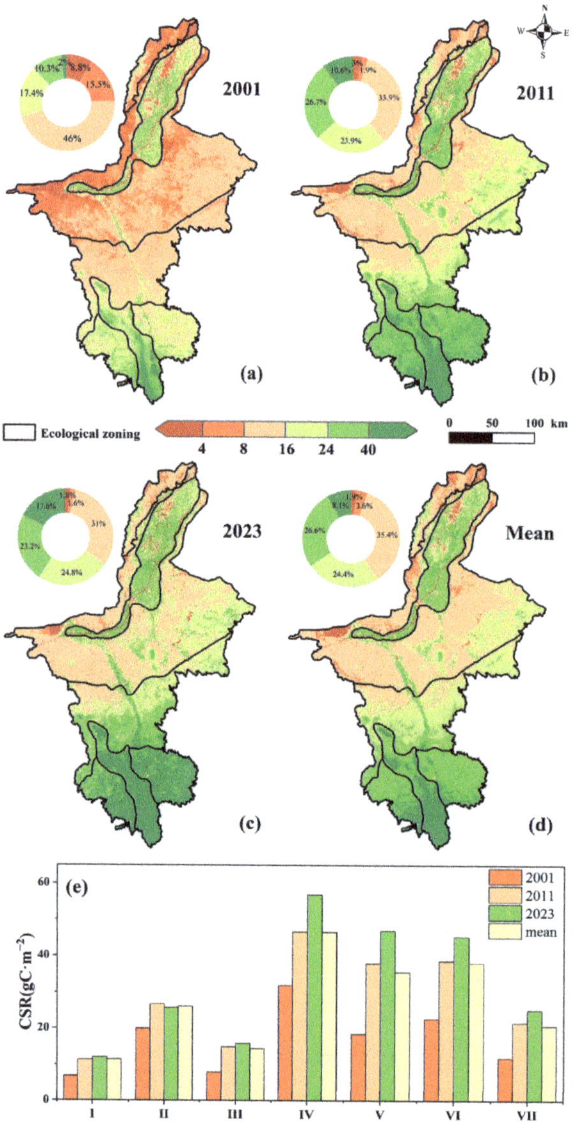

Figure 5. Spatial Distribution of CSR in Ningxia for 2001 (**a**), 2011 (**b**), and 2023 (**c**), and the average (**d**) and CSR statistics for different ecological zones (**e**).

According to Figure 6, the CSR changes in Ningxia and its different ecological zones exhibit significant spatial differentiation. Overall, the carbon sequestration rate in the southern regions of Ningxia has increased significantly, especially in zones V (the Loess hilly and gully residual tableland water erosion zone) and VI (the Loess hilly and gully water erosion zone). A large proportion of land in these regions showed significant increases in the CSR, with growth rates in the range of 0.7–1.2 gC·m^{-2}·a^{-1} and 1.2–7.9 gC·m^{-2}·a^{-1} accounting for about 90%. However, in the northern part of Ningxia, particularly in zones I (the Helan Mountain water erosion zone) and II (the Yinchuan Plain potential wind erosion zone), some areas showed a downward trend in the CSR. Specifically, in zone II, about 24% of the area showed negative growth in the CSR, indicating a weakening of the carbon sequestration capacity of the ecosystem in that region. Overall, the spatial distribution characteristics of the CSR in Ningxia indicate that the carbon sequestration capacity has significantly increased in the southern regions, while in some northern areas, particularly in the Yinchuan Plain, the carbon sequestration function has weakened.

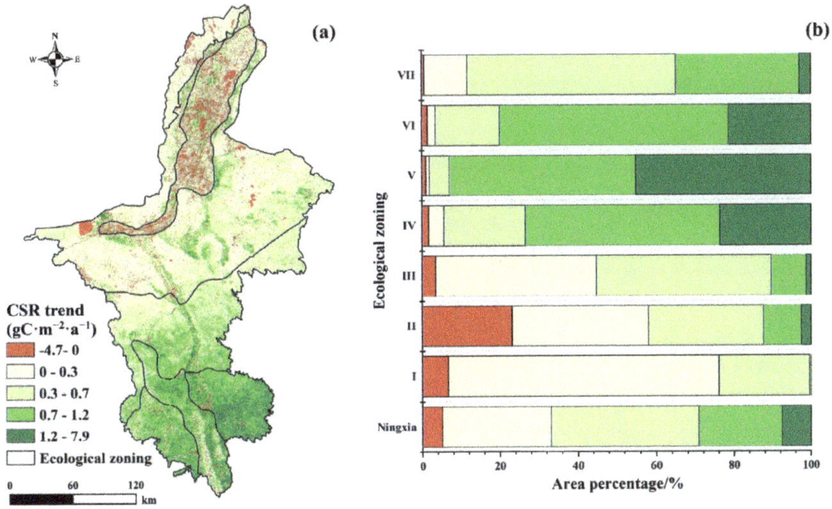

Figure 6. Spatial distribution of Ningxia CSR trend: (**a**) Spatial distribution of CSR trend; (**b**) Statistical proportions of different CSR trend levels within each ecological region.

3.3. *Temporal Changes in CSR (2001–2023)*

Between 2001 and 2023, the CSR in Ningxia and its seven different ecological zones showed an overall fluctuating increase, peaking in 2019. As shown in Figure 7, the carbon sequestration capacity of the overall Ningxia region has significantly increased, with an average annual growth rate of 0.53 gC·m^{-2}·a^{-1}. Among the ecological zones, the CSR growth rates vary significantly. Zone V (the water erosion area of Loess hilly and gully residual tableland zone) exhibited the most significant increase, with an average annual growth rate of 1.16 gC·m^{-2}·a^{-1}. The CSR variations in zones IV (the Liupan Mountain water erosion zone), VI (the Loess hilly and gully water erosion zone), and VII (the Loess hilly and gully water wind erosion staggered zone) were also relatively large, all exceeding the overall level in Ningxia. In contrast, the CSR changes in zones I (the Helan Mountain water erosion zone), II (the potential wind erosion area of Yinchuan Plain zone), and III (the hill platform arid grassland geomantic erosion staggered zone) were relatively stable.

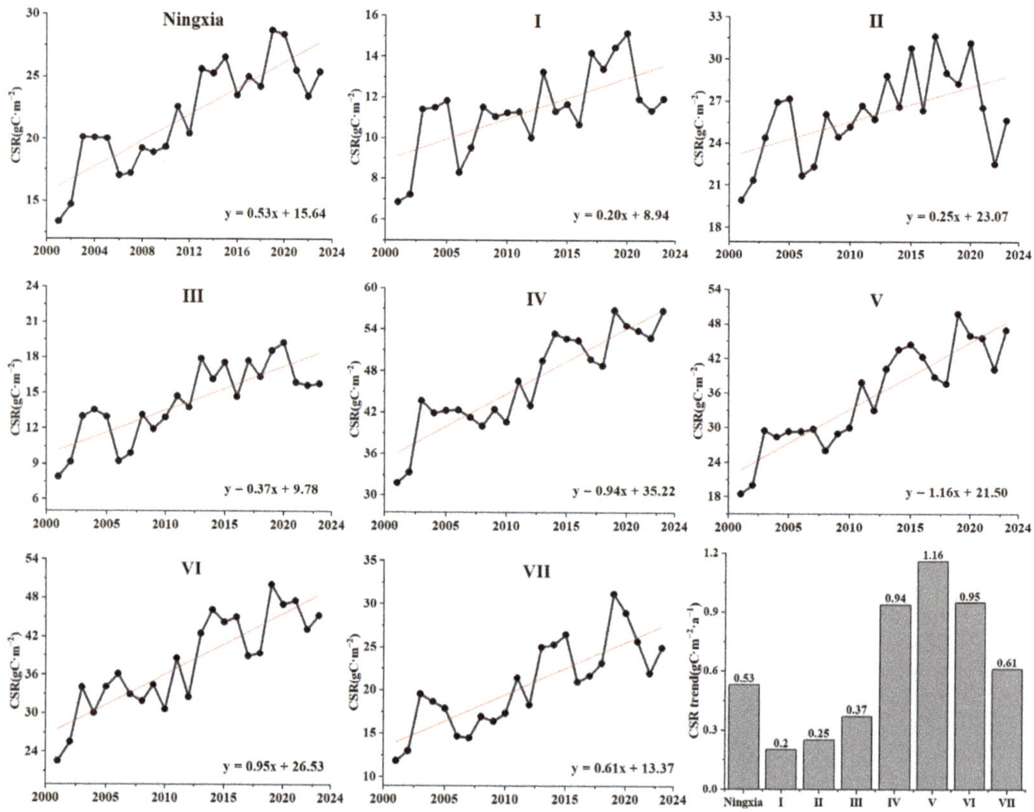

Figure 7. Interannual variation trends of CSR from 2001 to 2023 in Ningxia and its 7 ecological regions. I (the Helan Mountain water erosion zone), II (the potential wind erosion area of Yinchuan Plain zone), III (the hill platform arid grassland geomantic erosion staggered zone), IV (the Liupan Mountain water erosion zone), V (the water erosion area of Loess hilly and gully residual tableland zone),VI (the Loess hilly and gully water erosion zone), and VII (the Loess hilly and gully water wind erosion staggered zone).

3.4. Influence of Land Cover Types on Ecosystem CSR

Based on land cover products from 2000 to 2020, land use changes in Ningxia were analyzed, as shown in Figure 7. The results show that approximately 68.64% of land use remained unchanged, with arable land, grassland, barren, and forest areas remaining stable, occupying 32.26%, 28.17%, 7.13%, and 1.08% of the total area, respectively. It is noteworthy that the forest coverage rate is the lowest, mainly distributed in the southern Liupan Mountain water erosion zone of Ningxia. In contrast, barren land is primarily distributed in the western region of the Hill platform arid grassland geomantic erosion staggered area. Land use change accounts for approximately 30.83% of the total area, with the largest change being barren to forest–grassland (16.15%), followed by C-Fg (13.68%). The conversion rate of G-F is the lowest, at only 1%. This indicates that Ningxia has undergone significant land cover changes over the past two decades.

Based on the 2001–2023 CSR, its trend over the years, and land cover-type products, the CSR and carbon accumulation capacities (CSR trend) of different land cover types in Ningxia were calculated, as shown in Figure 8. According to Table 2, the average CSR ranking for unchanged land use types is as follows: forest > cropland > grassland > barren, indicating that forests have the highest carbon sequestration capacity, while barren land has the lowest. The average CSR for forests is 56.53 gC·m^{-2}, significantly higher than that of cropland, grassland, and barren land. The total carbon sequestration amount ranks as follows: cropland > grassland > barren > forest. Although forests have the highest per-unit carbon sequestration capacity, cropland has the highest total carbon sequestration due to its larger area, reaching 622.91 Gg C. The CSR growth trends are the highest for cropland and forests, with rates of 0.65 gC·m^{-2}·a^{-1} and 0.62 gC·m^{-2}·a^{-1}, respectively. In particular, cropland shows the highest carbon sequestration increase rate, reaching 13.92 Gg C·a^{-1}, indicating that the promotion of conservation tillage and organic agriculture in Ningxia has significantly enhanced cropland's carbon sequestration potential.

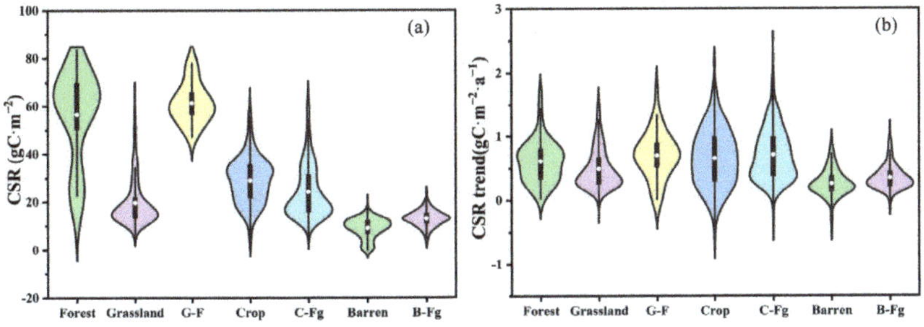

Figure 8. Chart of CSR (**a**) and CSR trend (**b**) for the land cover types.

Table 2. Statistical table of carbon sequestration capacity and carbon sequestration enhancement capacity for different land cover types.

Land Use	CSR/gC·m^{-2}	Carbon Sink /Gg C	CSR Trend /gC·m^{-2}·a^{-1}	CSR Trend /Gg C·a^{-1}	Enhancement Contribution Rate/%
Forest	56.53	40.54	0.62	0.44	1.26%
Grassland	19.70	368.49	0.49	9.17	26.04%
Crop	29.08	622.91	0.65	13.92	39.56%
Barren	9.09	43.03	0.26	1.23	3.50%
G-F	61.40	40.77	0.70	0.46	1.32%
C-Fg	24.34	221.09	0.71	6.45	18.32%
B-Fg	12.93	138.66	0.35	3.75	10.66%

The ranking of the CSR for land use change types is as follows: G-F > C-Fg > B-Fg. The ranking of the CSR amount is C-Fg > B-Fg > G-F. The total carbon sequestration amount for C-Fg is 221.09 Tg C, far higher than that of other types, while B-FG reaches 138.66 Gg C. Compared to grassland, cropland, and barren land, the conversion of G-F, C-Fg, and B-Fg can, respectively, enhance the carbon sequestration capacity by 42.9%, 9.2%, and 34.6%. After cropland is converted to forest–grassland, although the soil carbon sequestration capacity increases, the improvement is less pronounced than the conversion of grassland to forest–shrub due to the relatively higher carbon sequestration capacity of the cropland itself.

In summary, significant differences exist in the carbon sequestration and enhancement capacities across different land cover types in the Ningxia region. Through scientific land management measures and the optimization of the land-use structure, the carbon sequestration capacity in Ningxia can be effectively improved, contributing to sustainable ecological development.

3.5. Influencing Factors on CSR in Ningxia

Figure 9 shows the weight and interaction of the CSR's influencing factors from 2001 to 2023. A univariate analysis shows that the NDVI, with the highest correlation of 0.69, plays a dominant role in the carbon sequestration process. Other vegetation indices such as the EVI, NDMI, and RVI also show a significant influence, reflecting the importance of the vegetation condition on the CSR. Soil property factors such as the bulk density, clay content, sand, and slit also have relatively high q-values of 0.41, 0.36, 0.36, and 0.21, respectively, indicating that these soil characteristics also play a significant role in the CSR. The q-values for DEM, precipitation, and temperature are around 0.25, indicating that elevation and rainfall are also important influencing factors. The interactions between influencing factors exhibit characteristics of two-factor reinforcement or nonlinear amplification. The interaction between the EVI and bulk density has the strongest explanatory power at 0.78, followed by the interaction between the NDMI and bulk density at 0.77. This further confirms the core role of vegetation indices in the CSR.

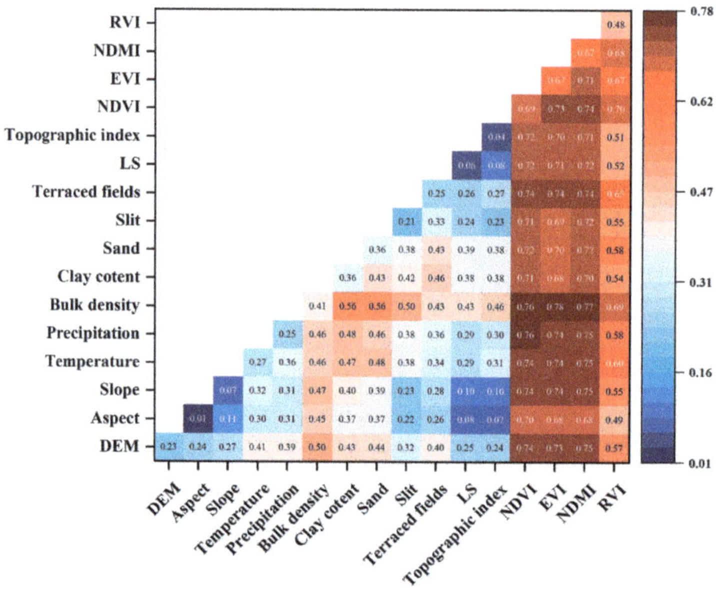

Figure 9. Weights and interactions of CSR influencing factors in Ningxia from 2001 to 2023.

4. Discussion

4.1. Uncertainty Analysis

Different methods and data sources in various studies lead to discrepancies in the CSR results. This study combines remote sensing inversion with ground observation data to more accurately assess the carbon sequestration capacity in Ningxia, using data with a relatively low spatial resolution to reduce estimation errors. However, this estimation still carries some uncertainty. First, the spatial and temporal resolution of remote sensing data limits the precise capture of small-scale carbon density changes, especially in areas with a complex topography. Related studies have shown that this uncertainty is not an isolated case. For example, Yu et al. (2024) [37] pointed out that remote sensing data can lead to biases in CSR estimates for local areas. Secondly, the insufficient distribution of ground observation points, particularly in the arid regions of northern Ningxia, limits the representativeness of the estimates for this area [38,39]. Therefore, future studies should consider incorporating higher-resolution remote sensing data and expanding the coverage of observation points to reduce these uncertainties [40,41].

Overall, the estimation of the carbon density using remote sensing technology still faces numerous challenges, including insufficient sample sizes, limitations in the accuracy of remote sensing data, constraints on the applicability of research methods, and uncertainties and biases in data sources. These issues significantly affect the accuracy and reliability of CSR estimates. Therefore, it is urgently necessary to expand the coverage and improve the precision of field observation data to provide more comprehensive and high-quality validation data. Simultaneously, enhancing the accuracy of remote sensing products, improving algorithms, and adopting advanced data fusion techniques are critical for reducing uncertainties. Additionally, optimizing simulation methods to enhance their applicability across different ecosystems and spatial scales is crucial. Future research should also focus on the potential of more advanced remote sensing technologies, the further integration of ground-based measurements and multi-source information, and an in-depth exploration of the mechanisms by which various variables influence CSR estimates. These advancements are of great significance for accurately characterizing the spatiotemporal dynamics of the CSR, uncovering its driving mechanisms and providing scientific support for ecosystem management and carbon neutrality policy formulation.

4.2. Comparison with Other MODIS Products

This study estimates a 30 m high-resolution CSR dataset for the Ningxia region and compares it with the net primary productivity (NPP) from the MODIS product MOD17A3HGF(V006), as shown in Figure 10. This comparison reveals the impact of different remote sensing data sources and resolutions on CSR estimation results. As shown in the figure, the 30 m resolution CSR data provide a higher level of spatial detail, offering the better capture of small-scale changes in the carbon sequestration process. This high resolution is particularly important for areas with a complex terrain and diverse vegetation types, as it more accurately reflects the characteristics of local ecosystems. While the 500 m resolution NPP has advantages in terms of the coverage and data acquisition frequency, its lower resolution may fail to capture micro-scale changes within the region, leading to reduced accuracy in carbon sequestration estimation. In conclusion, future research could consider combining both approaches to leverage their respective advantages, improving the accuracy and timeliness of carbon sequestration estimates and providing more comprehensive support for ecosystem management.

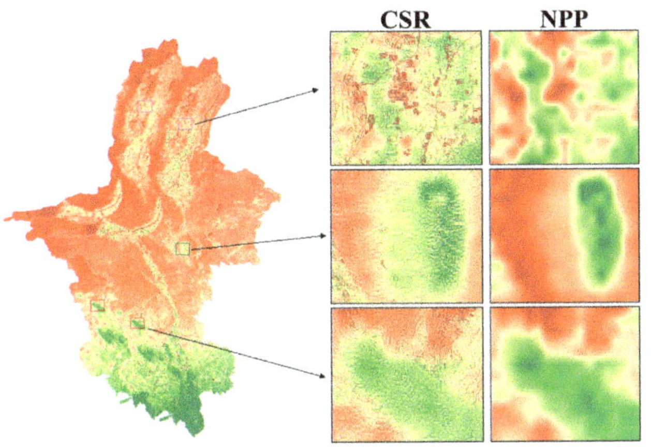

Figure 10. Comparison of Ningxia CSR dataset with details from other MODIS products.

4.3. Spatial Heterogeneity of CSR in Ningxia and Its Driving Factors

In recent years, Ningxia has implemented large-scale ecological conservation projects such as reforestation, grassland restoration, and protective farming practices, significantly increasing the vegetation coverage and enhancing the CSR function of the ecosystem [42]. The research results indicate that the CSR in Ningxia shows clear spatial heterogeneity, particularly a decreasing trend from south to north [4], which is closely related to regional precipitation, vegetation coverage, and soil fertility. Specifically, the CSR benefits in the southern, more humid regions are significantly higher than those in the northern, arid areas. Precipitation is a crucial climatic factor affecting the CSR, as the higher rainfall in the southern regions promotes plant growth and biomass accumulation, thereby enhancing the carbon storage capacity [43]. At the same time, the increase in vegetation coverage directly improves the soil's carbon storage capacity, creating a positive CSR effect [44].

Additionally, soil fertility is a key factor influencing the CSR's spatial heterogeneity [45]. In areas with good vegetation coverage, the soil fertility is typically higher, which aids in plant growth and carbon accumulation [46,47]. In contrast, in areas with poor or severely eroded soils, the CSR remains limited despite moderate precipitation, which further explains why the carbon sequestration capacity is lower in the northern arid regions.

Topographic factors also play an important role in the CSR. Areas with steeper slopes often experience severe soil erosion, leading to a decline in soil fertility and a reduction in the soil carbon storage capacity [48]. This topographic effect exacerbates the CSR differences between regions, suppressing the carbon sequestration capacity in certain areas. Therefore, the combined effects of the climate, topography, and vegetation indices contribute to the spatial heterogeneity of the CSR in Ningxia. Future research could further explore the interactions between these factors to gain a more comprehensive understanding of the driving mechanisms behind the regional carbon sequestration capacity. This will provide more effective management strategies to enhance the regional CSR capacity and promote the sustainable development of ecosystems. By integrating data from climate, vegetation, soil, and topography, targeted measures can be developed to better promote ecological restoration and the CSR in Ningxia.

5. Conclusions

This study, based on ground observation data and multimodal datasets, employs the EXT optimal machine learning model to invert a 30m resolution VTCD and SOCD dataset for Ningxia from 2000 to 2023. It further evaluates the spatiotemporal distribution characteristics of carbon sequestration rates (CSR) from 2001 to 2023 and reveals the associated influencing factors. The results indicate that:

(1) During 2001–2023, the CSR of Ningxia's ecosystems exhibited a spatial distribution characterized by higher values in the south and lower values in the north. The mean CSR was 21.95 $gC \cdot m^{-2}$, with an overall fluctuating upward trend and a growth rate of 0.53 $gC \cdot m^{-2} \cdot a^{-1}$.

(2) The CSR means significantly differ across different ecological regions. The Liupan Mountain water erosion area had the highest carbon sequestration capacity with a mean of 46.51 $gC \cdot m^{-2}$, while the Helan Mountain water erosion zone had the lowest CSR mean of 11.34 $gC \cdot m^{-2}$. The carbon sequestration rate in the Water Erosion Area of Loess Hilly and Gully Residual Tableland showed the most significant increase, with an annual growth rate of 1.16 $gC \cdot m^{-2} \cdot a^{-1}$.

(3) For land use types with unchanged coverage, the carbon sequestration capacity is ranked as forest > cropland > grassland > barren, while the enhancement capacity is ranked as cropland > forest > grassland > barren. In terms of land-use change types,

the CSR ranking is as follows: G-F > C-Fg > B-Fg. The enhancement capacity ranking is C-Fg > G-F > B-Fg.

Compared to grassland, cropland, and barren land, the transitions of G-F, C-Fg, and B-Fg can enhance carbon sequestration capacity by 42.9%, 9.2%, and 34.6%, respectively.

(4) Among the individual influencing factors, the NDVI is the primary driver of the spatiotemporal dynamics of the CSR in Ningxia's ecosystems. However, the two-factor interaction between the EVI and Bulk Density provides a more significant explanatory power for the CSR.

This study demonstrates that ecological restoration projects such as returning farmland to forest (grassland) and conservation tillage play a significant role in enhancing the regional carbon sequestration capacity. Future carbon-neutral policies for Ningxia should prioritize the implementation of vegetation restoration measures and further optimize the design and management of restoration projects across different ecological regions to maximize their carbon sequestration benefits.

Author Contributions: Conceptualization, Y.Z. and Z.W.; methodology, Y.Z. and C.C.; software, L.M. and C.C.; validation, H.H., Y.Z., and C.C.; formal analysis, Y.Z.; investigation, Y.Z. and C.C.; resources, Z.W.; data curation, H.H. and L.M.; writing—original draft preparation, Y.Z.; writing—review and editing, Z.W.; visualization, Y.Z.; supervision, Z.W.; project administration, Z.W.; funding acquisition, Z.W. All authors have read and agreed to the published version of the manuscript.

Funding: This research was funded by the National Key Research and Development Program of China (2023YFC3209303), the Science Foundation for Young Elite Talents of YRCC (HQK-202307), and the Henan Provincial Natural Science Foundation (242300421254).

Data Availability Statement: The shared vegetation and soil carbon density data used in this manuscript are available at http://www.cnern.org.cn/data/meta?id=40579 (accessed on 16 May 2024). However, our measured data are not available in a raw form due to laboratory policy or confidentiality agreements. We have fully described the experimental design, analyses, and results, as well as the process of the data analysis and processing. If editors and reviewers have questions about specific data, we will endeavor to provide more detailed explanations and descriptions. Readers interested in further information may contact the corresponding author. All data used in this thesis are publicly available through the websites provided in the text.

Conflicts of Interest: The authors declare no conflicts of interest.

References

1. Lv, Y.; Zhang, L.; Li, P.; He, H.; Ren, X.; Zhang, M. Ecological Restoration Projects Enhanced Terrestrial Carbon Sequestration in the Karst Region of Southwest China. *Front. Ecol. Evol.* **2023**, *11*, 1179608. [CrossRef]
2. Bu, X.; Cui, D.; Dong, S.; Mi, W.; Li, Y.; Feng, Y. Effects of Wetland Restoration and Conservation Projects on Soil Carbon Sequestration in the Ningxia Basin of the Yellow River in China from 2000 to 2015. *Sustainability* **2020**, *12*, 10284. [CrossRef]
3. Niu, Y.; Li, Y.; Liu, W.; Wang, X.; Chen, Y. Effects of Environmental Factors on the Carbon Fluxes of Semi-Fixed Sandy Land Recovering from Degradation. *Front. Ecol. Evol.* **2023**, *11*, 1178660. [CrossRef]
4. Wang, M.; Hu, Z.; Wang, X.; Li, X.; Wang, Y.; Liu, H.; Han, C.; Cai, J.; Zhao, W. Spatio-Temporal Variation of Carbon Sources and Sinks in the Loess Plateau under Different Climatic Conditions and Land Use Types. *Forests* **2023**, *14*, 1640. [CrossRef]
5. Khodakarami, L.; Pourmanafi, S.; Soffianian, A.R.; Lotfi, A. Modeling spatial distribution of carbon sequestration, CO_2 absorption, and O_2 production in an urban area: Integrating ground-based data, remote sensing technique, and GWR model. *Earth Space Sci.* **2022**, *9*, e2022EA002261. [CrossRef]
6. Misiura, K.; Dąbrowska-Zielińska, K.; Gurdak, R.; Grzybowski, P.; Kluczek, M. CO_2 Modelling from Eddy Covariance Measurements for Biebrza Wetlands. In Proceedings of the IEEE International Geoscience and Remote Sensing Symposium (IGARSS), Brussels, Belgium, 11–16 July 2021; p. 9554112. [CrossRef]
7. Goulden, M.L.; Munger, J.W.; Fan, S.M.; Daube, B.C.; Wofsy, S.C. Measurements of carbon sequestration by long-term eddy covariance: Methods and a critical evaluation of accuracy. *Glob. Chang. Biol.* **1996**, *2*, 169–182. [CrossRef]

8. Nathaniel, J.; Klein, L.; Watson, C.; Nyirjesy, G.; Albrecht, C. Above Ground Carbon Biomass Estimate with Physics-Informed Deep Network. In Proceedings of the IGARSS 2023—2023 IEEE International Geoscience and Remote Sensing Symposium, Pasadena, CA, USA, 16–21 July 2023. [CrossRef]
9. Abbas, S.; Wong, M.S.; Wu, J.; Shahzad, N.; Muhammad Irteza, S. Approaches of Satellite Remote Sensing for the Assessment of Above-Ground Biomass across Tropical Forests: Pan-tropical to National Scales. *Remote Sens.* **2020**, *12*, 3351. [CrossRef]
10. Tripathi, S.; Setia, R.; Ghosh, T.; Uniyal, P.; Kumar, A. Evaluation of Spectral Indices to Monitor the Vegetation Species in a Sanctuary for Carbon Sequestration in Soils Using Landsat Satellite Remote Sensing. *Int. J. Environ. Earth Sci.* **2022**, *10*, 357–365. [CrossRef]
11. Feng, X.; Fu, B.; Lu, N.; Zeng, Y.; Wu, B. How Ecological Restoration Alters Ecosystem Services: An Analysis of Carbon Sequestration in China's Loess Plateau. *Sci. Rep.* **2013**, *3*, 2846. [CrossRef] [PubMed]
12. Huang, L.; Liu, J.; Shao, Q.; Xu, X. Carbon Sequestration by Forestation across China: Past, Present, and Future. *Renew. Sustain. Energy Rev.* **2012**, *16*, 1291–1299. [CrossRef]
13. Sandi, S.; Rodriguez, J.; Saco, P.; Saintilan, N.; Riccardi, G. Accelerated Sea-Level Rise Limits Vegetation Capacity to Sequester Soil Carbon in Coastal Wetlands. *Earth's Future* **2021**, *9*, e2020EF001901. [CrossRef]
14. Aubrey, D.; Blake, J.; Zarnoch, S. From Farms to Forests: Landscape Carbon Balance after 50 Years of Afforestation, Harvesting, and Prescribed Fire. *Forests* **2019**, *10*, 760. [CrossRef]
15. Lal, R. Food Security in a Changing Climate. *Ecohydrol. Hydrobiol.* **2013**, *13*, 8–21. [CrossRef]
16. Mekonnen, Z.; Riley, W.J. Climate Change Will Increase Biomass Proportion of Global Forest Carbon Stocks. *Geophys. Res. Lett.* **2023**, *50*, e2023GL104612. [CrossRef]
17. Moisa, M.B.; Dejene, I.N.; Deribew, K.T.; Gurmessa, M.M.; Gemeda, D.O. Impacts of Forest Cover Change on Carbon Stock, Carbon Emission and Land Surface Temperature in Sor Watershed, Baro Akobo Basin, Western Ethiopia. *J. Water Clim. Chang.* **2023**, *14*, 323–334. [CrossRef]
18. Patton, N.; Lohse, K.; Seyfried, M.; Godsey, S.; Parsons, S. Topographic Controls of Soil Organic Carbon on Soil-Mantled Landscapes. *Sci. Rep.* **2019**, *9*, 6397. [CrossRef] [PubMed]
19. Zhang, M.; He, H.; Zhang, L.; Ren, X.; Wu, X.; Qin, K.; Lv, Y.; Chang, Q.; Xu, Q.; Liu, W.; et al. Increased Forest Coverage Will Induce More Carbon Fixation in Vegetation than in Soil during 2015–2060 in China Based on CMIP6. *Environ. Res. Lett.* **2022**, *17*, 105002. [CrossRef]
20. Sharma, S.; Jain, P.K.; Soloman, P. Carbon Storage Potential of Soil in Diverse Terrestrial Ecosystems. *Nature Environ. Pollut. Technol.* **2023**, *22*, 567–578. [CrossRef]
21. Xu, L.; He, N.; Yu, G. A Dataset of Carbon Density in Chinese Terrestrial Ecosystems (2010s). *Sci. Data Bank* **2019**, *4*. [CrossRef]
22. Beets, P.N.; Kimberley, M.O.; Oliver, G.R.; Pearce, S.H.; Graham, J.D.; Brandon, A. Allometric Equations for Estimating Carbon Stocks in Natural Forest in New Zealand. *Forests* **2012**, *3*, 818–839. [CrossRef]
23. Priya, M.V.; Kalpana, R.; Pazhanivelan, S.; Kumaraperumal, R.; Ragunath, K.; Vanitha, G.; Nihar, A.; Prajesh, P.; Vasumathi, V. Monitoring Vegetation Dynamics Using Multi-Temporal Normalized Difference Vegetation Index (NDVI) and Enhanced Vegetation Index (EVI) Images of Tamil Nadu. *J. Appl. Nat. Sci.* **2023**, *15*. [CrossRef]
24. Camps-Valls, G.; Campos-Taberner, M.; Moreno-Martínez, Á.; Walther, S.; Duveiller, G.; Cescatti, A.; Mahecha, M.; Muñoz-Marí, J.; García-Haro, F.J.; Guanter, L.; et al. Generalization of Vegetation Indices for Monitoring the Terrestrial Biosphere. In Proceedings of the EGU General Assembly Conference Abstracts, Vienna, Austria, 19–30 April 2021. [CrossRef]
25. Kumar, G.K.; Babu, M.; Mani, A.; Luther, M.; Rao, V. Spatio-Temporal Variation of Vegetation in Godavari Eastern Delta. *Curr. J. Appl. Sci. Technol.* **2019**, *38*, 1–13. [CrossRef]
26. Nejad, M.F.; Zoratipour, A. Assessment of LST and NDMI Indices Using MODIS and Landsat Images in Karun Riparian Forest. *J. Forest Sci.* **2019**, *65*, 59–67. [CrossRef]
27. Lu, Y.; Li, X.; Xin, L.; Song, H.; Wang, X. Mapping the Terraces on the Loess Plateau Based on a Deep Learning-Based Model at 1.89 m Resolution. *Sci. Data* **2023**, *10*, 115. [CrossRef]
28. Wang, Z.; Cheng, M.; Zhang, Y.; Peng, S.; Feng, F. Attributing trend in naturalized streamflow to temporally explicit vegetation change and climate variation in the Yellow River basin of China. *Hydrol. Earth Syst. Sci.* **2022**, *26*, 5291–5314. [CrossRef]
29. Liu, S.; Bond-Lamberty, B.; Hicke, J.A.; Vargas, R.; Zhao, S.; Chen, J.; Edburg, S.L.; Hu, Y.; Liu, J.; McGuir, A.D.; et al. Simulating the Impacts of Disturbances on Forest Carbon Cycling in North America: Processes, Data, Models, and Challenges. *J. Geophys. Res. Biogeosci.* **2011**, *116*, G00K08. [CrossRef]
30. de Blécourt, M.; Corre, M.D.; Paudel, E.; Harrison, R.D.; Brumme, R.; Veldkamp, E. Spatial Variability in Soil Organic Carbon in a Tropical Montane Landscape: Associations between Soil Organic Carbon and Land Use, Soil Properties, Vegetation, and Topography Vary Across Plot to Landscape Scales. *Soil* **2017**, *3*, 123–137. [CrossRef]
31. Belgiu, M.; Drăguţ, L. Random Forest in Remote Sensing: A Review of Applications and Future Directions. *ISPRS J. Photogramm. Remote Sens.* **2016**, *114*, 24–31. [CrossRef]
32. Breiman, L. Random Forests. *Mach. Learn.* **2001**, *45*, 5–32. [CrossRef]

33. Trichakis, I.C.; Nikolos, I.K.; Karatzas, G.P. Artificial Neural Network (ANN) Based Modeling for Karstic Groundwater Level Simulation. *Water Resour. Manag.* **2011**, *25*, 1143–1152. Available online: https://link.springer.com/article/10.1007/s11269-010-9628-6 (accessed on 5 November 2023). [CrossRef]
34. Pisner, D.A.; Schnyer, D.M. Support Vector Machine. In *Machine Learning*; Academic Press: Cambridge, MA, USA, 2020; pp. 101–121. [CrossRef]
35. Fan, J.; Ma, X.; Wu, L.; Zhang, F.; Yu, X.; Zeng, W. Light Gradient Boosting Machine: An Efficient Soft Computing Model for Estimating Daily Reference Evapotranspiration with Local and External Meteorological Data. *Agric. Water Manag.* **2019**, *225*, 105758. [CrossRef]
36. Shi, H.; Wang, P.; Zheng, J.; Deng, Y.; Zhuang, C.; Huang, F.; Xiao, R. A Comprehensive Framework for Identifying Contributing Factors of Soil Trace Metal Pollution Using Geodetector and Spatial. *Sci. Total Environ.* **2023**, *857*, 159636. [CrossRef]
37. Yu, Z.; Ciais, P.; Piao, S.; Houghton, R.A.; Lu, C.; Tian, H. Forest Expansion Dominates China's Land Carbon Sink Since 1980. *Nature Commun.* **2022**, *13*, 5374. [CrossRef] [PubMed]
38. Padrón, R.S.; Gudmundsson, L.; Liu, L.; Humphrey, V.; Seneviratne, S. Drivers of Intermodel Uncertainty in Land Carbon Sink Projections. *Biogeosciences* **2022**, *19*, 5435–5452. [CrossRef]
39. Scholze, M.; Kaminski, T.; Knorr, W.; Vossbeck, M.; Wu, M.; Ferrazzoli, P.; Kerr, Y.; Mialon, A.; Richaume, P.; Rodríguez-Fernández, N.; et al. Mean European Carbon Sink over 2010–2015 Estimated by Simultaneous Assimilation of Atmospheric CO_2, Soil Moisture, and Vegetation Optical Depth. *Geophys. Res. Lett.* **2019**, *46*, 14169–14178. [CrossRef]
40. Wang, Y.; Tian, X.; Chevallier, F.; Johnson, M.S.; Philip, S.; Baker, D.; Schuh, A.; Deng, F.; Zhang, X.; Zhang, L.; et al. Constraining China's Land Carbon Sink from Emerging Satellite CO_2 Observations: Progress and Challenges. *Global Chang. Biol.* **2022**, *28*, 4364–4382. [CrossRef]
41. He, W.; Jiang, F.; Wu, M.; Ju, W.; Scholze, M.; Chen, Z.; Byrne, B.; Liu, J.; Wang, H.; Wang, J.; et al. China's Terrestrial Carbon Sink over 2010–2015 Constrained by Satellite Observations of Atmospheric CO_2 and Land Surface Variables. *J. Geophys. Res. Biogeosciences* **2022**, *127*, e2021JG006644. [CrossRef]
42. Bu, X.; Dong, S.; Mi, W.; Li, F. Spatial-Temporal Change of Carbon Storage and Sink of Wetland Ecosystem in Arid Regions, Ningxia Plain. *Atmos. Environ.* **2019**, *204*, 89–101. [CrossRef]
43. Wei, X.; Yang, J.; Luo, P.; Lin, L.; Lin, K.; Guan, J. Assessment of the Variation and Influencing Factors of Vegetation NPP and Carbon Sink Capacity under Different Natural Conditions. *Ecol. Indic.* **2022**, *138*, 108834. [CrossRef]
44. Tu, H.; Jiapaer, G.; Yu, T.; Zhang, L.; Chen, B.; Lin, K.; Li, X. Effects of Land Cover Change on Vegetation Carbon Source/Sink in Arid Terrestrial Ecosystems of Northwest China, 2001–2018. *Remote Sens.* **2023**, *15*, 2471. [CrossRef]
45. Srinivasarao, C.; Kundu, S.; Rakesh, S.; Lakshmi, C.S.; Kumar, G.R.; Manasa, R.; Somashekar, G.; Swamy, G.N.; Mrunalini, K.; Jayaraman, S.; et al. Managing Soil Organic Matter under Dryland Farming Systems for Climate Change Adaptation and Sustaining Agriculture Productivity. In *Soil Organic Carbon and Feeding the Future*; CRC Press: Boca Raton, FL, USA, 2021; pp. 219–251. [CrossRef]
46. Ou, Y.; Rousseau, A.N.; Wang, L.; Yan, B. Spatio-Temporal Patterns of Soil Organic Carbon and pH in Relation to Environmental Factors—A Case Study of the Black Soil Region of Northeastern China. *Agric. Ecosyst. Environ.* **2017**, *245*, 22–31. [CrossRef]
47. Nie, X.; Yang, L.; Li, F.; Xiong, F.; Li, C.; Zhou, G. Storage, Patterns and Controls of Soil Organic Carbon in the Alpine Shrubland in the Three Rivers Source Region on the Qinghai-Tibetan Plateau. *Catena* **2019**, *178*, 154–162. [CrossRef]
48. Feyissa, A.; Raza, S.T.; Cheng, X. Soil Carbon Stabilization and Potential Stabilizing Mechanisms along Elevational Gradients in Alpine Forest and Grassland Ecosystems of Southwest China. *Catena* **2023**, *229*, 107210. [CrossRef]

MDPI AG
Grosspeteranlage 5
4052 Basel
Switzerland
Tel.: +41 61 683 77 34

Land Editorial Office
E-mail: land@mdpi.com
www.mdpi.com/journal/land

www.ingramcontent.com/pod-product-compliance
Lightning Source LLC
Chambersburg PA
CBHW061834210926
51850CB00006B/882
* 9 7 8 3 7 2 5 8 5 8 3 7 8 *